U0263369

Physical Science Under Microgravity: Experiments on Board the SJ-10 Recoverable Satellite

实践十号返回式卫星科学实验. Ⅰ 微重力科学

Wenrui Hu · Qi Kang

Editors

Physical Science Under Microgravity: Experiments on Board the SJ-10 Recoverable Satellite

实践十号返回式卫星科学实验. Ⅰ 微重力科学

Editors
Wenrui Hu
Chinese Academy of Sciences
Beijing, China

Qi Kang
Chinese Academy of Sciences
Beijing, China

ISBN 978-7-03-061942-6
Science Press, Beijing, China

Jointly published with Springer Nature Singapore Pte Ltd.
The printed edition is not for sale outside China Mainland. Customers from outside China Mainland please order the print book from: Springer Nature.
ISBN of the outside China Mainland edition: 978-981-13-1339-4

Published by Science Press
16 Donghuangchenggen North Street,
Beijing 100717, China

The two books *Physical Science Under Microgravity: Experiments on Board the SJ-10 Recoverable Satellite* and *Life Science in Space: Experiments on Board the SJ-10 Recoverable Satellite* are organized under the guidance of the Chief Scientist of SJ-10 Satellite Mission, Professor Wenrui Hu.

Foreword

The SJ-10 recoverable microgravity experimental satellite, the 24th recoverable satellite of China, was successfully launched on April 6, 2016, completely for the experiments of microgravity physics and space biology. The recoverable capsule was landed safely on the ground after 12 days of the launch and the experiments of cool combustion and fluid physics onboard the un-recoverable capsule continued for another week afterward. Scientific purpose of the SJ-10 mission is to promote scientific research in the space microgravity environment by operating the satellite at lower earth orbit. There are totally 28 space experiments, including 18 experiments in the field of microgravity physics (microgravity fluid physics 6, microgravity combustion 4, and space materials science 8) and 10 experiments in the field of space biology (radiation biology 3, gravitational biology 3, and space biotechnology 4). The SJ-10 mission was arranged as one of the four scientific satellites by the space science program of the Chinese Academy of Sciences in the "12th Five-Year-Plan" period (2011–2015).

Microgravity experiments for a long period, which could be performed only in the space infrastructures such as space station, space shuttle, and satellite, are essential for the development of microgravity science and space life science. The recoverable satellite is a useful and efficient tool for space experiments in the microgravity environment, and such kind of satellites have been launched firstly for remote sensing observation for many years, and then for microgravity experiments since the late 1980s in China. Many space experiments have been also arranged in China Manned Space Engineering Program consisting of the spaceships Shenzhou mission since the late 1990s, the space laboratory mission at the present time, and then the space station mission in the near future.

China National Space Administration (CNSA) organized an expert group for the microgravity research in the early twenty-first century, and seven panels were established under the expert group: microgravity fluid physics, microgravity combustion, space material sciences, space fundamental physics, radiation biology, gravitational biology, and space biotechnology. The program of SJ-8 recoverable satellite launched on September 9, 2006, was arranged jointly by CNSA and Chinese Academy of Sciences (CAS), the recoverable capsule was used for the breeding experiments as the main mission, and the un-recoverable capsule was used for the microgravity research (the results of microgravity research of SJ-8 satellite were published in a special issue of Microgravity Science And Technology, vol. 20, number 2, August 2008). The program of a microgravity satellite was originally organized by CNSA in the middle of 2000s. Ten experiments of microgravity science and 10 experiments of space life science were selected from more than 200 applications in late 2004 and early 2005. The mission proposals of space experiments (including two cooperation experiments in collaboration with the French Space Agency and one with European Space Agency) were reviewed in October 2005. The engineering proposal of satellite platform was reviewed in May 2006 by CNSA. Then, the demonstration working group on "recoverable satellite of scientific experiments for space environment utilization" was formally organized, and the mission was determined as SJ-10. Unfortunately, the demonstrative phase was stopped after one year due to the reform of CNSA and restarted when the government of China determined to move the national management of scientific satellite from CNSA to CAS in 2011. The restarted demonstration phase was completed at the end of 2012, and the engineering phase of program SJ-10 started since the beginning of 2013. According to the schedule, the satellites were purposed to be launched at the end of 2015, but it was delayed for 4 months.

A grant for research on the space experiments of SJ-10 missions was found jointly by National Nature Science Foundation of China (NSFC), the CAS in 2017. This grant supports the analysis of the space experiments, which the results are summarized mainly in the present two books: *Physical Science Under Microgravity: Experiments on Board the SJ-10 Recoverable Satellite* edited by Profs. Wenrui Hu and Qi Kang, and *Life Science in Space: Experiments on Board the SJ-10 Recoverable Satellite* edited by Profs. Enkui Duan and Mian Long. The first book contains the results of fluid physics, combustion, and materials science, and the second book consists of the radiation biology, gravitational biology, and bio-logical technology and biological facility studies.

I am grateful to the authors of these books for their respective contributions, to the book editors and mission organizers for their outstanding work, and Dr. Jian Li and June Tang of Springer and Yufeng Niu of Science Press for their kind assistance in publishing the books. Many thanks are due to the Chinese Academy of Sciences, National Space Center of CAS, and National Nature Science Foundation of China for their great support.

Beijing, China   Wenrui Hu
                                                  Chief Scientist of SJ-10 Satellite Mission

Contents

Introduction to Physical Sciences Experiments Completed in SJ-10 Recoverable Satellite

Wenrui Hu and Qi Kang

Abstract SJ-10 satellite is the recoverable scientific experiment satellite special for the space experiments of microgravity physics science and space life science. This mission was officially started on December 31, 2012, and the satellite was launched on April 6th, 2016. This chapter introduced in briefly the SJ-10 mission, the progress of SJ-10 engineering and the projects constitution of physical sciences experiments onboard SJ-10 satellite.

Keywords SJ-10 mission · Space experiment · Microgravity physical sciences · Microgravity fluid physics · Microgravity combustion sciences · Space material sciences

1 Introduction to SJ-10 Mission

"Shi-Jian Ten" (SJ-10) satellite is the second mission of the Strategic Priority Research Program (First-Stage) on Space Science, the Chinese Academy of Sciences (CAS). The first mission is Dark Matter Particle Explorer (DAMPE, also known as "Wukong", Launch date: December 18, 2015); the third mission is Quantum Experiments at Space Scale (QUESS, also known as "Mozi", Launch date: August 16, 2016); and the last mission is Hard X-ray Modulation Telescope (HXMT, also known as "Insight", Launch date: June 15, 2017).

On April 6th 2016, at 01:38:04, Long March 2D rocket was launched at Jiuquan Satellite Launch Center (see Fig. 1). 559 s later, SJ-10 recoverable scientific experiment satellite was successfully sent into the near-circular orbit at the height about 250 km. This is the 24th successful launch of China's recoverable satellite. In the past,

W. R. Hu · Q. Kang (✉)
Key Laboratory of Microgravity (National Microgravity Laboratory),
Institute of Mechanics, Chinese Academy of Sciences, No. 15 Beisihuanxi Road,
Haidian District, Beijing 100190, China
e-mail: kq@imech.ac.cn

School of Engineering Science, University of Chinese Academy of Sciences,
No. 19(A) Yuquan Road, Shijingshan District, Beijing 100049, China

© Science Press and Springer Nature Singapore Pte Ltd. 2019
W. R. Hu and Q. Kang (eds.), *Physical Science Under Microgravity: Experiments on Board the SJ-10 Recoverable Satellite*, Research for Development,
https://doi.org/10.1007/978-981-13-1340-0_1

Fig. 1 SJ-10 satellite was
launched successfully at
Jiuquan Satellite Launch
Center (6 April 2016, Photo
courtesy of Q. Kang)

we also carried out many microgravity science experiments with China's recoverable
satellites [1–3].

SJ-10 satellite is the recoverable scientific experiment satellite special for the space
experiments of microgravity physics science and space life science. This mission was
officially started on December 31, 2012 (see Fig. 2). With the long period of time
in the microgravity environment and the radiation condition in space provided by
SJ-10 satellite, a number of scientific and technological experimental studies on
the law of matter movement and the rule of life activity were carried out through
remote scientific and technological methods and sample recovery analytic techniques
of space experiments. These space experiments focused on some fundamental and
hot issues in microgravity physics science and space life science, especially for the
verification of the original physical model, the utilization of spacecraft technology
and space environment, and the important applications and theoretical breakthrough
in space science in the future. The purpose of this mission is to discover the law of
matter movement and the rule of life activity that cannot be discovered on the ground
due to the existence of gravity, and to know the acting mechanism on organisms by
the complex radiation of space that cannot be simulated on the ground.

SJ-10 engineering includes six systems: Satellite, Rocket, Launch, Control and
Recovery, Ground Support and Scientific Applications (as shown in Fig. 3). The
National Space Science Center of CAS is the general engineering management orga-
nization, and Wenrui Hu, the academician of Chinese Academy of Sciences, is the

Fig. 2 Launching ceremony of SJ-10 engineering mission (31 December 2012, Photo courtesy of S. Chen)

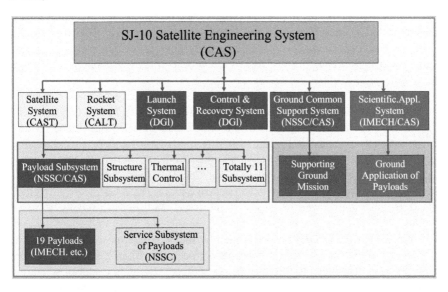

Fig. 3 Organization structure of SJ-10 engineering mission

principal investigator (PI) of SJ-10 mission. The Institute of Mechanics of CAS is in charge of the scientific application system of the mission. Their tasks include overall organization, coordination, operation, and management of the scientific research work, as well as completing the planning and guiding the implementation of scientific experiments, etc.

2 Introduction to SJ-10 Engineering

There are totally 28 research topics on microgravity physics and space life science aboard the SJ-10 satellite, and they are integrated into 19 projects of payloads. Among them, there are 10 payloads about microgravity physics sciences as listing in Table 1, and the research fields include 3 subjects: microgravity fluid physics (A1), microgravity combustion science (A2), and space material science (A3). The satellite fully inherited China's recoverable satellite technology, and it is divided into the orbit capsule and the reentry capsule. All the 8 projects in the orbit capsule are about experiments of microgravity physics sciences, including 5 experimental payloads of fluid physics (A1-1–A1-5) and 3 experimental payloads of combustion (A2), which all do not require sample recovery. In the reentry capsule, there are two experimental payloads about microgravity physics sciences: (1) synthetic furnace of multifunctional material in space (A3-1), (2) SCCO experimental payload of complex fluid (A1-6); and nine other experimental payloads are all for life science research. All experimental samples in the reentry capsule need to be analyzed furtherly in the ground laboratory after the recovery.

The research and development of the payload prototypes of SJ-10 were completed in September 2013. The development of engineering prototypes and flight prototypes was completed in December 2014 and January 2016 respectively. The total weight of the 10 experimental payloads of microgravity physics science is about 350 kg. Figure 4 shows the debugging and testing site of scientific experiment payloads and the satellite of SJ-10.

On April 6th, 2016, four hours after the satellite was launched into the orbit, the orbit capsule firstly started the research experiment of colloidal crystals (A1-5), then the experimental payloads in the orbit capsule carried out cycled rounds of scientific experiments in a serial operating mode (see Table 2). In the reentry capsule, the experiment of SCCO payload lasted for 270 h (A1-6), and the melting experiment on synthetic furnace of multifunctional materials accumulated totally 208 h with 6 working positions switched in turn (A3-1). The scientific experimental data from the satellite were received by the three satellite earth receive stations of CAS at Miyun, Sanya and Kashi, and delivered to the integrated operating and control center of the ground support system at the National Space Science Center of CAS in Beijing in real time for primary processing; Then the first level data packages were distributed to the scientific experiment operating center of scientific application systems at the Institute of Mechanics of CAS. Scientists interpreted and processed the space experimental data on line and adjusted space experimental parameters, operating mode and plans according to the space experimental status. Figure 5 shows the working scenes of Huairou integrated operating and control center of the ground common support system and Beijing scientific experiment operating center of the scientific application system. After 12 days in orbit, the reentry capsule separated from the orbit capsule and returned to Siziwangqi recovery area in Inner Mongolia, China. The payloads and samples were disassembled on site and taken

Table 1 Space experimental projects on board the SJ-10 satellite—the part of microgravity physics

No.	Subject	Project ID	Project	Executor	Location
1	A1 microgravity fluid physics	A1-1	Space experiment of evaporation and fluid interfacial effects	Institute of Mechanics, CAS	Orbit Capsule
2		A1-2	Phase separation and dynamic clustering in granular gas	Institute of Physics, CAS	Orbit Capsule
3		A1-3	Thermal dynamical behavior of vapor bubble during pool boiling	Institute of Mechanics, CAS; Xi'an Jiaotong University	Orbit Capsule
4		A1-4	Space experimental on surface wave of thermocapillary convection	Institute of Mechanics, CAS	Orbit Capsule
5		A1-5	Study on the colloidal assembling	Institute of Mechanics, CAS; Institute of Chemistry, CAS	Orbit Capsule
6		A1-6	Soret coefficients of crude oil (SCCO)	ESA Institute of Mechanics, CAS	Reentry Capsule
7	A2 microgravity combustion science	A2-1	Study on ignition, soot emission and smoke distribution of wire insulations by overload	Institute of Engineering Thermophysics, CAS	Orbit Capsule
8		A2-2/3	Investigation of the coal combustion and pollutant formation characteristics under microgravity	Tsinghua University; Huazhong University of Science & Technology	Orbit Capsule
9		A2-4	Ignition and burning of solid materials in microgravity	Institute of Mechanics, CAS	Orbit Capsule
10	A3 space materiel science	A3-1	Melt material science experiments	Institute of Semiconductors, CAS; etc.	Reentry Capsule

Fig. 4 The test of payloads
and satellite (Photo courtesy
of SJ-10 mission
department). **a** The desktop
test of payloads; **b** The
satellite test of payloads;
c SJ-10 satellite test

(a)

(b)

(c)

(a)

(b)

Fig. 5 **a** Huairou integrated operating and control center of payloads; **b** Beijing scientific experiment operating center (Photo courtesy of Q. Kang)

back to laboratories for post-processing. The orbit capsule continued to work 8 days in space to complete all the follow-up and expanded experiments.

3 Introduction to the Objectives of Microgravity Physical Sciences Projects

The research on microgravity physical sciences depends heavily on advanced space technologies. It promotes the development of space science and applications. It fully demonstrates a country's ability in science and technology. It is the important

Table 2 The procedure of space experiments on physics science in the orbit capsule

driven force for the development of science and technology in the world. Therefore, microgravity physical sciences naturally become the hot subject of research in the world (especially in power countries in spaceflight). In recent years, the International Space Station (ISS) has become the main platform for the research on microgravity physical sciences internationally. NASA, ESA, JAXA and Roskosmos/RKA established many special experimental racks for the research on microgravity physical sciences, and greatly improved the experimental platform in the space station to carry out studies on microgravity physical sciences. They are promoting the fast development of microgravity physical sciences at an unprecedented speed.

The study on microgravity physical sciences in China started at late 1980s. Since then, more than ten batches of space experiments on microgravity physical sciences have been carried out mainly on China's recoverable satellites and Shenzhou spaceships. Chinese scientists have obtained some valuable first-hand experiences in space research, which has laid a good foundation for the development of this subject. In recent years, microgravity science, as an important content in space science, has been deeply demonstrated and planned for a medium and long term by the Strategic Priority Research Program on Space Science of CAS and China Space Station Program. Among the pre-research projects of the Strategic Priority Research Program on Space Science and the first batch of science projects of China Space Station Program, dozens of microgravity physical science projects have been supported.

The space experimental research on microgravity physical science in SJ-10 focuses on some frontier subjects in the field of microgravity science in the world—microgravity fluid physics, microgravity combustion science, and space material science [4, 5]. It is based on the fact that, some important physical processes could be understood clearly only in the long period of time in the microgravity environment of space, and the research could increase people's knowledge of the laws of matter movement under the extreme condition of microgravity: the processes of convection, self-organizing and phase change, and the laws of heat and mass transfer; material ignition and combustion behavior, coal combustion mechanism under the microgravity condition; the growth and solidification processes of new material samples in the microgravity environment. It will improve and optimize engineering fluid and thermal power machinery as well as material processing technology on the ground and in space, obtain high-quality materials that are difficult to grow in the gravity field on the ground, and provide scientific basis and fundamental data for the safety of manned spacecraft in China and some major national requirements such as energy and carbon emission reduction, etc. [6–8].

The projects objectives of microgravity physical sciences are as follows:

(1) **Microgravity fluid physics**: to study the internal mechanisms, dynamic processes and instabilities of heat and mass transfer in the convection and phase change (evaporation and boiling), and discover new laws and verify independently developed physical models; to verity the such molecule gas-liquid separation theory of granular gas; using typical colloid system to study the establishment and evolution processes of the ordered phase driven by pure entropy, the establishment of liquid crystalline phase and the self-assembly

mechanism of metal nano-particles; to accurately measure the Soret coefficient of the samples including Chinese petroleum, and study the cross diffusion rule of multi-component medium. In this book, these studies are presented in Chap. 3 through Chap. 8.

(2) **Microgravity combustion science**: to discover the laws of ignition, combustion, flame spread, flue gas precipitation, soot emission and smoke distribution of typical non-metallic materials and wire insulations under microgravity condition; and to reveal the laws of pyrolysis, ignition, combustion and pollutant generation of typical coal of China under microgravity condition. To provide a theoretical basis and technical support for ground combustion and space safety by microgravity research results. In this book, these studies are presented in Chap. 9 through Chap. 11.

(3) **Space material science**: to discover the selective occupation law of dopant atoms, the morphogenesis and evolution mechanisms of the alloy structure during the crystal growth process, and understand in depth the interface dynamics in the formation of materials from the melt and develop relative theories; to realize the mass transport process dominated by diffusion, and obtain uniform and large scale semiconductor crystals, high-quality metal alloys and composite materials that are difficult to grow in the gravity field on the ground. In this book, only two related studies are presented in Chaps. 13 and 14.

The eleven space experiments on microgravity physical sciences will be introduced in detail in the following chapters. In addition, SJ-10 recoverable satellite platform and the synthetic furnace of multifunctional material will also be introduced in detail in Chap. 2 and Chap. 12 respectively.

Acknowledgements to All the Contributors The authors are grateful for all the individuals and institutions to implement the SJ-10 mission. They are (but not limited to) Prof. Hejun YIN and Prof. Bin XIANGLI as the Chief of mission, Prof. Ji WU as the Acting Chief of mission, Prof. Bochang TANG as the Chief Designer of mission, Prof. Xin MENG as the Deputy Chief of mission; Prof. Chengguang HUANG as the Chief Commander of scientific application system; Profs. Xiaohui ZHANG, Huiguang ZHAO, Jiawen QIU and Changbin XUE as the Director of SJ-10 engineer, satellite and payload. The authors also acknowledge all the colleagues, experts, engineers, administrators, and participants from the six systems of Satellite, Rocket, Launch, Control and Recovery, Ground Support and Scientific Applications for the SJ-10 satellite mission. Academician Wenrui HU and Prof. Qi KANG, as the chief scientist of mission and the chief engineer of scientific application system respectively, we are also particularly grateful to our scientist/payload teams. All the SJ-10 microgravity physical sciences projects are supported by the Strategic Priority Research Program of the Chinese Academy of Sciences (Grant No. XDA04020000) and United Funding from National Natural Science Foundation of China and Chinese Academy of Sciences.

References

1. Zhong XR, Lin LY (1997) GaAs single crystal growth in space. In: Hu WR (ed) Space Science in China. Gordon & Breach, New York, USA, 333

2. Hu WR (2008) Microgravity experiments on board the Chinese recoverable satellite. Microgravity Sci Tech 20(2):59–60
3. Hu WR, Long M, Kang Q, Xie JC, Hou MY et al (2009) Space experimental studies of microgravity fluid science in China. Chin Sci Bull 54(4035)
4. Hu WR et al (2014) Space program SJ-10 of microgravity research. Microgravity Sci Technol 26:159–169
5. Zhao JF, Kang Q (2016) Ground-based researches related to microgravity science experiments aboard SJ-10. Microgravity Sci Technol 28(2):79–188
6. Zhao HG, Qiu JW, Tang BC, Kang Q, Hu WR (2016) The SJ-10 recoverable microgravity satellite of China. J Space Explor 4(3)
7. Kang Q, Hu W (2016) Microgravity experimental satellite—SJ-10. Bull Chin Acad Sci 31(5):574–580 (In Chinese)
8. Hu WR, Tang BC, Kang Q (2017) Progress of microgravity experimental satellite SJ-10. Aeronaut Aerosp Open Access J 1(3)

System Design and Flight Results of China SJ-10 Recoverable Microgravity Experimental Satellite

Huiguang Zhao, Jiawen Qiu and Ying Wang

Abstract China recoverable satellite is a multi-utility, mid-scale satellite with the capability of capsule recovery. Twenty-five recoverable satellites have been developed successfully ever since 1970s. SJ-10 recoverable microgravity experimental satellite (SJ-10 satellite) is the 25th recoverable satellite in China recoverable satellite series and also the first China space microgravity experiment satellite. The main mission of SJ-10 satellites to carry out experimental researches on microgravity science and space life science by means of satellite space flight and capsule recovery. The experiments cover six disciplines, including microgravity fluid physics, microgravity combustion, space materials science, space radiation effect, biological effects of gravity, and space biotechnology. In order to satisfy the requirements of space experiments as well as under the constraints of schedule, funds and others, the new approach of satellite engineering development is promoted and carried out during the process of satellite development. The mission of SJ-10 satellite has been carried out successfully in 2016, and the flight results show that the satellite has absolute advantages in many aspects, such as the ability of recovery, better microgravity level and others. In this chapter, the technical characteristics and mission execution results of the recoverable satellite SJ-10 satellite are summarized. The further development ideas and the improvements of service performances of China recoverable satellite are also discussed briefly.

Keywords SJ-10 recoverable satellite · Microgravity experiment platform

This chapter is simultaneously published in "Physical Science under Microgravity: Experiments on Board the SJ-10 Recoverable Satellite" and "Life Science in Space: Experiments on Board the SJ-10 Recoverable Satellite". The editors feel that this chapter would be useful to the different audiences of both books to better understand the system design and flight test results of SJ-10 recoverable microgravity experimental satellite.

H. Zhao (✉) · J. Qiu · Y. Wang
China Academy of Space Technology, Beijing, China
e-mail: zhaohuiguang@spacechina.com

© Science Press and Springer Nature Singapore Pte Ltd. 2019
W. R. Hu and Q. Kang (eds.), *Physical Science Under Microgravity: Experiments on Board the SJ-10 Recoverable Satellite*, Research for Development,
https://doi.org/10.1007/978-981-13-1340-0_2

Abbreviations

CCSDS	Consultative Committee for Space Data Systems
DC	Direct Current
D-S theory	Dempster-Shafer evidence theory
FIR	Finite Impulse Response
FMEA	Failure Mode and Effects Analysis
g	Gravity
ISO	International Organization for Standardization
ISS	International Space Station
mg	Milli gravity
N	Newton
SA76	The U.S. Standard Atmosphere 1976
SJ-10 satellite	SJ-10 recoverable microgravity experimental satellite
TTC	Tracking, telemetry and command

1 Introduction

China recoverable satellite is a multiutility, mid-scale satellite with the capability of capsule recovery. Twenty-five recoverable satellites have been developed successfully ever since 1970s. SJ-10 satellite is the 25th recoverable satellite in China recoverable satellite series and also the first China space microgravity experiment satellite. The main mission of SJ-10 satellite is to carry out experimental researches on microgravity science and space life science by means of satellite space flight and capsule recovery. The experiments cover six disciplines, including microgravity fluid physics, microgravity combustion, space materials science, space radiation effect, biological effects of gravity, and space biotechnology. This Space Project was included in Strategic Priority Research Program of Chinese Academy of Sciences during the end of twelfth five-year plan. In September 2012, the scientific goals and payload configuration were reviewed and 28 experiments were confirmed. The SJ-10 satellite was approved officially and the engineering development was started at the end of 2012.

To satisfy the requirements of on-orbit experiments and sample recovery, the system design of SJ-10 satellite was upgraded significantly based on previous platform status. The performances such as microgravity level and data services, payload thermal control, remote-control ability for on-orbit experimental operation and prelaunch operation for life samples mounting into satellite, were all improved greatly. Many aspects have gained significant breakthrough, playing a major role in on-orbit experiments of SJ-10 satellite.

In order to be more efficient under the constraints of schedule, funds and others, the new approach of satellite engineering development was promoted and carried out during the process of satellite development. We tailored the ordinary satellite development process by means of reducing the initial system validations, adding the special and necessary validations, and breaking through the product assurance method of reusing the product to reduce the cost and shorten the development cycle. The product quality was ensured simultaneously. The mission of SJ-10 satellite has been carried out successfully in 2016, and the flight results show that the satellite has absolute advantages in many aspects, such as the ability of recovery, better microgravity level and others.

With the progress of science and technology, the pace of human exploration and utilization of space will be deepened. Future space science experiments, research and application of space environment utilization will have broad prospects. China recoverable satellite can be used as a customized platform with better performances for the research and development of space environmental utilization. This chapter summarizes the technical characteristics and mission execution results of the SJ-10 satellite, and further discusses the development ideas and approaches of the satellite system technology in view of the future applications of scientific experiments and others.

2 Development of Foreign and Domestic Platform

2.1 Brief Introduction

Different from the methods of drop tower, drop tube, sounding rocket and parabolic flight of aircraft, the method of orbital flight of spacecraft can provide high-level and longer time of microgravity environment. The comprehensive environment of microgravity and space radiation provide better space science research conditions. Various types of spacecraft have become important platforms for space science research. At present, the recoverable satellites and manned space stations are the best platforms for scientific application.

As a kind of experimental platform with high-precision and long duration microgravity, the prominent characteristics of the recoverable satellites are as follows: (1) Experimental samples can be returned to the ground. (2) Based on the orbital operation mode of unmanned spacecraft, the microgravity level of the experimental platform is higher, meeting the high microgravity level requirements of many experimental projects. (3) A satellite platform can provide customized service support for on-orbit parallel experiment tasks of multi-experiment projects, and has high adaptability. (4) Satellite engineering has a high cost-effectiveness ratio and is supposed to meet the needs of both fundamental science research and commercial applications at a low cost.

2.2   Development of Foreign Platform

In the international space research field, the main spacecraft used to support space microgravity science, life science experiments and new technology experiments are the Russian unmanned recoverable satellites and the International Space Station.

After more than 40 years of development, Russia's recoverable satellite spectrum has evolved into two series, namely Foton and Bion series. In many space flights and return missions of the Foton/Foton-M and Bion/Bion-M satellites, a large number of space science experiments from the United States, Europe, Russia, Japan and other countries have been carried out, and fruitful scientific achievements have been obtained.

Foton was developed in 1983–1999, and 12 satellites have been successfully launched. It carried out a large number of experiments in the semiconductor and optical materials, biological engineering, cell biology, molecular structure, crystal culture, and fluid physics.

After 2000, Russia upgraded the Foton satellite and formed a new generation of Foton-M recoverable satellite. The technical upgrade of the Foton satellite platform mainly includes: enhancing the energy supply capability of the payload and the tele-operation capability of on-orbit experimental payloads, and improving the thermal control system. From 2002 to 2015, Foton-M recoverable satellites have carried out 4 missions. The main technical parameters of Foton-M satellite are as follows:

- **Orbit**: circular orbit, altitude 205–550 km;
- **Inclination**: 62.8–64.9°;
- **Mass**: 6500 kg;
- **Payload mass**: 650 kg (max);
- **Power for load**: 500 W (normal), 700 W (peak);
- **No. of experiments**: 21 (Foton-M3);
- **Microgravity level**: 10^{-5} to 10^{-6} g_0;
- **Landing speed (vertical)**: ≤ 5 m/s;
- **Lifespan**: 12–44 days.

Bion serial recoverable satellite has successfully completed 11 launches from 1973 to 1996. In recent years, Russia upgraded the Bion series satellite platform and formed Bion-M recoverable satellite platform. Major improvements include the enhanced payload return capability and better power supply with solar cells. The Bion-M series completed its first flight in 2013. The main technical parameters of Bion-M satellite are as follows:

- **Orbit**: circular orbit, altitude 205–550 km;
- **Inclination**: 64.9°;
- **Mass**: 6300 kg;
- **Payload mass inside cabin**: 650 kg (max);
- **Payload mass outside**: 250 kg (max);
- **Power for Load**: 550 W;
- **No. of experiments**: 19;

- **Microgravity level**: 10^{-5} to 10^{-6} g_0;
- **Landing speed (vertical)**: ≤ 5 m/s;
- **Lifespan**: 30 days (6 months with solar array).

The International Space Station (ISS) is a large-scale, manned orbital space experiment platform. Its outstanding feature is that it can support space science experimental projects with long-term orbital microgravity environment requirements, and the experiments can be carried out under manned conditions, therefore it is suitable for many highly complex experiments which cannot be automatically controlled. Since ISS establishment and application, a large number of space science experiments and research projects has been completed in many disciplines, including materials, biology, medicine and other new technologies. Relatively inadequate is that, for scientific experiments with very high microgravity levels, it is necessary to develop and install ultra-quiet isolation platforms on the basis of cabins.

2.3 Development of Domestic Platform

At present, there are two types of spacecraft in China, namely the recoverable satellite and Shenzhou manned spacecraft, which can support the development of space microgravity scientific experiments. The Chinese Space Station is under construction, which will gradually provide more and more comprehensive space science experimental services. In view of the experimental requirements of space life science and microgravity science, both the recoverable satellite and Shenzhou spacecraft need to improve and enhance their technical capability and economy.

The key technology upgrades should include improving the payload return capability, and providing better microgravity level, radiation experimental conditions and other platform service performance. The payload capability of the recoverable satellite can only be improved by layout optimization because of the limited volume of the recovery capsule. The Shenzhou spacecraft and the Chinese Space Station project obviously have more enhanced downward return capability of the experimental payload, especially in the future. In terms of micro-gravity level, China's recoverable satellite running on-orbit has good rigidity, and there is no flutter disturbance of low-frequency from large flexible body. A better micro-gravity level can be obtained by adopting appropriate orbit height and restraining the vibration of moving parts on the satellite. For the space station, in order to obtain higher microgravity level, it is necessary to develop an ultra-quiet and vibration suppression device to isolate the various environmental disturbances of the space station. For the improvement of radiation conditions, except for the configuration of environmental data measurement function, all the radiation test conditions without shielding can only be provided by installing the payload outside the cabin. In terms of engineering economy, the total cost of the recoverable satellite project can be reduced by using reusable technologies.

The research and development of China's recoverable satellites began in the 1970s. Up to the SJ-10 satellite mission, 25 recoverable satellites have been successfully developed, forming a series of reliable and mature recoverable satellites. These satellite missions are mainly used in remote sensing and space science experiments. The main technical parameters of the satellite are as follows:

- **Orbit**: 260×340 km;
- **Inclination**: $63°$;
- **Mass**: 3600 kg;
- **Payload mass**: 250 kg (recovered), 300 kg (not recovered);
- **Power for payload**: 350 W;
- **Microgravity level**: 10^{-3} to 10^{-5} g_0;
- **Landing speed (Vertical)**: ≤ 13 m/s;
- **Lifespan**: 15–30 days.

3 Object of System Design

SJ-10 satellite is China's first microgravity science experimental satellite with return function, and the system design goal includes:

- Improving microgravity level in an all-round way to reach the best level of similar spacecraft in the world;
- To provide flexible and efficient operation control and management for multiple payloads on-orbit experiments. Payloads on board satellite supported by satellite-ground links can respond quickly to scientific experimental instructions on the ground, less than 1–2 flight tracks;
- To provide excellent environmental protection conditions for all types of experimental payloads, and to support large power consumption, heat recovery of the experimental load to carry out in-orbit test;
- Low-cost design of the system can be used as commercial microgravity experimental platform to promote the application.

4 System Design

4.1 Characteristics and Difficulties Analysis

SJ-10 satellite is a new type of recoverable satellite which is developed on the basis of Chinese traditional satellite platform with comprehensive adaptability improvement. The new space mission requires new technologies to achieve greater capability under low cost constraints. It brings new difficulties and challenges throughout satellite development:

(1) The system design is difficult. From the lunch of SJ-8 satellite in 2006 to now, SJ-10 satellite is the only recoverable satellite in this decade. Almost all on-board electric products need to be upgraded comprehensively, and all subsystem programs need to be redesigned except for structure and recovery subsystem. Especially for electronic system, the information architecture needs to be redesigned to adapt to the mission requirements of multiple payloads.

(2) Payload design and validation is difficult. SJ-10 satellite will support 28 scientific experiments (19 equipment included) with various types and quantity of payloads. Each payload is an individualized equipment with different scientific objective. Besides, the environmental condition, installation and layout, experiment process and time sequence control for each payload are also different. All of these differences increase the complexity of payload system, and therefore, the difficulties of information management, on-orbit experiment process design, fault isolation, system integration and verification arise significantly.

(3) The requirement of microgravity level is much higher than before. SJ-10 satellite is China's first space microgravity scientific experiment satellite, which is required to have the world-class leading microgravity level so as to support high-level experimental projects. One of the design goals is to realize an order of magnitude higher than the previous platform, which is a great challenge and need to improve the overall system design.

(4) The budget is limited. SJ-10 satellite devotes to build low-cost and affordable microgravity platform, and the cost is strictly controlled from the beginning of the project. From the system scheme to product, every design step must reduce cost through system optimization in the premise of guarantee of reliability.

4.2 System Composition and Architecture

SJ-10 satellite is composed of 11 subsystems, including payload, power supply and distribution, control, propulsion, telemetry and control, engineering parameter measurement, on-board data handling, structure, thermal control, antenna, and recovery subsystem (Fig. 1).

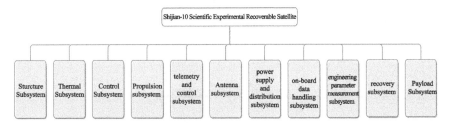

Fig. 1 System composition and architecture

4.3 System Technical Characteristics

- **Total weight**: not more than 3600 kg;
- **Recoverable/unrecoverable payload carrying capacity**: 236 kg/279 kg;
- **Satellite lifetime**: 20 days, reentry module returns in 12 days;
- **Quasi-steady microgravity level**: 10^{-6} g;
- **Micro-vibration level**: better than 1.5 mg, transient peak 7 mg;
- **Power supply capacity for payload**: long-term 350 W, short-term 500 W;
- **Recoverable/unrecoverable payload temperature**: 22 ± 2 °C/8–22 °C;
- **Scientific demand adjustment time**: 1–2 orbits;
- **Data download rate**: 150/300 Mbps;
- **Experimental sample disposal time before launch**: 5 h;
- **Pressure in sealed cabin**: 40–60 kPa;
- **Attitude stabilization**: ≤ 0.03°/s (3σ);
- **Attitude pointing accuracy**: ≤ 3° (3σ);
- **Landing speed of reentry module**: not more than 13 m/s.

4.4 Orbit Design

Inclination Design: The orbit inclination is designed close to the latitude of reentry module landing area to increase the number of return windows and improve the emergency return disposal capacity. Besides, it can reduce the lateral error of landing point caused by the low accuracy of orbit measurement and control for low orbit satellite. Due to the latitude of Siziwangqi landing area, the orbit inclination is chosen as about 43°.

Orbit altitude design: The altitude is selected mainly considering the requirement of returning maneuver. An over high orbit leads to a long return voyage, and the landing accuracy is difficult to guarantee for ballistic return. But a very low orbit cannot meet the requirements of microgravity, and the satellite lifetime and obit measurement and control accuracy cannot be guaranteed. The orbit altitude is chosen as about 250 km finally, and the orbit parameters are designed as follows:

- **Altitude**: 256 km;
- **Eccentricity**: 0;
- **Inclination**: 42.893°.

4.5 System Configuration and Layout Design

The configuration of SJ-10 is a combination of cylinder and cone, with a height of 5144 mm, a maximum diameter of 2200 mm, and a shell as the main bearing structure, as shown in Fig. 2. The satellite consists of reentry module and instrument module,

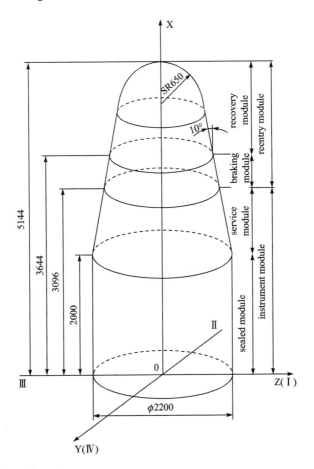

Fig. 2 Satellite configuration

in which the reentry module is composed of braking module and recovery module, and the instrument module is composed of sealed module and service module.

Recoverable life science payloads are installed in recovery module, and those physical science payloads unrecoverable are installed in sealed module, which provides pressure environment for experiments. The braking module has a braking system, which mainly provides power for the return, and the service module has service equipment.

4.6 Flight Procedure

SJ-10 flight process can be divided into four stages, that is, launch, on orbit, return and reentry and landing, as shown in Fig. 3 and specified below.

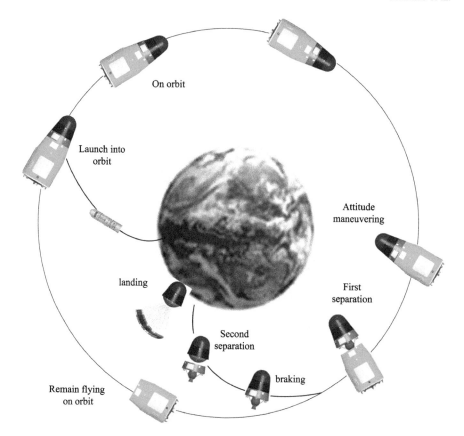

Fig. 3 Flight procedure of satellite

Launch: Launching satellite into orbit under the support of launch site system, telemetry and control system, and rocket system.

On orbit: After launching, carrying out experiments on the orbit.

Return and reentry: Attitude maneuvering before return, reentry module separation from the instrument module, firing the braking engines, recovery module separation from braking module, and recovery module reentry.

Landing: Recovery module reentry, opening the umbrella, and safe landing.

4.7 Scientific Experiment Flow Design

SJ-10 has 28 scientific experiments, 19 in recovery module and 9 in sealed module. Some projects have high power consumption, some generate large amounts of data, and two combustion payloads may produce a large number of exhaust gas needed to be exhausted outside of the module. All of these need special considerations in payload experimental time sequence design:

(1) Experimental time sequence must meet the requirements of scientific experiments;
(2) Projects with high power consumption should be scheduled non-simultaneously, in order to lower requirements for thermal control;
(3) Projects needed to exhaust gas should not be carried out simultaneously to simplify gas pipeline design and control;
(4) Carrying out experiments with large amounts of data should consider peak load shifting to guarantee data download in time.

Experimental time sequences for sealed module projects and recovery module projects are as shown in Table 2 of Chap. 1 and Fig. 4 respectively.

4.8 Design of the Micro-gravity Environment

Some payloads are so sensitive to microgravity environment that they would be affected by acceleration even down to 10^{-7} g. Improving the microgravity environment faces the following difficulties: (1) The structure and configuration of the recoverable satellite imposes restrictions on the improvement of the microgravity environment; (2) As ballistic reentry is adopted for recoverable satellite, and the orbit altitude cannot be too high. So it is difficult to reduce the atmosphere drag force; (3) There are lots of moving parts on the satellites that deteriorates the microgravity environment, especially the pumps, valves, stepper motor and others (Fig. 5). The system solutions need to be optimized to improve the microgravity environment over one order of magnitude.

Microgravity environment is composed of quasi-steady acceleration, oscillatory acceleration and transient acceleration. The methods of eliminating or reducing these three kinds of acceleration are different. It is necessary to optimize and synthesize them from the system level based on the effect and cost of the corresponding measures. Methods adopted for reducing the microgravity acceleration is shown in Table 1.

4.8.1 Improvement of the Quasi-steady Micro-gravity Environment

Quasi-steady acceleration represents the microgravity varies little over long time period, typically with frequency lower than 0.1 Hz. Quasi-steady acceleration exists mainly due to rotation of the satellite, the gradient of the earth gravitational field and drag of the atmosphere. Additionally, the recoverable satellite uses thrusters for attitude control. Quasi-steady acceleration of the satellite will be disturbed transiently while the thrusters work.

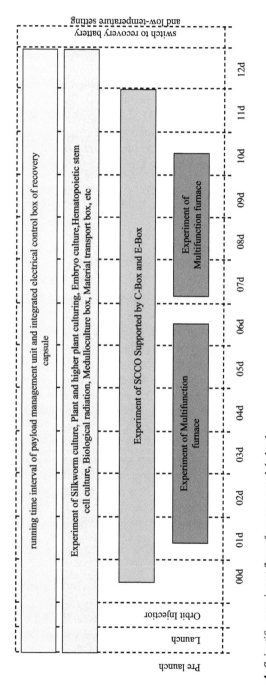

Fig. 4 Scientific experiment flow of recovery module load

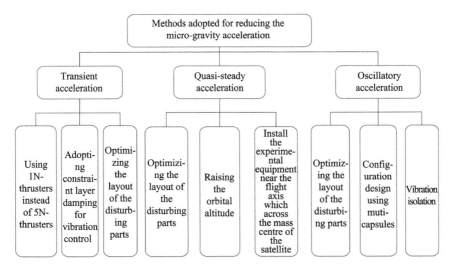

Fig. 5 Keep of the micro-gravity environment

Table 1 Methods adopted for reducing the microgravity acceleration

Quasi-steady acceleration	Oscillatory acceleration	Transient acceleration
Raising the orbital altitude	Optimizing the layout of the disturbing parts	Using 1N-thrusters instead of 5N-thrusters
Install the experimental equipment near the flight axis which across the mass center of the satellite	Configuration design using multi-capsules	Optimizing the layout of the disturbing parts
Using 1N-thrusters instead of 5N-thrusters	Vibration isolation	Adopting constraint layer damping for vibration control
Shut off sensitive experiments during certain period when the microgravity environment deteriorates		

(1) Measures to reduce the atmosphere drag force

The atmosphere drag force which depend on the atmospheric density and the ratio between windward area and mass of the satellite acts on the center of mass of the recoverable satellite. The area-mass ratio of the satellite has already reached 0.0011 m²/kg and is difficult to further minimize. The atmospheric density is the function of orbital height. So, reduction of the atmosphere drag force can be realized by raising the orbit of the satellite. The aerodynamic acceleration caused by the atmosphere drag can be expressed as follow [1]:

$$|\vec{g}_{s1}| = \left|\frac{\vec{F}_k}{m}\right| = \frac{0.5C_D A\rho V^2}{m} \tag{1}$$

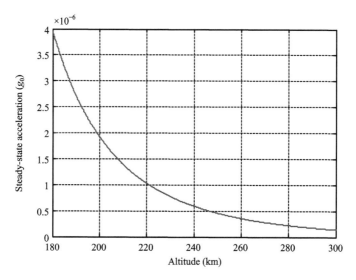

Fig. 6 Quasi-steady acceleration induced by atmosphere drag

in which $\vec{F}_k$ is the atmosphere drag force, A is the windward area of the satellite, ρ is the density of atmosphere, V is the velocity of the spacecraft with respect to the atmosphere, C_D is the aerodynamic drag coefficient, m is the mass of the recoverable satellite.

Using the typical atmosphere model SA76, the quasi-steady acceleration induced by atmosphere drag is presented in Fig. 6, in which it can be clearly seen that the acceleration reduces rapidly from 180 to 250 km.

(2) Measures to reduce the tidal force

The rest of major contributions to the quasi-steady acceleration are the tidal force and the rotation of the satellite, which acts on the satellite except for the mass center. For any position on the spacecraft deviated from center of mass, there exists quasi-steady acceleration caused by tidal force and the rotation of the satellite, which can be expressed as follows [2]:

$$|\vec{g}_{s2x}| \approx \frac{3\Omega^2 xy}{R} = \frac{3 \times 0.0012^2 \times 1.1 \times 1.1}{6378137} \approx 0$$
$$|\vec{g}_{s2y}| \approx 3\Omega^2 y = 3 \times 0.0012^2 \times 1.1 = 4.6 \times 10^{-7} g_0$$
$$|\vec{g}_{s2z}| \approx -\Omega^2 z = 0.0012^2 \times 1.1 = 1.5 \times 10^{-7} g_0 \tag{2}$$

in which Ω is the average rotation velocity of the spacecraft, which is equal to the rotation velocity around the center of the earth. The main parameters that affect the tidal force are y and z, which represent the distance between the analyzed location and flying axis on y and z respectively. The nearer the payload installed to the flying axis, the smaller the tidal force is.

(3) Measures to reduce the disturbance on the quasi-steady acceleration from propulsion system

Transient disturbance on the quasi-steady acceleration of the traditional recoverable satellite of China was caused by the propulsion system, which contains one orbital control thruster (20N) and twelve attitude control thrusters (5N). All the thrusters were installed on the bottom of the recoverable satellite. In order to lower the transient disturbance from propulsion system, 5N-thrusters were replaced by 1N-thrusters.

The acceleration induced by 1N-thrusters can be calculated as follows [2]:

$$|\vec{g}_{t1z}| = \left| \frac{d\vec{\omega}_z}{dt} \times \vec{r} + \vec{\omega}_z \times (\vec{\omega}_z \times \vec{r}) \right| \leq 7.34 \times 10^{-5} g_0$$

$$|\vec{g}_{t1y}| = \left| \frac{d\vec{\omega}_y}{dt} \times \vec{r} + \vec{\omega}_y \times (\vec{\omega}_y \times \vec{r}) \right| \leq 1.46 \times 10^{-4} g_0$$

$$|\vec{g}_{t1x}| = \left| \frac{d\vec{\omega}_x}{dt} \times \vec{r} + \vec{\omega}_x \times (\vec{\omega}_x \times \vec{r}) \right| \leq 9.82 \times 10^{-5} g_0 \tag{3}$$

in which $\vec{g}_{t1x}$, $\vec{g}_{t1y}$ and $\vec{g}_{t1z}$ represents the acceleration along x-, y- and z-axis respectively, $\vec{\omega}_x$, $\vec{\omega}_y$ and $\vec{\omega}_z$ are the angular velocity about x-, y- and z-axis respectively.

It has to be noted that the recoverable satellite is considered as a rigid-body in the above evaluation of the acceleration. The above results are the angular acceleration of the satellite generated by the propulsion system. Disturbance on the quasi-steady acceleration also depends on the actual wave of the thrust, duration of the thrust and interval between two adjacent thrusts.

The 20N-thruster which is installed along the x-axis was used for orbital control. The acceleration disturbance caused by orbit control will reach $10^{-4} g_0$ magnitude level. During the 15-day on-orbit mission of the recoverable satellite, the 20N-thruster only worked for very short period, during which sensitive experiments were shut off.

4.8.2 Improvement of the Oscillatory Environment

Oscillatory accelerations represent microgravity which is periodic in nature with a characteristic frequency. The characteristic frequency of oscillatory acceleration on the recoverable satellite is in the sub-Hertz to hundreds of Hertz range. The source of oscillatory accelerations on the recoverable satellite includes pumps of the fluid loop, air fans for the thermal control and the infrared earth sensor. It has to be noted that the payloads for carrying out space experiments may also contain several disturbance sources which could deteriorate the microgravity environment.

The ground vibration measurement tests show that disturbance from infrared earth sensor is negligible compared to pumps and air fans. Acceleration caused by the pumps of the fluid loop is below 500 mg (milli gravity) at the location where the pump is installed, with frequencies of 146 and 292 Hz. Acceleration caused by the air fan is about 70 mg at the location where it is installed.

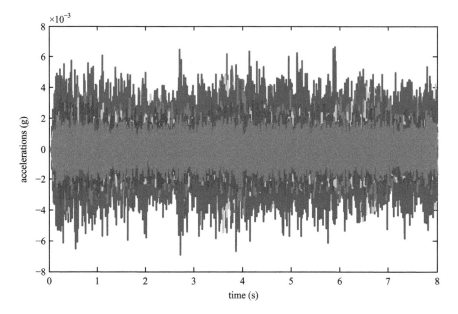

Fig. 7 Comparison of acceleration disturbance caused by fluid loop pump before and after vibration reduction

Therefore, primary measures for reducing the oscillatory acceleration is to carry out vibration isolation for the pumps. The pumps are installed far from the payloads. After translating over several mechanical joints, the oscillatory acceleration induced by the pumps can be reduced by more than two orders of magnitude. Figure 7 shows the comparison of acceleration disturbance caused by fluid loop pump before and after vibration reduction of connected structures.

4.8.3   Improvement of the Transient Acceleration Environment

Transient acceleration has wide range of exciting frequency and usually lasts for less than 1 s. The source of transient acceleration disturbance in recoverable satellite includes electric relay movement, valve movement, jet thrusters and others. Using low force thrusters and optimizing layout of disturbance source can improve the transient acceleration environment. In addition, constrained damping approach is used for thrusters' mounting bracket to decrease the wide band vibration. All the thrusters of recoverable satellite are installed on lower bottom which consists of shell structure. For the shell structure, installing constrained damping layer on it is an effective method of suppressing vibration. The constrained damping layer consists of 0.3 mm damping layer and 2 mm restraint layer, which can reduce transient vibration speed of payload installation site to about 50%.

4.9 High Payload-to-Total-Weight Ratio Design

The main structure of SJ-10 satellite inherited from traditional reentry design. 28 experiments were performed in 19 installations—515 kg total—where 19 of them required recovery. SJ-10 satellite is the first recoverable satellite that applied fluid circuit system. The large amount of payloads, heavy weight, large equipment volume and irregularity and high assembly requirements bring difficulties to payload bay design. The high payload-to-total-weight ratio is realized by system design in not only structure, configuration and layout optimizations, but also system function optimization, sequence optimization, effective energy management and precise control of the center of mass. The design outline mainly includes:

(1) Top level optimization. The functions of each segment of the satellite are optimized. Deorbiting and recovery system is kept in minimum scope. 400 kg from retrorockets is thus removed since required total impulse halved. Only one monopropellant system is installed instead of traditional configuration of one monopropellant system plus one cool gas system, reducing 30 kg from the propulsion system.

(2) Effective energy management. Batteries are used for energy supply on SJ-10 satellite instead of solar panels. The power of the reentry capsule is supplied by Li-Thionyl chloride batteries in the sealed capsule instead of Silver-Zinc batteries that only have lower specific energy, removing 140 kg from batteries. The power supply bus is in parallel. All subsystems share one batch of batteries while payloads share another. Power waste is well avoided by compensations between each subsystem through the bus instead of independent power supply on each subsystem.

(3) Structure lessening design. Structure lessening is realized by material selection, optimization and crafting improvement: (1) Material selection. Materials are selected for different applications so that the material matches functional requirements. Crucial bearing parts are made of titanium for high strength and specific stiffness. The seal capsule is made of 5A06 for good mechanical property, weldable and corruption resistance. Skin is made of 2A12 for high strength and heat resistance. Bearing frames apply 2A14 for high strength. Mounting brackets are magnesium alloy for light weight. (2) Optimization. Optimization is performed by analysis software and mathematical tools. The instrument disk and the twin platform, for example, loss 20% weight through topology configuration optimizations. (3) Crafting improvement. Dead weight is reduced by manufacturing improvement. For instance, solder consumption is reduced by laser welding.

(4) Effective space utilization configuration. Optimizations are conducted on recovery capsule: (1) Separation design. Payload and functional equipment are separated, avoiding coupling and connections of cables or pipelines. (2) Conformal design of payload bay within recovery capsule. The payload combination is conformal with the capsule, increasing space utilization. (3) Some of the payload instrument is deformed in order to further improve space utilization.

(5) Precise control of center of mass. The spinning reentry lays strict requirement on mass property of the recovery capsule. 1 mm precision of lateral center of mass without counter mass is acquired by model-based automatic configuration and detailed design in capsule structure, instrument arrangement and pipeline layout.

The payload-to-total-weight ratio of SJ-10 satellite was improved to 15.3% (3370 kg gross weight with 515 kg payloads) by applying these systematic designs.

4.10 Temperature Control System Design of the Payload of Recovery Capsule

Payloads to carry out on-orbit experiments need the platform of satellite to provide the appropriate temperature environment. This is because the recovery capsule surface of satellite is wrapped with a thick layer of ablative heat protection structure and the heat produced in the high-power load working process cannot dissipate effectively, which becomes a technical bottleneck that it hinders the traditional return satellite to support the high power load experiment.

SJ-10 satellite is facing a more serious situation: there are 11 scientific loads arranged in the recovery capsule, and the average power is about 400 W, which is 10 times of heat consumption of the traditional return type satellite recovery capsule. In addition, unlike the traditional return type satellite, the SJ-10 satellite recovery capsule continues to operate within 3 h after loads unblocking, re-entry into the atmosphere, and landing. The ambient temperature of the load equipment continues to rise, which brings a lot of pressures to thermal and energy of recovery capsule.

In order to solve the thermal control problems in the on-orbit and return phase of the recovery capsule, the system takes into account the factors such as satellite load management, energy management and thermal control management. A scheme was proposed to manage the heat energy inside the recovery capsule with the single-phase fluid circuit as the core subsystem. The details are as follows:

Two cold plate of approximately 1 m in diameter was installed for recovery capsule which is used to collect heat of the experimental load. The mechanical pump drives the fluid circuit to transfer heat to the radiator on the surface of sealed cabin, and to dissipate the heat into the cryogenic space. The fluid circuit is divided into internal circulation and external circulation. The thermal control unit adjusts the opening of thermostatic valve according to the set temperature, and controls the flow proportion of internal circulation to the external circulation to realize the precise control on the load temperature. With the above technical scheme, the on-orbit data shows that the temperature of recovery capsule can be controlled at 22 ± 2 °C.

4.11 Design of Multi-payload Space Laboratory Management System

There are many kinds of payloads onboard satellite, such as microgravity fluid physics, microgravity combustion, space material science, space material science, space radiation effect, microgravity biological effect and space biotechnology. The working pattern is diverse and the timing control is complex. According to the on-orbit experimental results of the payload experimental device, scientists need to adjust experimental parameters flexibly and quickly to plan new test requirements. It takes one to two orbital cycles to transform the new experimental requirements proposed by scientists into uplink instruction injection for satellites. The fault isolation problem of the experimental device is also a key issue. For multi-payload parallel experiment tasks, any individual failure or abnormal problem of the experimental payload must not affect the rest of the experimental payloads, including the safe operation of the satellite platform. Parallel experiments with multiple payloads may lead to resource constraints such as high heat consumption and large amount of data in specific time periods, and thus it is necessary to optimize the experimental timing.

The main controller is payload manager for recovery and sealed capsule. It is made of 11 recovery capsule load test device and 8 sealed capsule load test device that is independent of each other but also can be interactive information collection and distribution of space experimental control system. Payload system and 19 payload test devices are combined to form the distribution of space experimental control system by the configuration of payload support subsystem (as seen in Fig. 8 and specified below).

(1) By the reference of SIOS standard published by CCSDS, for the different needs of the payload, payload subsystem configured data bus architecture and a variety of special communication interface to meet the stand-alone load control requirements and data management requirements.
(2) Using master-slave system bus architecture, see Fig. 9 for topology. The character includes: RS485 bus bases physical layer and UART serial communication link layer protocol, supports automatic identification for site address, and customizes high-reliability application layer protocol. Low speed data management is realized for 11 test devices in recovery capsule and 8 test devices in sealed capsule, and also, cost and development cycle is reduced because of the easy of protocol realization.
(3) Payload support subsystem uses a variety of interfaces (RS422, LVDS and others) to achieve low- and high-speed data acquisition and storage functions.
(4) Data merging and data multiplexing function of adaptive low- and high-speed data seamless switching are realized. The sealed capsule payload manager receives the LVDS high-speed image data and the RS422 low-speed real-time engineering parameters of each load in the cabin, and then packs the data to the storage unit. After the satellite enters, RS422 real-time engineering parameters of the recovery capsule and sealed capsule, and the LVDS playback data

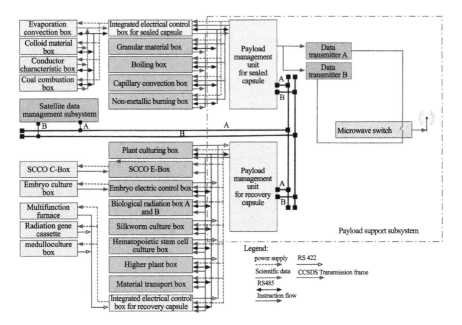

Fig. 8 Distributed control system of payload

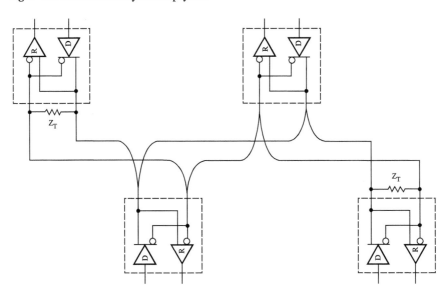

Fig. 9 Low cost master-slave system bus architecture

input from the large-capacity storage unit are multiplexed. Finally, LDPC code is transmitted to the digital transmission transmitter.

(5) The on-orbit health management and fault isolation are realized. The load manager acquire important engineering parameters for critical product in real time, according to the default or injection judgment on-orbit to perform automatic judgment, so that products can achieve automatic shutdown and fault isolation abnormal condition.

(6) Rapid adjustment of scientific experiment processing and parameters in orbit is achieved. In order to ensure that the experimental devices fulfill the scientific objects and even find more unknown physical phenomenon, SJ-10 satellite perform on-orbit autonomous management by event table from payload manager. Firstly, the simulation and analysis of the power and energy consumption of the multi-load simultaneous working mode under various complex conditions are carried out. Under the condition of satisfying the safety of the whole power supply, the flight process is designed according to each scientific target demand and in-orbit configuration requirement. At the same time, the rapid generation, transmission and confirmation strategy of the dedicated configuration instruction is realized by using the terrestrial network, which satisfies all the requirements of quick setting and parameter adjustment of all 28 scientific experiments in the limited transit time of the satellite. The configuration period is optimized from 4 tracks to 1–2 tracks.

By designing a distributed load control system which is independent and efficient with the reclaiming payload manager of recoverable module and sealed module as the hub, it can solve the problem of complex time sequence control, fault isolation and scientific demand change. The on-orbit implementation satisfies all the requirements of fast reconstruction and parameter adjustment that related to 28 scientific experiments. The configuration period is optimized from 4–8 tracks to 1–2 tracks (1.5–3 h).

4.12 Aerospace Products Reuse Technology

SJ-10 satellite project propose strict limit to development costs. The most efficient technical means of cutting the cost is making use of recovery products from other retrievable spacecraft. This is the first project of China which uses large scale of recovery products and of high risk. No experience can be used for reference. We have great challenges on product state assessment and product assurance. The solution of assuring adaptability and reliability of recovery products is to propose the principle of using, assess the healthy state, and innovate product assurance mode and state management by classification.

Technical Program

Carding reuse principle: by carding product state and proposing reuse guiding ideology, we summarize and refine the technology adaptability principles, entirely assess the state principle, and minimize change principle and full test principle.

In aspect of assessing the healthy state of recovery products: we make use of complete machine test data as the main criterion, and at the mean time lead into biographical analysis as another important criterion. Combining the two criterions, we get the final assessment result and employ the healthy assessment method which is based on D-S theory and test data analysis. Product biographical analysis makes use of reliability budget, expert assess and empirical formula. We assess the influence on product healthy state from design, technological level, test experience and storage experience.

In order to analyze the machine healthy state, we first define key parameter by analyzing the test data of key parameter from first assembly to reuse, and get the belief function and finally the assess results which can guide choose and state control of the following products.

In aspect of product state control: we perform the category management according to assess results. Environment test must be carried out, according to the recovery product from former retrievable satellite, for analyzing the reason of property drift, periodic inspection, multiple test to assure product state. For the good state product, the emphasis is interpreting the uniformity of test data, ensuring that the test result is stable. For the product which does not need to rewrite the software, environment test can be cancelled by making use of the like products to confirm the state as a side witness.

In aspect of product assurance: assurance methods such as replacing the component independent management with complete machine management, reusing software specific analysis, and performing technology materials specific check can be used. Making special provisions to software, component, craftwork, material, data pack, file management and security, all these provision can make sure the reuse product meet the reuse requirements.

SJ-10 satellite is the first spacecraft of China which makes great scale use of recovery aerospace product on platform. Reuse rate of information system is 70% which save cost as high as 7.6%. Through this practice, we get a new method and technical norms of reusing the recovery products, and make a benefit attempt for the future reusable spacecraft.

4.13 Low Cost Design

On the one hand, the demand of low-cost system development comes from the pressure of funding constraints; on the other hand, from the development process of spacecraft, reducing the cost is the eternal theme of spacecraft development, especially for spacecraft with respect to the space microgravity scientific experiments.

The cost is of great significance to improve the market competitiveness and enhance the vitality of the spacecraft itself.

Technical approach of SJ-10 low-cost development includes.

4.13.1 Product Reuse

The most effective way to reduce costs is product reuse. SJ-10 used seven devices from the previous satellites to complete information management functions. List of equipment is shown in Table 2.

4.13.2 System Integration of Products Among Fields and Systems

In order to realize the cost control, SJ-10 satellite have chosen the products from many fields, such as recoverable satellite, manned spacecraft, transmission remote sensing, deep space exploration and small satellite platform. Through the design and optimization of electronic system architecture and information system architecture, a good solution comes out for the problem of multi-system products compatible matching and system integration testing.

4.13.3 Using Commercial Modules and Components

For module parts, in order to reduce development costs, satellite non-critical equipment selected commercial products, such as sealed cabin convection fan 4414F/2 from Ebmpapst company as the core component for convection in sealed cabin. On the above basis, secondary development is performed for electronic control module. To ensure the products meet the requirements of in orbit working, performance test,

Table 2 List of Equipment

No.	Product	Function	Quantity
1	Remote control demodulator	Demodulation and distribution of direct remote command and injection data	1
2	Telecontrol central control acquisition unit	Framing and modulation of satellite LLC data	1
3	Landing search beacon	Data providing	1
4	Pulse transponder (band C)	Signal tracking providing in return leg	1
5	CTU	Center of satellite data management	1
6	RTUa	Indirect command source	1
7	RTUb	Indirect command source	1

mechanical and thermal environmental testing and EMC test are performed to ensure the products meet the launch and in-orbit requirements.

For components parts, most of the payload used industrial-grade components and modules, and even commercial-grade products. By checking the working principle and design scheme of payloads, system level characteristic analysis and FMEA analysis of each load, critical products and key module of each product are identified, and the functional redundancy, especially the heterogeneous functional redundancy design is emphasized. Moreover, we formulate the criteria of low-grade components and modules within the product: the product of critical items should use quality components that has strict accordance with the overall satellite product assurance requirements, the products of non-critical are graded according to the failure severity of functional unites, the non-single-point failure can use low-level components and modules, and the single-point failure should use components and modules that has passed the upgraded screening or stricter test assessment. In addition, system test syllabus and rules are developed to ensure the test coverage of product redundant design. As a result of in-orbit operation, low-level components and modules did not appear to clap, working in good condition.

4.13.4   Development Procedure Optimization

Development process of SJ-10 satellite is carried out a number of optimization to simplify the system work, shorten the development cycle, reduce system risk and reduce the repeated work. Also, these acts will provide significant contribution for system development costs reducing. Optimization of development technical process is mainly reflected in the following aspects.

(1) There is no structural model and thermal control model in preliminary design phase. The satellite mechanical and thermal control system design directly goes into critical design phase by the means of design and simulation, recheck and recalculation, virtual test, and others.
(2) Parallel work is taken through manufacturing, assembly, testing and others according to the characteristics of four modules of the satellite. At the same time, technological approaches about satellite fast manufacturing, assembly and testing are explored to compress the development cycle and save the money.
(3) Optimize work order, save the money of crafts development and reduce AIT projects.

According to the approaches above, total cost is reduced by 15%.

5   Evaluation of Flight Test Results

The flight test results are evaluated by the implementation of system technical indexes, the assessment of microgravity level condition and the conduct situation of in-orbit scientific experiments.

5.1   Evaluation of System Technical Characteristics

The comparisons of key technical indexes are shown in Table 3. All the functional performance parameters satisfy the task requirements.

5.2   Evaluation of Microgravity Environment

On-orbit evaluation of microgravity environment was carried out based on the on-orbit measured microgravity data. Qualifications of SJ-10 micro gravity admeasuring apparatus are illustrated in Table 4.

5.2.1   On-Orbit Evaluation of the Quasi-steady Microgravity Environment

There are two microgravity sensors which are made up of quartz accelerometers installed on the satellite. One is installed in the re-entry capsule and the other is installed in the sealed capsule.

Although quartz flexibleness accelerometer of microgravity admeasuring apparatus can measure the DC component (the frequency of acceleration is 0 Hz), the offset and resolution of microgravity admeasuring apparatus can not satisfy the requirement for measurement of quasi-steady acceleration on satellite. Then quasi-steady acceleration caused by atmospheric drag and tidal force can be got from analytic technique. Considering that the analytical error of atmospheric drag and tidal force is small, which cannot cause greater impact to the evaluation of quasi-steady microgravity environment, the theoretical analysis results can explain the order of quasi-steady microgravity.

At the same time, because satellite attitude control thrust is big and the quasi-steady microgravity fluctuation of recoverable satellite mainly depends on the turbulence of attitude control thrusters, the evaluation focus of quasi-steady microgravity is to evaluate the low-frequency turbulence of attitude control thrusters.

In the process of SJ-10 satellite attitude maneuver, the attitude angular velocity is about 0.5°/sand the microgravity sensors are installed about 2.5 m away from

Table 3 Satisfaction situation of system technical indexes

Content	Flight test result	Technical index	Compare result
The weight of whole satellite	3368 kg	≤3600 kg	Satisfied
Carrying capacity of return payload	236 kg	236 kg	Satisfied
Carrying capacity of non-return payload	279 kg	279 kg	Satisfied
Flight time of re-entry capsule	12 days and 14.5 h	≥12 days	Superior to index
Flight time in-orbit	8 days	≥3 days	Superior to index
Quasi-steady Acceleration	Reaches 10^{-6} g order	Superior to 10^{-3} g	Superior to index
Micro vibration level	Superior to 1.5 mg for long-term, 7 mg for transient peak	–	–
Power supply capacity for payload	350 W for long-term, 500 W for short-term	320 W for long-term, 450 W for short-term	Superior to index
Payload temperature in re-entry capsule	20.6–23.6 °C	5–35 °C	Superior to index
Payload temperature in sealed cabin	8–22 °C	0–35 °C	Superior to index
Experimental sample handling time before launch	5 h before launch	24 h before launch	Superior to index
Response time of in-orbit adjustment for scientific demand	1–2 orbits	4 orbits	Superior to index
Re-entry capsule searching time	40 min	≤3 h	Superior to index
sealed cabin pressure	47.55–55.6 kPa	40–60 kPa	Superior to index
Scientific data download rate	300/150 Mpbs optional	150 Mpbs	Superior to index
Satellite attitude stabilization	≤0.01°/s (3σ)	<0.03°/s (3σ)	Superior to index
Attitude pointing precision	≤2° (3σ)	<3° (3σ)	Superior to index
Landing accuracy	39 km from landing site in southwest	Landing ellipse: 190 × 110 km	Superior to index
Landing velocity	12.5 m/s	≤13 m/s	Superior to index

Table 4 Qualifications of microgravity admeasuring apparatus

Item	Qualification
Measuring range	(1) X1: (−250 to +250) mg (2) X2, Y, Z: (−20 to +20) mg
Resolution	(1) Wide range: ≤35 μg (2) Narrow range: ≤narrow

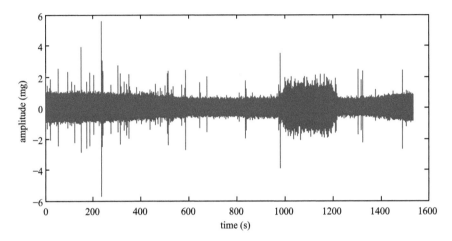

Fig. 10 Time course of microgravity measurement in the process of attitude maneuver

barycenter, the centrifugal acceleration of this position can be calculated theoretically below:

$$\omega^2 r = 2.5 \times (0.5/57.3)^2 = 1.9 \times 10^{-5}\,g \tag{4}$$

In the process of attitude maneuver, the results of microgravity measurements are shown in the Fig. 10.

Making use of FIR low pass filter to process the signal in Fig. 10, the window function is KASIER, the cut-off frequency of pass band is 0.1 Hz, and the attenuation of pass band is less than 1 dB. Given that the cut-off frequency of stop band is 0.2 Hz and the attenuation of stop band is more than 80 dB, the vibration acceleration in the process of attitude maneuver after filtering is shown in Fig. 11. It is obviously that the quasi-steady acceleration in the process of attitude maneuver changes about 1.75×10^{-5} g, which is in agreement with the theoretical analysis.

When attitude control thrusters are implementing attitude maintenance, the quasi-steady turbulence caused by air injection can be extracted by low pass filter. Fig. 12 intercepts the biggest influence of quasi-steady microgravity level caused by attitude control thrusters when on-orbit. Five jumps working one by one bring transient turbulence to quasi-steady microgravity environment. It is clear that, when attitude control thrusters are working, their influence to the quasi-steady microgravity ranges from 10^{-6} to 10^{-5} g.

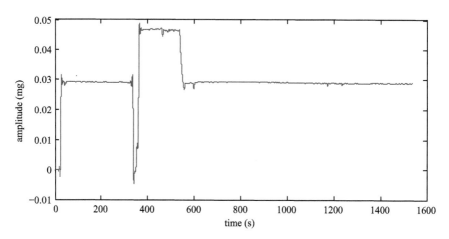

Fig. 11 Time course of microgravity measurement after filtering in the process of attitude maneuver

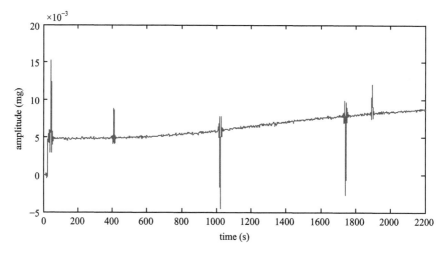

Fig. 12 Jumps of quasi-steady microgravity caused by attitude control thrusters

5.2.2 Micro-vibration Environment On-Orbit Evaluation

(1) Influence of fluid loop pump

As mentioned before, fluid loop pump has the greatest impact on micro vibration environment of satellite platform. Here we focus on the on-orbit micro vibration evaluation result of fluid loop pump.

When the satellite is on-orbit operating, we take shutting down action to fluid loop pump, and then analyze the time-domain curve before and after shutting down. The results are shown in Figs. 13, 14 and 15. From the comparison, after the fluid loop is

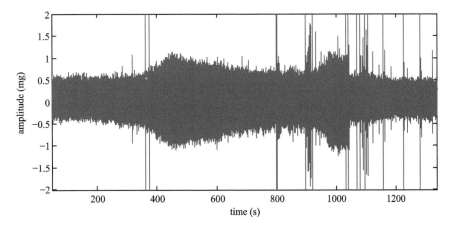

Fig. 13 Directly collected curve before and after fluid loop pumps shut down (X direction)

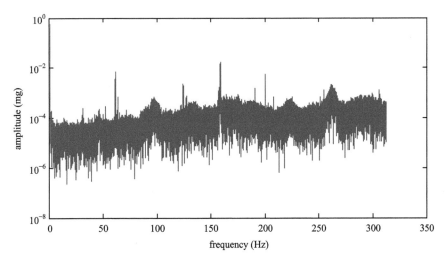

Fig. 14 Amplitude frequency diagram after fluid loop pumps shut down (X direction)

closed, two peaks near 146 and 292 Hz are disappeared, implying that the influence of fluid loop pump mainly concentrates upon 146 and 292 Hz which is consistent with the result of ground experiment.

Making use of band pass filter to process these data and extracting the components near 146 and 292 Hz (Figs. 16 and 17), we can see that the greatest influence of fluid loop pump is about 0.6 mg on recovery capsule and about 0.7 mg on sealed module. Moreover, through on-orbit data analysis, the micro vibration peak value caused by fluid loop pump changes with angle of temperature-control valve and speed of fluid, but the frequency remains stable. In summary, the greatest impact of fluid loop pump is less than 2 mg under different operating mode.

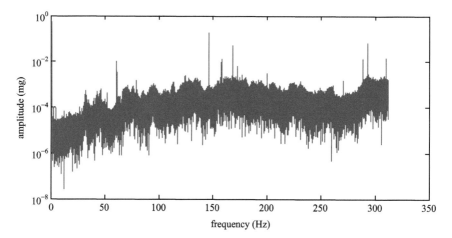

Fig. 15 Amplitude frequency diagram before fluid loop pumps shut down (X direction)

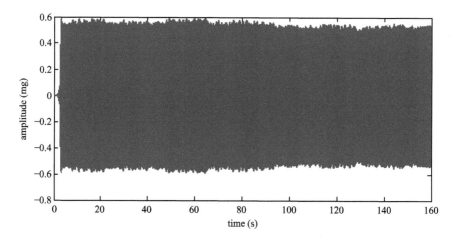

Fig. 16 Influence of fluid loop pump on recovery capsule (Y direction near 146 Hz after filtering)

(2) Influence of other equipment

For the source of disturbance that generates small amplitude of vibration, it is difficult to extract the frequency-amplitude information through the measured data which has massive noises. Thus, we measured the microgravity acceleration of each disturbance source one by one with payloads stop working.

With each disturbance source working alone, oscillatory vibration of temperature control valve, spinning top and fan can be measured respectively though special experiment of microgravity. With the analysis results, the influence of these devices can be excluded from the data measured on orbit, and therefore, further extraction of

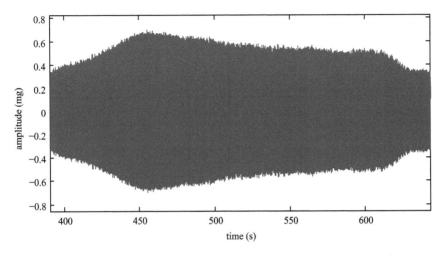

Fig. 17 Influence of fluid loop pump on sealed module (X direction near 146 Hz after filtering)

Table 5 Micro vibration environment influenced by equipment on the SJ-10 platform

Direction	Infrared earth sensor(mg)	Gyro(mg)	Pump(mg)	Air fan(mg)
X	About 0.3	About 0.05	0.2–1.8	0.3
Y	About 0.2	About 0.05	0.3–2	0.25
Z	About 0.25	About 0.05	0.4–2	0.08

the micro vibration disturbance of other equipment can be obtained. Table 5 shows that micro vibration environment influenced by equipment on the SJ-10 satellite platform.

5.2.3 Evaluation of Transient Vibration on Orbit

Transient vibration is caused by relays, dynamos and others, which usually last for less than 1 s. The peak value of transient vibration can be measured directly.

Typical transient vibration on orbit can be seen in Fig. 18. According to measurement data, maximum value of transient acceleration is approximately 30 or 40 mg in respective sealed or reentry modules, in which most transient vibration is less than 10 mg except for very few peaks. The maximum transient acceleration are coursed by certain movements such as pump turning on/off. Sensitive experiments were shut off when these movements showed up.

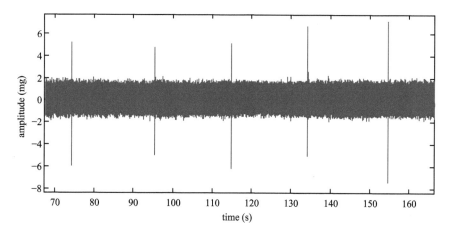

Fig. 18 Curve of measured micro vibration on orbit

Table 6 Microgravity environment of SJ-10 satellite

	Sealed module	Re-entry module
Platform micro vibration environment (0.1–300 Hz)	Approximately 0.6–2 mg frequency of peak value (approximately 146 Hz)	approximately 0.4–2 mg frequency of peak value (approximately 146 Hz)
Transient micro vibration environment	10^{-3} to 3×10^{-2} g	10^{-3} to 4×10^{-2} g
Quasi steady state microgravity environment (0–0.1 Hz)	Orbit control: ≈ 1 mg	
	Attitude adjustment: better than 4×10^{-5} g	
	Quiet state: 10^{-6} g; centroid: 10^{-7} to 10^{-6} g	
	Quasi steady state microgravity influenced by attitude adjustment thruster: 10^{-6} to 10^{-5} g	

5.2.4 Evaluation of Microgravity Environment

Microgravity level of SJ-10 satellite can be seen in Table 6. Flight results showed that quasi-steady microgravity level of SJ-10 satellite was about 10^{-6} g, micro vibration acceleration influenced by equipment was about 2×10^{-3} g, and transient micro vibration was less than 40 mg.

5.3 Evaluation of In-orbit Scientific Experiments

The scientific experiments carried out on-orbit are summarized as follows:

- Microgravity experiments: Among 67 samples and 78 experiment stations, 430 experiments are successful out of total 478 experiments, and the cumulative experiment time is 740 h. The success rate of the experiments is 98.2%, the success rate of the experiment station is 100%, and the success rate of the samples is 77%.

- Space life experiment: Among 28 samples and 90 culture mediums, the cumulative test is 3064 h for all the 11 experiments. The process and recovery success rate of the samples is 100%.

The payloads successfully implemented 28 scientific experiments in accordance with the experiment process, and a lot of scientific experiment data and clear images and video of experiment process were obtained. A number of experiments presented groundbreaking scientific results: the results of liquid evaporation and phrase change heat transfer in microgravity were measured for the first time, and mammalian embryo development was first realized in space. With the future research on the ground, these 28 scientific experiments will continue to bring a large number of scientific and technological achievements with independent intellectual property rights.

6 Future Development Outlook of China Recoverable Satellite

Based on the experience of 25 China recoverable satellite missions, we can summarize several service requirements needed to be enhanced in foregoing China recoverable satellite platform. These items and its technical roads can be generalized as follow:

(1) More recoverable payloads with more chances for orbit access

For scientists, it is still difficult to get a chance to carry out space experiment up to now. The payload mass for recovery and orbit running is still limited. The China recoverable satellite can only provide about 300 kg payloads for recovery and 250 kg payloads is unable to recover. Next we can enlarge the recovery module to improve its return capability based on the principle of aerodynamic similarity. At the same time, we are considering the satellite to adopt modular design and shelf storage to provide shuttle bus space mission services for scientists.

(2) Better performance and service for space experiments

Space science experiments using recoverable satellite require, in particular, better microgravity levels and radiation conditions. At the same time, it requires the platform to provide convenient and sufficient service capability for instruction, data and communication. Good microgravity levels need to consider the overall performance of quasi stationary and micro vibration frequencies. So in order to construct better microgravity environment, the ways will include the selection of orbit altitude, smaller drag area along the direction of flight, and disturbance reduction of moving parts dwell on satellite platform. If necessary, active or passive vibration isolation measures should be taken to ensure that some bands have ultra clean micro-vibration environment. Satellite platform already have several technical means to ensure adequate instructions, data, power supply and other services for experimental projects, which can fully meet this demand.

(3) Lower cost

Space science experiments using spacecraft space flight methods are still a costly means. Recoverable satellites are relatively inexpensive compared to space stations, but current costs still need to be further improved greatly. The main solution is to carry out comprehensive low-cost design, development and testing. In particular, the whole satellite should adopt the technologies, such as non-destructive recovery and reuse technology, low-cost materials and structural technology, and high-integration, commercial devices used on-board electric systems. And it is also required to consider using the global business aerospace TTC network and the internet based experimental control to carry out the satellite operation.

7 Conclusion

Focusing on high performance and low cost system design, several key technologies such as high microgravity level security, high load ratio system design, thermal control technologies for high thermal consumption loads, space experiment management for multi-loads, massive reuse of aerospace products, low cost system development were studied and solved in SJ-10 satellite. The satellite performance was greatly improved, and the cost was well controlled. The in-orbit flight test shows that the technical indexes of the satellite meet the scientific experiment requirements very well. The satellite can provide a high performance, high reliability and low-cost space experiment platform for domestic and foreign scientific and engineering users, and has broad application prospects.

Acknowledgements We credit numerous people who made the SJ-10 satellite a success. During the development of the satellite, we have done a lot of design and integration verification work, and here we would like to thank my team of designers particularly. It was their hard work over the past few years for supporting and ensuring the successful implementation of China's first microgravity scientific satellite mission. System designer, Ying Wang and her partners overcame many difficult problems. Our eleven sub-system chief designers also did great favors to success of the mission. As general directors, I and Jiawen Qiu thanks them from our heart. We also want to give our great thanks to Bochang Tang, our project engineering chief designer and our project office partners. Everybody did his best to our mission success. Thank them all sincerely.

References

1. Curtis HD (2005) Orbital mechanics for engineering students, 1st edn. Elsevier Butterworth-Heinemann, Oxford
2. Sidi MJ (1997) Spacecraft dynamics and control: a practical engineering approach. Cambridge University Press, London

Granular Clustering Studied in Microgravity

Meiying Hou, Wenguang Wang and Qian Jiang

Abstract Energy dissipation is one of the most important characteristics of granular gas, which makes its behavior different from that of molecular gas. In this chapter we first show the investigations on the freely-cooling evolution of granular gas under microgravity in a drop tower experiment, the granular segregation in a two-compartment cell, known as Maxwell's Demon phenomenon in granular gas, is then shown, which is investigated in SJ-10 satellite. DEM simulations of both investigations are given. The simulation results on granular freely-cooling evolution support Haff's law that the kinetic energy dissipates with time t as $E(t) \sim (1 + t/\tau)^{-2}$, with a modified τ, of which the friction dissipation during collisions has to be taken into account. The DEM simulation on Maxwell's Demon with and without gravity shows that the segregation quantified by parameter is non-zero in zero gravity. However, not like the case with gravity, it does not depend on the excitation strength. The waiting time τ, in zero gravity, depends strongly on: higher the, lower the waiting time. The simulation results are confirmed by the SJ-10 satellite microgravity experiments. The major content of this chapter is adopted from our previous papers (Wang et al. Chin Phys B 26:044501, 2017 [1]; Chin Phys B 27:084501, 2018 [2]).

Keywords Granular gas · Clustering · Freely cooling · DEM simulation · Microgravity · Segregation

1 Freely-Cooling Evolution of Granular Gas

Granular gas is a dilute ensemble of grains interacting by dissipative collisions. This dissipative nature of particle interactions determines its ensemble properties, and distinguishes it from molecular gas. Most prominent feature is granular cooling that system loses kinetic energy permanently, with no external energy input. From a

M. Hou (✉) · W. Wang · Q. Jiang
Key Laboratory of Soft Matter Physics, Beijing National Laboratory
for Condensed Matter Physics, Institute of Physics, Beijing 100190, China
e-mail: mayhou@iphy.ac.cn

© Science Press and Springer Nature Singapore Pte Ltd. 2019
W. R. Hu and Q. Kang (eds.), *Physical Science Under Microgravity: Experiments on Board the SJ-10 Recoverable Satellite*, Research for Development,
https://doi.org/10.1007/978-981-13-1340-0_3

homogeneous excited-state, the system enters into an initial period of homogeneous energy loss, and later grains can spontaneously cluster [3–19]. Theoretical modeling [20–31] and simulation investigations [22–24, 26, 30, 32–41] are based on simplifications and assumptions of grain properties. Quantitative experiments are very much needed for better understanding of fundamental features of such ensembles [42–45]. In this section our investigations on the freely-cooling evolution of granular gas under microgravity in a drop tower experiment are given, and the molecular dynamics (MD) simulation is conducted for comparison [2].

In order to investigate the energy loss due to grain-grain collisions in steady excitation and during freely cooling, experiments need to counter balance the gravitational force to float the particles [16, 42]. In 2008, Maaß et al. used magnetic forces to make the granular particles float in a container [42]. In their experiment, diamagnetic particles were chosen. The total energy of a particle in a magnetic field B is $U = -\chi VB^2/2\mu_0 + mgz$ [46], where U is the potential energy of the particle, χ is the magnetic susceptibility, V is the volume of the particle, μ_0 is vacuum permeability, m is the mass of particle, g is the gravity, and z is the position of particle in the z direction. Since $\mathbf{F} = -\nabla U$ and the force balance requires $\mathbf{F} = 0$, $-\nabla U = -mg\mathbf{e}_z + \chi V\mathbf{B}\nabla\mathbf{B}/\mu_0 = 0$ and $\mathbf{B}\nabla\mathbf{B} = \mu_0\rho g/\chi$, where $\rho = m/V$ is the density of particle. Since $\nabla^2 U = -\chi V\nabla^2\mathbf{B}^2/2\mu_0$ and $\nabla^2\mathbf{B}^2 > 0$, for diamagnetic particles $\chi < 0$, $\nabla^2 U > 0$, therefore the diamagnetic particles can float stably in such an external magnetic field. Maaß et al. used this method to make the particles float. They have successfully observed two stages. At early time, the evolution of the kinetic energy dissipation of granular gas follows Haff's law. At a later time, the granular gas clusters and behaves like a single particle motion, which depends on the size of the container.

Tatsumi et al. reported in 2009 the first microgravity experimental investigations on freely cooling granular gas system. Their microgravity experiment was performed during parabolic flight. They studied the kinetics of both freely cooling and steadily driven granular gas system in quasi-two-dimensional cells under micro-gravity [43]. Due to the g-jitter of the parabolic flights, Tatsumi et al. could observe only within one second of the cooling process. Their granular temperature decays as $T_g = T_0(1 + t/\tau)^{-2}$, which is consistent with the Haff's law that $E(t) \sim (1 + t)^{-2}$.

Very recently Kirsten Harth et al. studied freely cooling of a granular gas of rod-like particles in microgravity. For rod-like particles they found that the law of $E(t) \sim t^{-2}$ is still robust [45]. A slight predominance of translational motions, as well as a preferred rod alignment in flight direction were also found.

These studies concerned the form of kinetic energy dissipation with time. Recently, a Drop Tower micro-gravity experiment was conducted and the results were compared with simulations. Some detailed factors (such as restitution coefficient, number density, particles size) which affect Haff's law were shown. Taking into consideration the collision friction, the characteristic decay time τ in the model is modified. Experimental and simulation results show that the average speed of all particles in granular gas decays with time t as $v \sim (1 + t/\tau)^{-1}$ and the kinetic energy decays with time t as $E(t) \sim (1 + t/\tau)^{-2}$ as predicted in Haff's law. However, the clustering mode, as reported in Maaß's et al. work [42], was not observed in this recent work [2] due to its short (5 s) cooling observation time.

1.1 Haff's Freely Cooling Model

Haff's theory about freely cooling granular is based on the classical thermodynamics [20, 21]. There are several assumptions to simplify the analysis of freely cooling process of dilute granular gas. Firstly, the restitution coefficient is constant for each collision among particles. Secondly, the granular gas system must be homogeneous, so the mean free path is meaningful in the analysis and can be used to calculate the collision frequency. Thirdly, only collisions among particles contribute to the energy dissipation in the theory, while the part due to friction is not taken into account. Fourthly, the size of particles is relatively small compared with the mean free path.

The mean free path of particles in dilute and homogeneous granular gas is $\lambda = 1/(n\sigma)$, where n is the number density and $\sigma = \pi(2R)^2$ is the cross section of particles respectively. Here R is the particles' radius. The kinetic energy E of one-unit volume of granular gas is proportional to $nm\bar{v}^2/2$, namely $E \propto nm\bar{v}^2/2$, where $\bar{v}$ is the average speed of particles. When a particle collides with another one, the average kinetic energy loss of a particle is proportional to $(1 - e^2)m\bar{v}^2/2$. The collision frequency of a single particle, is $\bar{v}/\lambda$. The collision frequency of whole system will be $n\bar{v}/\lambda$. But a single collision between two particles is counted twice in a whole system, the actual collision frequency of the system is $n\bar{v}/2\lambda$. The total kinetic energy loss per unit time of one-unit volume of granular gas is

$$\Delta E = -\frac{(1 - e^2)nm\bar{v}^3}{4\lambda}. \tag{1}$$

The change of kinetic energy per unit time equals the loss of the energy per unit time due to collisions between particles,

$$\frac{d}{dt}\left(\frac{1}{2}nm\bar{v}^2\right) = -(1 - e^2)\frac{nm\bar{v}^3}{4\lambda}. \tag{2}$$

The solution of Eq. (2) gives

$$\bar{v}(t) = \frac{\bar{v}_0}{1 + t/\tau}, \tag{3}$$

$$\tau = \frac{4}{(1 - e^2)\bar{v}_0 n\sigma}, \tag{4}$$

where $\bar{v}_0$ is the initial average speed of the granular particles. We notice that Eq. (4) is different from that used in Maaß's research [42] by a factor of 2. Our simulation results support Eq. (4).

The kinetic energy E is proportional to $\bar{v}^2$, the evolution of kinetic energy, thus, is

$$E = \frac{E_0}{(1 + t/\tau)^2}. \tag{5}$$

After a duration of τ since the granular gas begins to cool freely, the average velocity of the system will drop by half as given in Eq. (3). This parameter has also been used in Brilliantov and Pöschel's paper [31, 43] that is,

$$\tau^{-1} = 2\sqrt{\frac{T_0}{\pi}}(1 - e^2)\frac{1 - (7/16)\phi}{(1 - \phi)^2}\frac{\phi}{d},\tag{6}$$

where T_0 is the initial granular temperature, d is the diameter of particle, and ϕ is the volume friction. In order to compare with their work, we also study the dependence of characteristic decay time τ on material parameters (the restitution coefficient e and the particle size R), and also the initial velocity of the system, $\bar{v}_0$. The τ dependence shows the complex property of granular matter—the historical dependency.

Most previous researches mainly concerned the form of Eq. (3), and the exact expression of τ is not yet unique. Hence, we pay more attention to the characteristic decay time τ here, and our numerical results support the form of Eq. (4).

1.2   Experiment

To study the granular gas cooling behavior on ground, gravity has to be counteracted by an external force, like buoyancy in liquid, electric or magnetic field force. For example, Maaß et al. used magnetic gradient to make the diamagnetic particle float [42]. Soichi Tatsumi et al. were the first to attempt microgravity experiment studying the freely cooling granular gas behavior in the parabolic flights [43]. However, due to the g-jitter during the flights, their cooling observations were no longer than 1.6 s. In the present work, we achieve microgravity in Bremen Drop Tower. The Bremen Drop Tower, with a height of 146 m, providesup to 10^{-6} g microgravity condition with a duration as long as 9.3 s.

In the experiments, the catapult operation mode with a weightless duration of 9.3 s was adopted. By this means, a drop capsule Fig. 1a, b was thrown from the bottom of the tower to the top and back into a buffer container filled with foam pellets. Before the freely cooling phenomenon is observed, the particles had to move at a certain initial velocity. In the initial 3 or 4 s, the granular gases were driven by two pistons which were controlled by two linear motors separately Fig. 1c. Then the motor was powered off, the process of freely cooling started. A high-speed camera was used to process and analyze the particle trajectories.

The size of the sample cell is $150 \times 50 \times 10$ mm. Three sub cells are separated by two pistons as is shown in Fig. 2. The numbers of particles in each subcell from left to right in Fig. 2 are 171, 286 and 64, respectively. The diameters of particles in each subcell from left to right are 2.5 mm, 2 mm and 2.5 mm, respectively. Particle motions are initiated by the two pistons, whose central positions, vibration frequencies and amplitudes are controlled. The pistons are stopped after 3–4 s for the fast camera to take images of grain motion during cooling. The image data in the middle and the right subcell are processed and analyzed. Two drop experiments are conducted. The key parameters in each experiment can be found in Table 1 of reference [2].

(a) (b) (c)

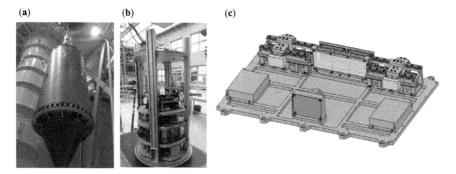

Fig. 1 **a** The drop capsule; **b** Inside the capsule: power system and experimental device; **c** Sketch of the experimental module

Fig. 2 Sample cell where two pistons are marked by red lines

1.3 Experimental Results

The experimental images are captured by a high-speed camera of 500 fps. Each frame of the image is of 512 × 512 pixels. The resolution of experimental images is about 0.3 mm/pixel, therefore the particle sizes with diameters 2 and 2.5 mm are about 6–7 pixels and 8–9 pixels, respectively. However, since we have only one camera, we can observe only a two-dimensional projection of the particle three-dimensional motion. Some trajectories of two overlapped particles may not be identified. Over exposure in one of the cell corners also hinders tracing particles in that area. Despite of all these, we can trace more than 95% of the total particles.

Fig. 3a shows the trajectories of all the traceable particles in the freely cooling processes. Not all the particles can be traced at all times. The average end speeds of particles in these four sets of experiments are about 4–5 mm/s. The average speed decreases by an order of magnitude in 4–5 s. Figure 3b shows that many particles move in a local area before they cool down.

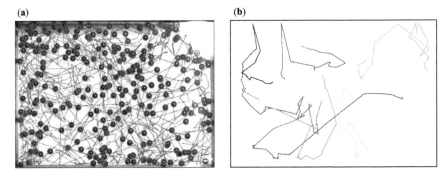

Fig. 3 a Traces of all particles in freely cooling process; **b** Demo of few traces of particles (adapted from Wang et al. [2])

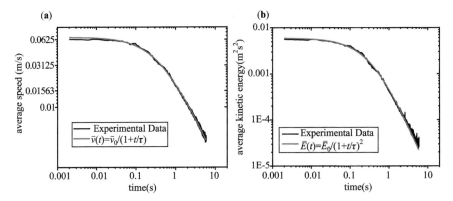

Fig. 4 a Average speed and **b** average kinetic energy of particles as a function of time (adapted from Wang et al. [2])

The speed of each particle v_i is calculated from $v_i = \sqrt{v_{ix}^2 + v_{iy}^2}$, where v_{ix} is the ith particle velocity along the excitation direction, and v_{iy} is the velocity of the particle along the perpendicular y direction. The $\bar{v}$ denotes the average velocity of all particles. Here we set the mass of a particle $m = 1$, so the average kinetic energy is $\bar{E} = <v_i^2>$. In Fig. 4 we fit the average speed and the average kinetic energy of particles by Haff's law. It is seen that Haff's equation fits well, although it does not take into consideration the inelastic collisions with the rigid walls nor frictions in the particle-particle collisions. The values of the two fitting parameters in Haff's law $\bar{v}_0/(1 + t/\tau)$, i.e., $\bar{v}_0$ (the initial average speed) and τ (the decay time), in the four sets of experiments are given in Table 2 of reference [2].

In our experiments, collisions with the two pistons provide particles with initial velocities in the x direction. Figure 5 shows time evolutions of both $\bar{v}_x$ and $\bar{v}_y$, which are consistent with those from the Haff's law. Since particles gain velocities in the y

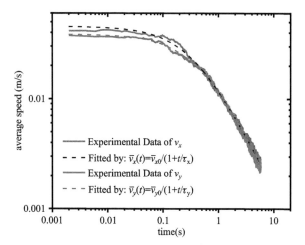

Fig. 5 Average speeds in vertical (x, black curves) direction and horizontal (y, red curves) direction decaying with time (adapted from Wang et al. [2])

direction through particle-particle collisions, the initial $\bar{v}_y$ is seen to be smaller than that of $\bar{v}_x$ due to the dissipation caused by the inelastic collisions. However, we see in Fig. 5 that although initially $\bar{v}_x$ is greater than $\bar{v}_y$, it takes about the same time (approximately 0.5 s) for them to reduce to the same velocity. This again is predicted by Haff's Law. As we can see, when the time t is large enough compared with the characteristic decay time τ, Eq. (3) becomes

$$\bar{v}(t) \cong \frac{\bar{v}_0}{t/\tau}. \tag{7}$$

Substituting Eq. (4) into Eq. (7), we have

$$\bar{v}(t) = \frac{4}{\left(1 - e^2\right)n\sigma t}. \tag{8}$$

Equation (8) shows that when $t \gg \tau$, the average speeds do not depend on the initial velocities, and the velocities of different directions converge into the same value. The average velocity depends on the restitution coefficient and the particle number density.

The kinetic energy E is proportional to the square of the average speed of particles $\bar{v}^2$. The kinetic energy of each dimension in freely cooling state obeys the energy equipartition theorem.

1.4 Simulation Results

Considering the limitation of the experimental opportunities, we perform simulations based on discrete elements method (DEM) [1, 47]. The cell size in the simulation is fixed at $10 \times 10 \times 10$ cm. In Haff's theory, the contribution of the energy dissipation caused by friction during collision between particles is not considered. But in reality, friction exists. In this simulation, the influence of friction is considered.

Figure 6 shows the decays of the simulated average speed with and without considering friction. It is seen that it decays faster when friction is considered. Both can be fitted to Haff's law, but with different values of τ. Haff's freely cooling law is rather robust even though several assumptions in the theory are not satisfied. In the following section, the influence of friction on τ will be discussed.

1.4.1 Effects of Restitution Coefficient and Friction on Characteristic Time τ

Figure 7 shows the simulation results for the τ versus restitution coefficient e when friction is not taken into account. Green line is plotted by using Eq. (4). The inset in Fig. 7 shows the linear relationship between τ and $(1-e^2)^{-1}$. When the restitution coefficient e is close to 1, the τ approaches to infinity.

Figure 8 shows the simulated τ versus restitution coefficient e with taking rotation and friction into account. The inset of Fig. 8 still shows the good linear dependence of τ on $(1-e^2)^{-1}$, however, the intersection shifts. We consider the shift to be due to rotation and friction by adding a γ term $(1-e^2 + \gamma)^{-1}$. The total kinetic energy loss after each collision on average is $\Delta E = (1-e^2 + \gamma)E$. Equation (4) is thus modified into

Fig. 6 Average speeds decaying with time t with (blue dot) and without (green dot) considering friction (adapted from Wang et al. [2])

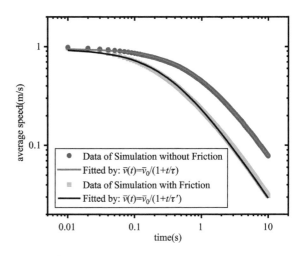

Fig. 7 Simulated τ versus the restitution coefficient e with no friction. Inset shows linear relationship between $(1-e^2)^{-1}$ (adapted from Wang et al. [2])

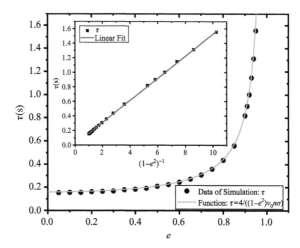

Fig. 8 Simulated τ versus restitution coefficient e with friction. Inset shows linear relationship between $(1-e^2 + \gamma)^{-1}$ and τ (adapted from Wang et al. [2])

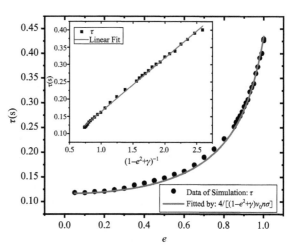

$$\tau = \frac{4}{\left(1 - e^2 + \gamma\right)\bar{v}_0 n\sigma}. \tag{9}$$

The red curve in Fig. 8 is fitted by Eq. (9), and the inset figure shows a good linear relationship between τ and $(1-e^2 + \gamma)^{-1}$. The friction coefficient in this group of simulations is taken to be 0.3, and the corrective frictional term γ fitted by Eq. (9) equals 0.37. Figure 9 shows that the $(1-e^{-2}) \sim 1/\tau$ straight lines with friction and without friction are almost parallel. The difference between the intercepts with x-axis of the two lines is 0.38, which is very close to the fitted value 0.37. The above result indicates that our assumption and modification are reasonably good when the influence of friction during the collision among particles is considered.

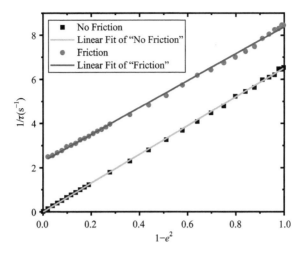

Fig. 9 Plots of $(1-e^{-2}) \sim 1/\tau$ are both straight lines with and without considering friction. Straight line (blue) with considering friction shifts upward with respect to straight line (green) without considering friction (adapted from Wang et al. [2])

1.4.2 Effects of Number Density and Particle Size on Characteristic Decay Time τ

From Eq. (4), it can be seen that $1/n$ and $1/R^2$ have a linear relationship with τ respectively. The inset in Fig. 10 shows that the effect of the number density on τ obeys Haff's law while other parameters hold constant, $e = 0.8, R = 1$ mm. The inset in Fig. 11 shows that the effect of the size of particles on τ nearly follows Eq. (4), but not strictly, while other parameters hold constant, $e = 0.8$ and $n = 2$ cm^{-3}. We notice that the slope of the curve in the inset becomes upward gradually. This means that the τ will be shorter with a larger size of particles. We recall that an assumption in Haff's law is that the size of particles must be relatively small compared with the mean free path λ of the granular gas system. Actually, when the size of particles is relatively large compared with λ, we must take into account the size of particles. The collision frequency is accurately related to the average separation s between neighboring particles, which depends on a surface-to-surface distance. A corrective term r_c must be subtracted from the mean free path λ to obtain an average separation s, namely $s = \lambda - r_c$. The corrective term is close to $1.654R$ according to Opsomer's work [48]. The collision frequency Z is calculated from $Z = \bar{u}/s = \bar{u}/(\lambda - 1.654R)$, where $\bar{u}$ is the average relative velocity during collisions among particles. The corrective term will contribute more to the collision frequency when the size of particles turns larger. Hence, the curve $1/\tau \sim R^2$ is a little concave as shown in the inset of Fig. 11.

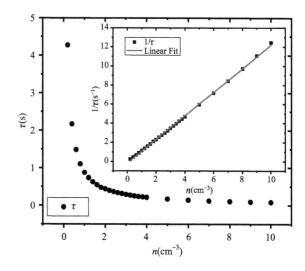

Fig. 10 Decay time τ as a function of number density n. The inset shows linear relationship between n and $1/\tau$ (adapted from Wang et al. [2])

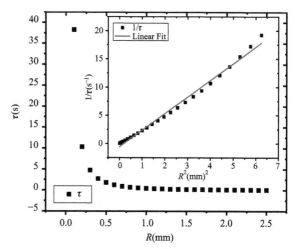

Fig. 11 Decay time τ as a function of particle radius R. Inset shows $R^2 \sim 1/\tau$ curve slopes upward gradually with respect to linear fitted line (adapted from Wang et al. [2])

1.4.3 Evolution of Velocity Distribution with Time

For a conservative classical ideal gas system in equilibrium state, the velocity distribution of each dimension follows the normal distribution. But it is still not widely accepted that what law the velocity distribution of a dissipative system like granular gas obeys. Figure 12a shows that our numerical simulation results of the velocity distribution at different times during the freely cooling obey their corresponding normal distribution. The scalar speed of each particle is set to be 1 m/s at the initial time, but the direction of velocity is randomly set. Although the velocity distribution does not satisfy the normal distribution at the initial time, it evolves into the normal distribution after a while. The standard deviation w represents the width of

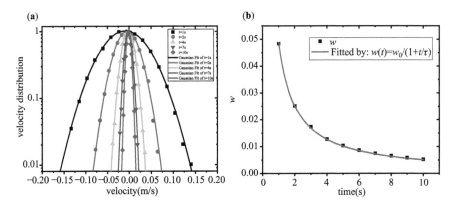

Fig. 12 **a** Evolutions of velocity distribution in one dimension with time. The number of particles is 10,000. The velocity distribution is standardized. **b** Standard deviation w of the normal distribution varying with time (adapted from Wang et al. [2])

the normal distribution curve. The width is related to the fluctuation of the average speed of particles and is defined as granular temperature. It is verified in Fig. 12b that standard deviation w evolving with time t obeys $w \sim (1 + t/\tau)^{-1}$ similar to the average speed of particles.

We recall that the state of energy equipartitions for freely cooling granular gas will be reached even if the symmetry is broken at initial time. Our numerical results show that the freely cooling granular gas not only follows Haff's law but also has two intrinsic properties, namely the energy equipartition and the normal velocity distribution.

2 Maxwell's Demon in Zero Gravity

Granular materials' extremely rich dynamical behaviors have attracted attention of physicists of different fields in recent years [3–5]. Examples are the heap formation of a granular bed [49–51] and the size segregation of a granular system with grains of various sizes under vertical vibration known as the Brazil Nut effects [52, 53]. In these phenomena, energy is being injected continuously into the system by the oscillating boundaries and propagated into the bulk by the inelastic collisions of the grains. A steady state of the whole system is reached when the dissipation of the system is balanced by the input of the energy. Since both the energy input and dissipation depend crucially on the configurations of the system, many intriguing steady states and even oscillatory states can be created.

Borrowing concepts from molecular gas system, low density granular system can be treated as a granular gas. The granular gas systems reach a steady state when input and loss of energy are balanced. They are not in thermal equilibrium and the laws of thermodynamics for molecular gases do not apply to these systems. For example, the thermodynamically impossible phenomenon such as the Maxwell's demon [54, 55] has been observed and successfully explained. In such an experiment, granular gas confined in a compartmentalized system can be induced to segregate into one of the compartments by lowering the vibration amplitude of the system. In this latter case, a decrease of the configurational entropy of the system takes place spontaneously; as if the second law of thermodynamics is violated. In fact, other similar intriguing segregation [56] and ratchet effects [57] have also been reported in compartmentalized granular gases.

Granular Maxwell's Demon phenomenon has been studied in simulation [58–68], theoretical modeling [69–74], and by experiments [75–85] in recent years extensively. The segregation phenomenon relies on the existence of a non-monotonic flux, which determines the number of particles per unit time flows between the two compartments [55]. The flux function is derived from the equation of gas state and therefore depends on the gravity. If applicable, the phenomenon can be used to transport granular materials in space [86, 87]. In this section, simulation study is given to find the condition for possible segregation in two-compartment granular gas system in an environment of zero gravity.

2.1 Model

The simulation is based on discrete element method, in which each particle is treated as a discrete element [88, 89]. The motion of each particle obeys the Newton's Second Law. The interaction between them is considered if and only if two particles collide. Taking the rotation into consideration, the particles' equations of motion are as follows:

$$m_i \ddot{\vec{r}}_i = \sum_{j, j \neq i} \vec{F}_{ij}^C + \vec{F}_i^O, \tag{10}$$

$$I_i \dot{\vec{\omega}}_i = \sum_{j, j \neq i} \vec{R}_i \times \vec{F}_{ij}^C + \vec{M}_i^O, \tag{11}$$

where m_i, I_i, $\vec{r}_i$ and $\vec{\omega}_i$ are the mass, the moment of inertia, the displacement and the angular velocity of particle i, respectively. $\vec{F}_{ij}^C$ is the contact force that the particle i is applied by the particle j. $\vec{R}_i$ is the radius vector of particle i. $\vec{F}_i^O$ and $\vec{M}_i^O$ are other forces and torques applied to the particle i, such as the gravity force. Generally, $\vec{F}_{ij}^C$ can be decomposed into two parts: a normal component and a tangential one,

$$\vec{F}_{ij}^C = \vec{F}_{ij}^n + \vec{F}_{ij}^t. \tag{12}$$

The normal and the tangential contact forces between two particles are handled separately.

Two particles interact with each other only during contact. The overlap in the normal direction between particles i and j with radii R_i and R_j, respectively, is

$$\delta = (R_i + R_j) - (\vec{r}_i - \vec{r}_j) \cdot \vec{n}, \tag{13}$$

where δ is a positive number when the two particles interact, $\vec{n}$ is the unit vector of the relative displacement $\vec{n} = (\vec{r}_i - \vec{r}_j)/|\vec{r}_i - \vec{r}_j|$. Considering the dissipative effect, the equation of the normal contact force is

$$F_{ij}^n = k_n \delta + \gamma_n v_n, \tag{14}$$

where k_n and γ_n are the spring stiffness and the dissipative coefficient, respectively, and v_n is the normal component of the relative velocity $\vec{v}_{ij} = \vec{v}_i - \vec{v}_j$. This is called Linear Spring Dashpot (LSD) model. Why do we select a linear mode other than the Hertz contact mode (a non-linear mode)? The reason is that the Hertz mode is based on an assumption that the loads applied to the particles are static [36]. The coefficient of restitution of two particles in the Hertz model is dependent on the relative velocity during colliding. When the clustering occurs in granular gas, the distribution of velocity of particles has a larger range. This will cause a larger variation of the coefficient of restitution. It's not correspondent with practice. For a dense packing system, the Hertz model is more suitable.

The tangential contact force is determined by the tangential component of the relative velocity and the normal contact force

$$F_{ij}^t = \min(\gamma_t v_{ij}^t, \mu F_{ij}^n), \tag{15}$$

where γ_t is a constant, μ is the friction coefficient and v_{ij}^t is the tangential component of $\vec{v}_{ij}$. Figure 13 shows the mechanism of the contact model in this work.

According to the LSD model, the collision process between two particles acts like a damped harmonic oscillator. We can get the contact duration t_C, half of the period of the damped harmonic oscillator, analytically,

$$t_c = \pi / \sqrt{k_n/m_{ij} - \gamma_n^2/4m_{ij}^2}, \tag{16}$$

Fig. 13 The mechanism of the contact model (adapted from Wang et al. [1])

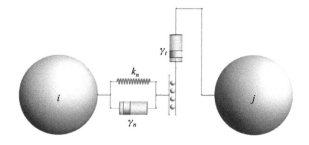

where m_{ij} is the reduced mass $m_{ij} = m_i m_j / (m_i + m_j)$. Using the oscillation half period, we obtain the coefficient of restitution by the ratio of the final velocity and the initial velocity,

$$e = v_{final} / v_0 = \exp(-\gamma_n t_c / 2m_{ij}). \qquad (17)$$

By solving Eq. (17), we can get the dissipative coefficient γ_n via the restitution coefficient e,

$$\gamma_n = -2\sqrt{k_n m_{ij}} \ln(e) / \sqrt{\ln^2(e) + \pi^2}. \qquad (18)$$

Using LSD model, the coefficient of restitution is a constant independent of the relative velocity of two particles during collision. Additionally, in order to guarantee the simulation's accuracy, the time duration per step Δt is required to be much less than t_C, i.e. $\Delta t \ll t_c$. In the simulation the step duration is taken as $\Delta t \approx t_C / 10$.

2.2 Experimental

Figure 14a is a photo of the two-compartment sample cell used in the experiment of the recoverable satellite "SJ-10". Grains in the cell are driven by four independently moving pistons, which are used as the end walls of the two compartments as plotted in Fig. 14b. The two compartments are connected by a window of 12 mm wide. The total length of the cell is 100 mm, as seen in Fig. 14b, and the individual piston size is 25 × 25 mm. The four pistons are driven by two linear motors. Each motor controls either left- or right-end two pistons, so that the two pistons at the same side vibrate synchronously. The piston's position, vibration frequency and amplitude are pre-programmed. In order to prevent the electrostatic effect, metallic particles are used and the cell is made of metal and has been grounded. The fused silica window plates are coated by conductive thin films. Particles are enclosed between the left and right pistons. The volumes of the two cells can then be controlled by the positions of the pistons.

2.3 Numerical

2.3.1 Double-Cell

The numerical parameters of the granular cell system are taken same as that in the experiments shown in Fig. 14b. In the simulation, the particle radius R is 0.5 mm. The density ρ of particles is 4500 kg/m³. The coefficient of restitution e for grain-grain collision is taken as 0.8. The coefficient of friction μ is 0.3. The total number

(a)   **(b)**

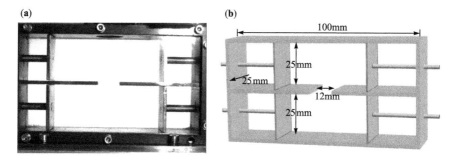

Fig. 14 a A photograph of the two-compartment sample cell used in the experiment of the recoverable satellite "SJ-10"; **b** Sketch of the setup (adapted from Wang et al. [1])

of particles in the connected two-compartment cell is taken as N. The number in one (say upper one) cell is N_1, and the number in the other (lower) cell is $N_2 = N - N_1$. To start with, we put particles evenly in each compartment. The initial velocity is arbitrarily set to be 0.1 m/s with random direction.

2.3.2   Single-Cell

Simulations of particles in single cell are performed to investigate in zero gravity the particle distribution along the vibration direction. With gravity, using the equation of gas state, Eggers [55] sets up a flux function to model the segregation phenomenon. In space under microgravity we need to know the profile of the particle distribution to set up a flux function in order to find the segregation condition. To know the number of particles flow instantaneously through the window per unit time, we count the number of particle-wall collisions at the virtual window of the cell [66]. The size of the single-cell is $80 \times 25 \times 25$ mm. Three rectangular shades along the side wall are shown in Fig. 15 to indicate the virtual windows at different positions. We record

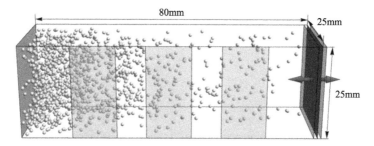

Fig. 15 Schematic of the virtual-window single cell. Particles are excited by the right-end moving wall. The virtual window at three different locations are indicated by rectangular shades at the side wall (adapted from Wang et al. [1])

how many times particles collide with a given virtual window at a time interval for a sufficiently long time. The flux profile can then be given by the counts of collisions per unit time across a given virtual window.

2.4 Results

2.4.1 Distribution of Particles in Single Cell

Under gravity in Eggers' model [55] the equation of gas state is conveniently used for the density distribution of particles along the vibration direction. In microgravity the particle distribution is determined by the geometry of the cell and the wall excitation condition. We, therefore, perform a numerical study in a same cell geometry configuration and wall excitation condition to obtain the zero-gravity particle distribution. One set of the simulation results at condition that frequency $f_r = 10$ Hz and shaking amplitude $A_r = 3$ mm, is shown in Fig. 16. Dividing the cell into 20 equal parts along the vibration direction, particle counts in each division are averaged for different total number N. Except for situation in Fig. 16a when $N = 500$, particle distributions in steady state for N greater than 1000 show exponential distribution along x-axis, which means similar distribution as the situation with gravity is achievable. For N is less than 1000, the distribution is relatively homogeneous. With greater number N, particles gather denser to the cool end. When the number N is 2000, nearly 40% of particles cluster in 5% of the total volume.

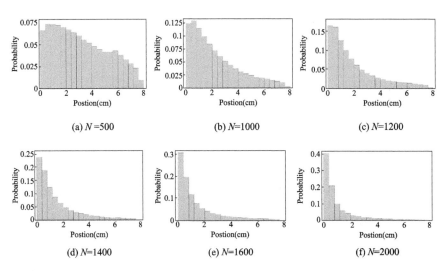

Fig. 16 Density distribution of particles in single cell with different numbers of particles (adapted from Wang et al. [1])

2.4.2 Flux Function Obtained in Virtual-Window Single-Cell Simulation

Finding flux function is necessary to understand the occurrence of the segregation. The segregation is governed by the following equation:

$$\frac{dN_i(t)}{dt} = -F(N_i(t)) + F(N - N_i(t)), \quad i = 1, 2, \tag{19}$$

where the $N_1(t)$ and $N_2(t)$ are the counts of particles in the two compartments at time t, and $F(N_i)$ is the number of particles per unit time at time t that flow from a compartment containing N_i particles to another compartment. Once knowing the function $F(N_i)$, $N_i(t)$ can be obtained from Eq. (19).

In order to characterize the segregation of particles in the two cells, a dimensionless parameter ε_i is introduced:

$$\varepsilon_i = (N_i - N/2)/N, \quad i = 1, 2. \tag{20}$$

Figure 17 shows a plot of simulation result how ε_i changes with time. It takes some time for the population $N_i(t)$ to reach a steady state. We call this time as the waiting time τ. At any time $\varepsilon_1(t) = -\varepsilon_2(t)$, and the absolute value of ε_i at the steady state is indicated as ε. $\varepsilon = 0$ means particles are equally populated in the two cells, and $|\varepsilon| = 0.5$ means all the particles gather in one compartment, i.e., fully segregated.

The flux function obtained by the virtual-window single-cell simulation method [66] is shown in Fig. 18a for three locations of the window. A non-monotonic function $F(N_i)$ guaranteed the occurrence of the segregation. A weakly position dependence of the flux function shown in Fig. 18a tells us that flux value is greater when the window position is moving away from the moving wall. Numerically solving the Eq. (19), from $N_i(t)$ we get the value $|\varepsilon_i(t)|$ and compare it with the simulation results, as shown in Fig. 18b. The comparison is qualitatively well except the time

Fig. 17 One example of simulation result of $\varepsilon_i(t)$ (adapted from Wang et al. [1])

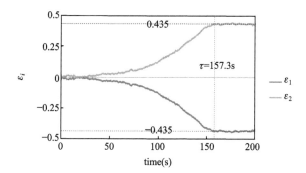

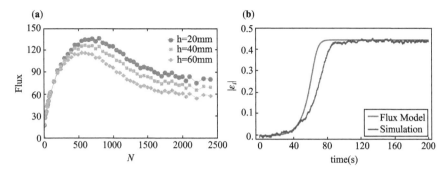

Fig. 18 **a** Flux as a function of the number N; **b** Comparison between the theoretical result from the flux model and the simulation result (adapted from Wang et al. [1])

during 60–80 s. The flux mode describes the process of the segregation in granular gas theoretically. But in numerical simulation or experiment, the fluctuation in the granular system may promote or delay the occurrence of Maxwell's demon.

2.4.3 The Effect of Total Number N on the Segregation ε

In the SJ-10 experiment, the segregation is observed at total particle number $N = 2400$, cell length $= 80$ mm, and with right-end shaking at frequency $f = 7$ Hz, amplitude $A = 2$ mm, (equivalently acceleration $\Gamma = 0.38g$). The vibration intensity is characterized by the acceleration $\Gamma = A_r (2\pi f_r)^2/g$, where g is the acceleration of gravity. In the simulation, we set amplitude A to be fixed and change frequency of the right piston to change the shaking acceleration Γ. Figure 19a–d are snapshots captured in the experiment at four different times. It shows the distribution segregation is fully developed at the time $t = 176$ s.

Using similar parameters, a simulation is performed and a comparison shows consistent results as seen in Fig. 19e–h. Initially particles are distributed equally in

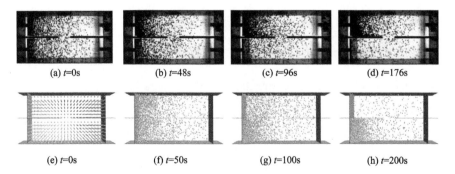

Fig. 19 **a–d** Snapshots captured in the experiments at different time; **e–h** Snapshots in the simulations taken at similar time (adapted from Wang et al. [1])

Fig. 20 Flux as a function of time (adapted from Wang et al. [1])

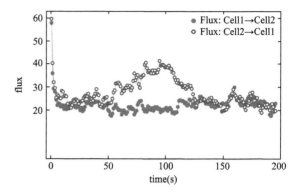

the two cells. Segregation appears after about 100 s, and is fully developed 200 s afterwards.

Figure 20 shows the dependence of the flux, i.e. the counts per unit time of particles flowing from one cell to the other, on time. The red (blue) dots in Fig. 20 show the flux from cell 1(2) to cell 2(1) along time. In the initial few seconds, the flow rates from cell 1(2) to cell 2(1) are quickly decreasing at about the same speed. Numbers of particles in two cells are almost equal, and the flux from one cell to the other is nearly equal to the flux of the opposite direction, namely the net flux is nearly zero. At some instant if cell 2 happens to have more particles flowing out than flowing in, less particles will be in cell 2 and less inelastic collisions will cause particles in cell 2 move faster on average than particles in cell 1. There will be a positive feedback to induce even more particles flow out from cell 2 to cell 1 until most of the particles are in cell 1 and almost no more particles can flow from cell 2 to cell 1. At this time the flux from either cell will be the same again, and the net flux will be zero again.

Next, we take the advantage of computer simulation to study the effect of total number of particles in the segregation conditions under no gravity.

Next we change the total number N from 800 to 1500. Figure 21a gives us the results how $\varepsilon_i(t)$ changes with N. When the total number N is less than or equal to 1100, the asymmetry ε fluctuate around 0. Segregation does not occur. Only when N exceeds a critical value (here the critical value is about 1100) the segregation occurs. Figure 21b shows that the asymmetry parameter ε grows from 0 and approaches 0.5 with increasing N.

2.4.4 Effect of Excitation Acceleration on the Segregation

Figure 22a gives the simulation results of the asymmetry parameter $|\varepsilon_i|$ changing with time under different Γ. For higher acceleration Γ, $|\varepsilon_i|$ reaches steady state value ε faster.

Since in microgravity experiments the operational time is normally limited, the waiting time τ is important to know in advance, which tells us how long to wait for seeing the effect. As shown in Fig. 22b, the rate $1/\tau$ seems linearly proportional to

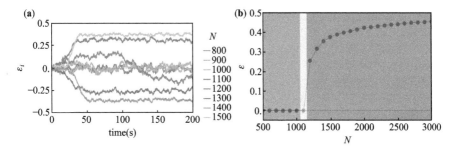

Fig. 21 **a** Variation of the asymmetry parameters ε under different N as a function of time; **b** Variation of the asymmetry parameter ε in the steady state as a function of N (adapted from Wang et al. [1])

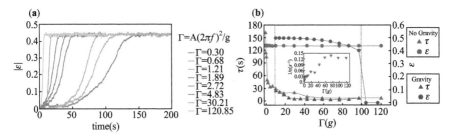

Fig. 22 **a** The asymmetry parameter $|\varepsilon|$ reaches the steady state after waiting time τ; **b** Effects of the vibration intensity on the asymmetry parameter ε and the waiting time τ under gravity and no gravity (adapted from Wang et al. [1])

Γ at low Γ. Only when Γ is high enough the rate $1/\tau$ reaches a maximum value. For the Γ value available in our microgravity experiment (normally less than 1 g), Fig. 22b tells us the time duration τ shall be of the order of 100 s. This result helps us in determining the experimental design.

In Fig. 22b, it shows that the waiting time τ does not go to zero, there is a minimum time required for particles to go to the steady segregation state no matter how strong the vibration intensity goes. In our case the value is of the order of 10 s.

The above numerical results are quite different with the experiments in gravity. With gravity particles act like gas molecules, and distribute in both cells. No segregation of the granular gas is observed at high Γ when particle random collisions dominate. When Γ is below a certain threshold, no segregation is observed either as the gravitational force dominates. With no gravity, the segregation ε becomes a constant and does not depend on Γ, as seen in Fig. 22b, but the waiting time τ depends strongly on Γ.

2.4.5 Effect of the Position of the Opening on the Segregation

The effect of the window position is investigated by changing the piston location. We change the length (volume) of the cell and the distance of the window from either side of the pistons. The window connecting the two compartments is located at a position x_l from the position of the left piston. The right piston is accelerated at 4.83 g (where $f_r = 20$ Hz and $A_r = 3$ mm). The effects on both the waiting time τ and the asymmetry parameter ε are shown in Fig. 23. The segregation efficiency is better when the window is closer to the cool-end window. As is shown in Fig. 23b, the value of ε changes from 0.44 to 0.36 when x_l changes from 10 to 35 mm. It means the segregation is more efficient when the window is closer to the cool-end. However, to reach the steady segregation we need to wait longer time when the window is closer to the cool-end as the time τ increases, as is seen in Fig. 23b.

The previous researches show that the window's position has influence on the occurrence of the segregation under gravity, namely that only when the window's position is closer enough to the bottom, the segregation can appear. Our numerical results in Fig. 24 also show the same result. When gravity does not play a role, the position effect to the segregation becomes weaker.

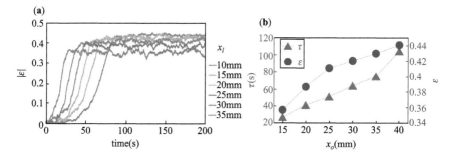

Fig. 23 **a** Variation of the asymmetry parameter $|\varepsilon_i|$ as a function of time; **b** The waiting time τ (red triangular dots) and the asymmetry parameter ε (blue dots) as a function of window position (adapted from Wang et al. [1])

Fig. 24 The asymmetry parameter ε as a function of opening position with and without gravity (adapted from Wang et al. [1])

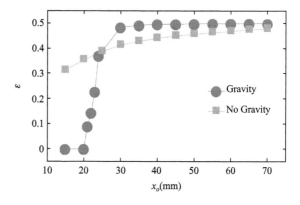

3 Summary

The microgravity-experimental and simulation results on the granular gas freely cooling process support Haff's law that the kinetic energy dissipates with time t as $E(t) \sim (1 + t/\tau)^{-2}$. The effects of particle number density and the particle size on τ, due to the rotation and friction during collision, are studied in simulation. A modified τ is given to take into account the dissipation due to frictions. The collision frequency affected by the number density and particle size is discussed. From the standard deviation of the velocity distribution we also verify the energy dissipation law, showing its good consistence with the Haff's Kinetic energy dissipation.

The investigation in two-connected-cells clustering shows granular segregation is achievable at zero gravity with a different condition from the case under gravity. With gravity, the segregation depends on Γ and can be divided into three regimes: one at low Γ when the gravitational force is dominant that no segregation appears and ε is zero; a second regime when Γ is at an intermediate value that segregation appears and ε becomes non-zero; and the third regime when Γ is high enough that random particle collisions dominate and ε becomes zero again. Under zero gravity, segregation ε does not depend on Γ. It is a constant weakly dependent on the position of the window. A minimum excitation time is necessary to observe the phenomenon with or without gravity. This time τ, however, depends strongly on the value Γ. With the affordable acceleration Γ in satellite condition, the excitation time is on the order of a few minutes, which means no segregation can be seen in drop tower or parabolic experiments [90] which offer microgravity condition only for a short time (3–22 s).

References

1. Wang W, Zhou Z, Zong J, Hou M (2017) Dem simulation of granular segregation in two-compartment system under zero gravity. Chin Phys B 26:044501
2. Wang W, Hou M, Chen K, Yu P, Sperl M (2018) Experimental and numerical study on energy dissipation in freely cooling granular gases under microgravity. Chin Phys B 27:084501
3. Jaeger HM, Nagel SR, Behringer RP (1996) Granular solids, liquids, and gases. Rev Mod Phys 68:1259
4. de Gennes PG (1999) Granular matter: a tentative view. Rev Mod Phys 71:S374
5. Kadanoff LP (1999) Built upon sand: theoretical ideas inspired by granular flows. Rev Mod Phys 71:435
6. Goldhirsch II, Zanetti G (1993) Clustering instability in dissipative gases. Phys Rev Lett 70:1619
7. McNamara S, Young WR (1994) Inelastic collapse in two dimensions. Phys Rev E 50:R28
8. Falcon É, Wunenburger R, Évesque P, Fauve S, Chabot C, Garrabos Y, Beysens D (1999) Cluster formation in a granular medium fluidized by vibrations in low gravity. Phys Rev Lett 83:440
9. Luding S, Herrmann HJ (1999) Cluster-growth in freely cooling granular media. Chaos 9:673
10. Painter B, Dutt M, Behringer RP (2003) Energy dissipation and clustering for a cooling granular material on a substrate. Phys D 175:43
11. Miller S, Luding S (2004) Cluster growth in two- and three-dimensional granular gases. Phys Rev E 69:031305

12. Efrati E, Livne E, Meerson B (2005) Hydrodynamic singularities and clustering in a freely cooling inelastic gas. Phys Rev Lett 94:088001
13. Meerson B, Puglisi A (2005) Towards a continuum theory of clustering in a freely cooling inelastic gas. Europhys Lett 70:478
14. Aranson IS, Tsimring LS (2006) Patterns and collective behavior in granular media: theoretical concepts. Rev Mod Phys 78:641
15. Fingerle A, Herminghaus S (2006) Unclustering transition in freely cooling wet granular matter. Phys Rev Lett 97:078001
16. Yu P, Frank-Richter S, Börngen A, Sperl M (2014) Monitoring three-dimensional packings in microgravity. Granul Matter 16:165
17. Chen Y, Pierre E, Hou M (2012) Breakdown of energy equipartition in vibro-fluidized granular media in micro-gravity. Chin Phys Lett 29:074501
18. Mei Y, Chen Y, Wang W, Hou M (2016) Experimental and numerical study on energy dissipation in freely cooling granular gases under microgravity. Chin Phys B 25:084501
19. Wang Y (2017) Granular packing as model glass formers. Chin Phys B 26:014503
20. Haff PK (1983) Grain flow as a fluid-mechanical phenomenon. J Fluid Mech 134:401
21. Brito R, Ernst MH (1998) Extension of Haff's cooling law in granular flows. Europhys Lett 43:497
22. Luding S, Huthmann M, McNamara S, Zippelius A (1998) Zippelius, homogeneous cooling of rough, dissipative particles: theory and simulations. Phys Rev E 58:3416
23. Garzó V, Dufty J (1999) Homogeneous cooling state for a granular mixture. Phys Rev E 60:5706
24. Huthmann M, Orza JAG, Brito R (2000) Dynamics of deviations from the gaussian state in a freely cooling homogeneous system of smooth inelastic particles. Granul Matter 2:189
25. Zaburdaev VY, Brinkmann M, Herminghaus S (2006) Free cooling of the one-dimensional wet granular gas. Phys Rev Lett 97:018001
26. Hayakawa H, Otsuki M (2007) Long-time tails in freely cooling granular gases. Phys Rev E 76:051304
27. Meerson B, Fouxon I, Vilenkin A (2008) Nonlinear theory of nonstationary low mach number channel flows of freely cooling nearly elastic granular gases. Phys Rev E 77:021307
28. Santos A, Montanero JM (2009) The second and third sonine coefficients of a freely cooling granular gas revisited. Granul Matter 11:157
29. Kolvin I, Livne E, Meerson B (2010) Navier-Stokes hydrodynamics of thermal collapse in a freely cooling granular gas. Phys Rev E 82:021302
30. Mitrano PP, Garzo V, Hilger AM, Ewasko CJ, Hrenya CM (2012) Assessing a hydrodynamic description for instabilities in highly dissipative, freely cooling granular gases. Phys Rev E 85:041303
31. Brilliantov NV, Pöschel T(2000)Velocity distribution in granular gases of viscoelastic particles. Phys Rev E 61:5573
32. McNamara S, Luding S (1998) Energy nonequipartition in systems of inelastic, rough spheres. Phys Rev E 58:2247
33. Ben-Naim E, Chen SY, Doolen GD, Redner S (1999) Shock like dynamics of inelastic gases. Phys Rev Lett 83:4069
34. Nie X, Ben-Naim E, Chen S (2002) Dynamics of freely cooling granular gases. Phys Rev Lett 89:204301
35. van Zon JS, MacKintosh FC (2004) Velocity distributions in dissipative granular gases. Phys Rev Lett 93:038001
36. Shinde M, Das D, Rajesh R (2007) Violation of the porod law in a freely cooling granular gas in one dimension. Phys Rev Lett 99:234505
37. Shinde M, Das D, Rajesh R (2009) Equivalence of the freely cooling granular gas to the sticky gas. Phys Rev E 79:021303
38. Shinde M, Das D, Rajesh R (2011) Coarse-grained dynamics of the freely cooling granular gas in one dimension. Phys Rev E 84:031310
39. Villemot F, Talbot J (2012) Homogeneous cooling of hard ellipsoids. Granul Matter 14:91

40. Pathak SN, Das D, Rajesh R (2014) Inhomogeneous cooling of the rough granular gas in two dimensions. Europhys Lett 107:44001
41. Pathak SN, Jabeen Z, Das D, Rajesh R (2014) Energy decay in three-dimensional freely cooling granular gas. Phys Rev Lett 112:038001
42. Maaß CC, Isert N, Maret G, Aegerter CM (2008) Experimental investigation of the freely cooling granular gas. Phys Rev Lett 100:248001
43. Tatsumi S, Murayama Y, Hayakawa H, Sano M (2009) Experimental study on the kinetics of granular gases under microgravity. J Fluid Mech 641:521
44. Burton JC, Lu PY, Nagel SR (2013) Energy loss at propagating jamming fronts in granular gas clusters. Phys Rev Lett 111:188001
45. Harth K, Trittel T, Wegner S, Stannarius R (2018) Competitive clustering in a bidisperse granular gas. Phys Rev Lett 120:214301
46. Berry MV, Geim AK (1997) Of flying frogs and levitrons. Eur J Phys 18:307
47. Luding S (2008) Cohesive, frictional powders: contact models for tension. Granul Matter 10:235
48. Opsomer E, Ludewig F, Vandewalle N (2012) Dynamical clustering in driven granular gas. Europhys Lett 99:40001
49. Clément E, Duran J, Rajchenbach J (1992) Experimental study of heaping in a two-dimensional sand pile. Phys Rev Lett 69
50. Jia LC, Lai PY, Chan CK (1999) Empty Site models for heap formation in vertically vibrating grains. Phys Rev Lett 83:3832
51. Lai PY, Jia LC, Chan CK (2000) Symmetric heaping in grains: a phenomenological model. Phys Rev E 61:5593
52. Hong DC, Quinn PV, Luding S (2001) Reverse Brazil nut problem: competition between percolation and condensation. Phys Rev Lett 86:3423
53. Breu APJ, Ensner HM, Kruelle CA, Rehberg I (2003) Reversing the Brazil-Nut effect: competition between percolation and condensation. Phys Rev Lett 90:014302
54. Schlichting HJ, Nordmeier V (1996) Math. Naturwiss. Unterr. 49323 (in German)
55. Eggers J (1999) Sand as Maxwell's Demon. Phys Rev Lett 83:5322
56. van der Weele K, van der Meer D, Versluis M, Lohse D (2001) Hysteretic clustering in granular gas. Europhys Lett 53:328
57. van der Meer D, Reimann P, van der Weele K, Lohse D (2004) Spontaneous ratchet effect in a granular gas. Phys Rev Lett 92:184301
58. van der Meer D, van der Weele K, Lohse D (2001) Bifurcation diagram for compartmentalized granular gases. Phys Rev E 63:061304
59. Meerson B, Pöschel T, Sasorov PV, Schwager T (2004) Giant fluctuations at a granular phase separation threshold. Phys Rev E 69:021302
60. Costantini G, Paolotti D, Cattubo C, Marconi UMB (2005) Bistable clustering in driven granular mixtures. Phys A 347:411–428
61. Lambiotte R, Salazar JM, Brenig L (2005) From particle segregation to the granular clock. Phys Lett A 343:224–230
62. Liu R, Li Y, Hou M (2007) Van der Waals-like phase-separation instability of a driven granular gas in three dimensions. Phys Rev E 75:061304
63. Liu R, Li Y, Hou M (2009) Oscillatory phenomena of compartmentalized bidisperse granular gases. Phys Rev E 79:052301
64. Hou M, Li Y, Liu R, Zhang Y, Lu K (2010) Oscillatory clusterings in compartmentalized granular systems. Phys Status Solidi A 207:2739–2749
65. Li Y, Hou M, Evesque P (2011) Directed clustering in driven compartmentalized granular gas systems in zero gravity. J Phys: Conf Ser 327:012034
66. Li Y, Liu R, Shinde M, Hou M (2012) Flux measurement in compartmentalized mono-disperse and bi-disperse granular gases. Granular Matter 14:137–143
67. Li Y, Liu R, Hou M (2012) Gluing bifurcation and noise-induced hopping in the oscillatory phenomena of compartmentalized bidisperse granular gases. Phys Rev Lett 109:198001

68. Shah SH, Li Y, Cui F, Evesque P, Hou M (2012) Irregular oscillation of bi-disperse granular gas in cyclic three compartments. Chin Phys Lett 29:034501
69. Bray JJ, Moreno F, García-Rojo R, Ruiz-Montero MJ (2001) Hydrodynamic Maxwell Demon in granular systems. Phys Rev E 65:011305
70. Lipowski A, Droz M (2002) Urn model of separation of sand. Phys Rev E 65:031307
71. Marconi UMB, Conti M (2004) Dynamics of vibrofluidized granular gases in periodic structures. Phys Rev E 69:011302
72. Mikkelsen R, van der Meer D, van der Weele K, Lohse D (2004) Competitive clustering in a bidisperse granular gas: experiment, molecular dynamics, and flux model. Phys Rev E 70:061307
73. Leconte M, Evesque P (2006) Maxwell demon in granular gas: a new kind of bifurcation? The hypercritical bifurcation. Physics 0609204
74. Lai P, Hou M, Chan C (2009) Granular gases in compartmentalized systems. J Phys Soc Jpn 78:041001
75. Mikkelsen R, van der Meer D, van der Weele K, Lohse D (2002) Competitive clustering in a bidisperse granular gas. Phys Rev Lett 89:214301
76. van der Meer D, van der Weele K, Lohse D (2002) Sudden collapse of a granular cluster. Phys Rev Lett 88:174302
77. Jean P, Bellenger H, Burban P, Ponson L, Evesque P (2002) Phase transition or Maxwell's Demon in granular gas. Pourders Grains 13:27–39
78. Mikkelsen R, van der Weele K, van der Meer D, van Hecke M, Lohse D (2005) Small-number statistics near the clustering transition in a compartmentalized granular gas. Phys Rev E 71:041302
79. Miao T, Liu Y, Miao F, Mu Q (2005) On cyclic oscillation in granular gas. Chin Sci Bull 50:726–730
80. Viridi S, Schmick M, Markus M (2006) Experimental observations of oscillations and segregation in a binary granular mixture. Phys Rev E 74:04130
81. Hou M, Tu H, Liu R, Li Y, Lu K (2008) Temperature oscillations in a compartmentalized bidisperse granular gas. Phys Rev Lett 100:068001
82. Hou M, Liu R, Zhai G, Sun Z, Lu K, Garrabos Y, Evesque P (2008) Velocity distribution of vibration-driven granular gas in knudsen regime in microgravity. Microgravity Sci Technol 20:73–80
83. Isert N, Maaß CC, Aegerter CM (2009) Influence of gravity on a granular Maxwell's Demon experiment. Eur Phys J E 28:205–210
84. Shah SH, Li Y, Cui F, Hou M (2012) Directed segregation in compartmentalized bi-disperse granular gas. Chin Phys B 21:014501
85. Zhang Y, Li Y, Liu R, Cui F, Evesque P, Hou M (2013) Imperfect pitchfork bifurcation in asymmetric two-compartment granular gas. Chin Phys B 22:054701
86. Opsomer E, Noirhomme M, Vandewalle N, Ludewig F (2013) How dynamical clustering triggers Maxwell's Demon in microgravity. Phys Rev E 88:012202
87. Noirhomme M, Opsomer E, Vandewalle N, Ludewig F (2015) Granular transport in driven granular gas. Eur Phys J E38
88. Luding S (2008) Cohesive, frictional powders: contact models for tension. Granular Matter 10:235–246
89. Dintwa E, Tijskens E, Ramon H (2007) On the accuracy of the hertz model to describe the normal contact of soft elastic spheres. Granular Matter 10:209–221
90. Wang H, Chen Q, Wang WG, Hou MY (2016) Experimental study of clustering behaviors in granular gases. Acta Phys Sin 65:014502

Thermal Dynamics of Growing Bubble and Heat Transfer in Microgravity Pool Boiling

Wangfang Du, Jianfu Zhao, Huixiong Li, Yonghai Zhang, Jinjia Wei and Kai Li

Abstract Boiling heat transfer realizes the high-performance heat exchange due to latent heat transportation, and then there are extensive industrial applications on Earth and many potential applications in space. Microgravity experiments offer a unique opportunity to study the complex interactions without external forces, and can also provide a means to study the actual influence of gravity on the pool boiling by comparing the results obtained from microgravity experiments with their counterparts in normal gravity. It will be conductive to revealing of the mechanism underlying the phenomenon, and then developing of more mechanistic models for the related applications both on Earth and in space. The present chapter summarize the up-to-date progress on the understanding of pool boiling phenomenon based on the knowledge obtained from microgravity experiments, focusing particularly on the thermal dynamics of growing bubble and heat transfer in microgravity pool boiling. The gravity scaling behavior, as well as the passive enhancement of heat transfer

W. Du · J. Zhao (✉) · K. Li
CAS Key Laboratory of Microgravity (National Microgravity Laboratory/CAS),
Institute of Mechanics, Chinese Academy of Sciences (CAS), 15 Beisihuanxi Road,
Haidian District, Beijing 100190, China
e-mail: jfzhao@imech.ac.cn

J. Zhao · K. Li
School of Engineering Science, University of Chinese Academy of Sciences (UCAS),
19A Yuquan Road, Shijingshan District, Beijing 100049, China

H. Li · J. Wei
State Key Laboratory of Multiphase Flow in Power Engineering, Xi'an Jiaotong University,
28 Xianning Xilu, Xi'an 710049, China

H. Li
School of Energy and Power Engineering, Xi'an Jiaotong University,
28 Xianning Xilu, Xi'an 710049, China

Y. Zhang · J. Wei
School of Chemical Engineering and Technology, Xi'an Jiaotong University,
28 Xianning Xilu, Xi'an 710049, China

© Science Press and Springer Nature Singapore Pte Ltd. 2019
W. R. Hu and Q. Kang (eds.), *Physical Science Under Microgravity: Experiments on Board the SJ-10 Recoverable Satellite*, Research for Development,
https://doi.org/10.1007/978-981-13-1340-0_4

73

performance of nucleate pool boiling on flat plates by using micro-pin-finned surface, is presented and discussed in detail. Based on the outcome of the current trends in pool boiling research, some recommendations for future work are also proposed.

Keywords Microgravity · Pool boiling · Bubble dynamics · Heat transfer · Gravity scaling law

1 Introduction

Boiling heat transfer realizes the high-performance heat exchange due to latent heat transportation, and then there are extensive industrial applications on Earth and many potential applications in space. It is, however, also a very complex and illusive process because of the interrelation of numerous factors and effects. Such factors and effects include the nucleate process, the growth of the bubbles, the interaction between the heater's surface with liquid and vapor, the evaporation process at the liquid-vapor interface, the transport process of vapor and hot liquid away from the heater's surface, and so on. Furthermore, adding to the complexity is the randomness of the distribution and the configuration of the activated nucleation sites, around which bubbles continue to form, grow up, depart off and move on. Some macro-scale statistical average parameters are then commonly used in the boiling study to fit the needs of engineering endeavors rather than to focus upon the physics of the boiling process. As a result, our present knowledge on boiling phenomenon has been built with the aid of numerous meticulous experiments in normal gravity on Earth where gravity is a dominant factor because of large density difference between the liquid and vapour phases. The literature on boiling research has then been flooded with empirical correlations and semi-mechanistic models involving several adjustable, empirical parameters. These empirical correlations and/or semi-mechanistic models can provide quick input to design, performance, and safety issues and hence are attractive on a short-term basis. However, the usefulness of them diminishes very quickly as parameters of interest start to fall outside the range of physical parameters for which the empirical correlations and/or semi-mechanistic models were developed. In particular, although many empirical correlations and/or semi-mechanistic models include gravity as a parameter, they usually fail when extended beyond the range of gravity levels, usually the sole $1g_0$ (here g_0 denotes the normal gravity on Earth) condition, they were based on. Thus, the physics of the boiling process itself is not properly understood yet, and is poorly represented in the most of empirical correlations and/or semi-mechanistic models, despite almost seven decades of boiling research.

This chapter focuses upon the so-called pool boiling phenomenon, in which the liquid is essentially quiescent and, in normal gravity on the ground, vapor bubbles rise as a result of buoyancy forces induced by gravity. It is, thus, well known that gravity strongly affects pool boiling phenomenon in the environment of normal gravity by creating forces in the systems that drive motions, shape boundaries, and compress fluids. Buoyancy dominates the bubble dynamics, which undermines the phase change heat transfer process and local convection feature near the heater surface, particularly in the vicinity of the liquid-vapor-solid three-phase contact line, which restricts the theoretical development of boiling heat transfer. Advances in the understanding of pool boiling phenomenon have been greatly hindered by masking effect of gravity.

Microgravity experiments offer a unique opportunity to study the complex interactions without external forces, such as buoyancy, which can affect the bubble dynamics and the related heat transfer. In microgravity, gravity effect is greatly weakened and even disappears, and some processes related to surface or interface which are ever undermined in normal gravity become very prominent. The driving force is weakened to make bubbles depart from the heater surface, and the effect of flow and heat transfer near the heater surface becomes much more prominent than that in normal gravity. The micro flow and heat transfer feature become clear near the liquid-vapor-solid three-phase contact line by excluding the buoyancy effect, which is convenient for studying the heat transfer mechanism deeply. Therefore, pool boiling in microgravity has become an increasing significant subject for investigation.

In addition, comparing the results obtained from microgravity experiments with their counterparts in normal gravity, it can also provide a means to study the actual influence of gravity on the pool boiling. Therefore, the microgravity researches will be conductive to revealing of the mechanism underlying the phenomenon, and then developing of more mechanistic models for the related applications both on Earth and in space.

Research on pool boiling heat transfer in microgravity has a history of more than 50 years with a short pause in the 1970s, and then has been advanced with the development of various microgravity facilities and with increased experimental opportunities, especially in the last three decades. On the progress in this field, many comprehensive reviews and monographs are available now. Among many others, Straub [1], Di Marco [2], Ohta [3], Kim [4, 5], and Zhao [6] summarized the experimental and theoretical works all over the world. Here, we focus particularly on the thermal dynamics of growing bubble and heat transfer of pool boiling in microgravity, and discuss in detail on the gravity scaling of bubble behaviour and heat transfer. The enhancement of heat transfer performance of nucleate pool boiling on flat plates with a passive method by using micro-pin-finned surface is also presented and discussed.

2 Pool Boiling Curve

A number of investigators have observed different regions of pool boiling heat transfer. The common-accepted pool boiling curve following Nukiyama [7] is shown, for example, in Fig. 1 for saturated water at atmospheric pressure. It incorporates a number of additional features that have been identified by later investigators. The heat flux q'' from the heater surface to the working fluid is plotted against the wall superheat $\Delta T_W = T_W - T_{SAT}$, where T_W and T_{SAT} denote the temperature at the

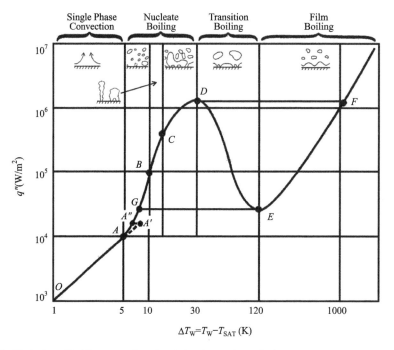

Fig. 1 Typical pool boiling curve for saturated water at atmospheric pressure

heater surface and the saturation temperature at the system pressure P, respectively. The nature of bubbles or a vapor film surrounding the heater in different regions is also depicted in Fig. 1.

In the first region $O-A/A'$ shown in Fig. 1 with a small ΔT_W, heat is transferred by single phase convection, or natural convection of liquid phase in normal gravity environment. The heat fluxes in the region of single phase convection are drastically reduced in microgravity compared with those in normal gravity due to great weakening of the gravity effect.

At location A' corresponding to a certain value of $\Delta T_{W, ONB}$, bubble nucleation is initiated on the cavities present on the heater surface, which is called onset of nucleate boiling (ONB). With the inception of nucleation, the heater surface temperature drops to A'' for a given imposed heat flux, which is termed the hysteresis effect of the onset of nucleate boiling. Beyond the ONB point, the slope of the curve increases at higher ΔT_W values as the bubbles grow and depart more rapidly in the partial nucleate boiling region (region $A/A''-B$). As heat flux is increased, more nucleation sites become active and fully developed nucleate boiling ensues (regions $B-C$). The slope of the boiling curve becomes to decrease and the heat transfer coefficient begins to decrease because the intense evaporation under the overcrowded bubble at higher heat fluxes leads to periodic dry patches on the heater surface that can be rewetted by the surrounding liquid. As the wall superheat increases further, liquid is unable to rewet the heater surface, causing a sudden formation of a dry patch that eventually covers a large region of the heater surface. A thin film of vapor eventually separates the liquid from the heater surface and leads to a large temperature excursion of the heater surface (drastic reduction in the heat transfer coefficient). The so-called the critical heat flux (CHF), q''_{CHF}, corresponding to this condition at location D represents the maximum heat flux sustained under the nucleate boiling condition. This is an unstable condition and the heater temperature jumps to point F under any small increase of the heat flux. The ensuing mode of heat transfer in which the heater is blanketed with a thin vapor film is called film boiling.

The region D–E following the CHF represents the transition boiling region, in which the heat transfer is associated with formation followed by rewetting of the dry patches in rapid successions. The transition boiling region is not accessible with the heat flux-controlled method, such as the heating with an electrical heater that imposes a constant heat flux boundary condition. It can only be traced under stable conditions by employing a constant temperature boundary condition on the heated wall. As the heater surface temperature is increased, eventually the rewetting cannot be sustained at the so-called Leidenfrost condition represented by E, and the heater is surrounded by a stable vapor film, resulting in the transition to the film boiling. The heat flux at E is called the minimum heat flux (MHF), q''_{MHF} under the film boiling condition.

Finally, the region E–F is the film boiling region in which heat is transferred by combined radioactive and convective modes across the thin vapor film. With the heat flux-controlled method, another hysteresis effect, shown as the curve $GBCDFEG$, can also be observed.

There are several factors, such as the system pressure P, the bulk liquid temperature T_L or the subcooling $\Delta T_{SUB} = T_{SAT} - T_L$, the gravity, and so on, affecting greatly the characteristics of heat transfer and then the boiling curves in pool boiling phenomenon. The major progress on the understanding of pool boiling, particularly relevant to the influence of the gravity on nucleate pool boiling, will be presented and discussed in the following sections.

3   Onset of Nucleation Boiling

As mentioned above, the onset of nucleate boiling (ONB) refers to the transition of heat transfer mode from the single-phase liquid convection to a combination of convection and nucleate boiling. In pool boiling, it is identified by the formation of vapor bubbles on the heated wall in a pool of liquid, which is termed "heterogeneous nucleation". Oppositely, the formation of a vapor bubble completely inside a superheated bulk liquid mass is termed "homogeneous nucleation". Theoretically, the upper limit on the superheat for homogeneous nucleation within a liquid mass at constant pressure is very high and equal to the spinodal limit that results from thermodynamic consideration [8].

The theoretical value of the superheat required for facilitating the heterogeneous nucleation from an atomically smooth surface has been estimated to be very high,

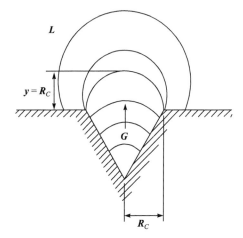

Fig. 2 Growth of a vapor bubble nucleating from a cavity

and often close to the homogeneous nucleation limit. Most experiments, however, reported a significantly lower value. This is usually explained by the fact that the heater surfaces used in practical applications are far from atomically smooth and/or the fluid is contaminated, resulting in that the surfaces are not completely wet by the liquid and there is always some entrapment of vapor/gas in the cavities or around the contaminant [9].

Small cavities trap vapor/gas and act as nucleation sites for bubbles on the heater surface. As the heater surface temperature exceeds the saturation temperature of the working fluid at the corresponding system pressure, a bubble may grow inside the cavity and appear at its mouth, as shown in Fig. 2 where L and G denote the liquid and vapor/gas phases, respectively. A certain value of the wall superheat is needed to activate the cavities depending on various factors: (1) cavity size and shape; (2) fluid properties, including surface tension and contact angles; and (3) temperature profile in the liquid immediately surrounding the heater surface. It is generally expected that the inception superheat or the superheat of ONB is independent of the gravity.

Griffith and Wallis [10] proposed firstly that incipient superheat for boiling from pre-existing nuclei corresponds to the minimum radius of curvature of the interface. The minimum radius of curvature of the interface was assumed to be equal to the radius of the cavity mouth. By replacing the pressure difference between the vapor bubble (no gas) and liquid with the liquid superheat through the use of the Clausius-Clapeyron equation, they obtained an expression for inception superheat as

$$\Delta T_{\text{W, ONB}} = \frac{2\sigma T_{\text{SAT}}}{\rho_V h_{LV} R_c} \tag{1}$$

here σ, ρ_V, h_{LV}, and R_c denote the surface tension, the vapor density, the latent heat of liquid-vapor phase change, and the cavity mouth radius, respectively. It should be noted that Eq. (1) includes neither the effect of contact angle on inception superheat nor the effect of temperature gradient that exists near a heated wall. Subsequently, Hsu and Graham [11] and Hsu [12] studied the effect of temperature profile adjacent to the heated surface on the minimum superheat needed for nucleation. In developing his model, Hsu [12] proposed that the top of a bubble embryo should be covered with warm liquid before it can grow. Since vapor in the embryo must be at saturation temperature corresponding to the pressure of vapor in the bubble (which is higher than the pool pressure by $2\sigma/R$), the liquid surrounding the bubble must be superheated to maintain the thermal equilibrium. If the required superheat does not exist, the heat transfer to colder liquid will cause the bubble embryo to shrink. Because heat is transferred from the wall, the liquid temperature decreases with distance from the wall, and the above criterion is satisfied everywhere around the embryo, if the temperature of the liquid at the tip of the embryo is equal to the saturation temperature corresponding to pressure in the bubble.

A general criterion of the range of the active nucleation cavities, according to Davis and Anderson [13], can be deduced as

$$\{R_{c,\,min},\, R_{c,\,max}\} = \frac{f_1\delta_t}{2f_2}\left(1 \mp \sqrt{1 - \frac{8\sigma T_{SAT}f_2}{\rho_v h_{LV}\delta_t\Delta T_W}}\right) \tag{2}$$

In Eq. (2), the minimum and maximum cavity radii $R_{c,\,min}$ and $R_{c,\,max}$ are obtained from the negative and positive signs of the radical, respectively. $f_1 = \sin(\theta + \alpha_c)$ and $f_2 = 1 + \cos(\theta + \alpha_c)$, in which θ and α_c denote respectively the contact angle and the cavity mouth angle, while δ_t denotes the thickness of the thermal boundary layer before ONB.

Different investigators have used different models to relate the bubble radius R_b or height y_b to the cavity radius R_c and to the location where the liquid temperature is determined. Hsu and Graham [11] and Hsu [12] assumed that $y_b = 2R_c$, which effectively translates into a contact angle of $\theta = 53.1°$. Bergles and Rohsenow [14] and Sato and Matsumura [15] considered a hemispherical bubble at the nucleation inception with $y_b = R_c$, namely $\theta = 90°$.

Based on Eq. (2), a graph showing the active range of nucleation cavities for different thermal boundary layer thicknesses at saturated temperature is plotted for water under atmospheric pressure in Fig. 3. It is evident that a certain value of superheat is required before any cavity becomes active, corresponding to ONB (black dots marked in Fig. 3).

For a given cavity radius R_c of the nucleate site, the criterion, in term of the inception superheat, can be re-written as [12]

$$\Delta T_{W,\,ONB} = \frac{2f_1\sigma T_{SAT}}{\rho_v h_{LV}R_c}\bigg/\left(1 - \frac{f_2R_c}{f_1\delta_t}\right) \tag{3}$$

For R_c much smaller than δ_t, the wall superheat varies inversely with size of a nucleating cavity, as was the case for Eq. (1). As an alternative to Hsu's criterion, Wang and Dhir [16] proposed that the instability of vapor nuclei in a cavity determines the inception superheat, namely

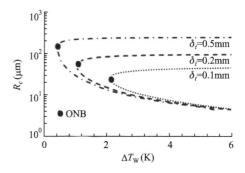

Fig. 3 Active range of nucleation cavities for different thermal boundary layer thickness during saturated boiling of water under atmospheric pressure

$$\Delta T_{W,\,ONB} = \frac{2\sigma T_{SAT}}{\rho_V h_{LV} R_c} K_{max} \qquad (4)$$

where $K_{max} = 1$ for $\theta < 90°$, while $K_{max} = \sin\theta$ for $\theta > 90°$. The above expression is obtained under the assumption that cavity radius, R_c, is much smaller than the thermal layer thickness. Thus, in this limit, Eq. (4) suggests that f_1 in Eqs. (2) and (3) is unity for $\theta < 90°$ and is equal to $\sin\theta$ for $\theta > 90°$. Or, for non-wetted surfaces, the required superheat is smaller than that given by Eq. (1) of Griffith and Wallis [10]. Wang and Dhir [16], from their experiments on surfaces with different contact angles, have shown the general validity of Eq. (4).

Straub [1] studied systematically the onset of boiling both in normal and microgravity conditions using wires as heating elements. The study demonstrates, as is generally expected, that the inception superheat is more or less independent of the gravity and of the subcooling if the overheating due to the saturation state is regarded. The inception superheat depends on the saturation state and decreases with increasing system pressure, which is consistent with the predictions of nucleation theory, for example, Eqs. (1), (3), and (4) mentioned above. Similar conclusions have also been obtained experimentally both for wire heater [17] and for flat plate heater [18]. Therefore, Zhang et al. [19, 20] and Li et al. [21] adopted Eq. (3) as the criterion for determining the beginning of the subsequent bubble cycle in the numerical simulation of single bubble pool boiling.

Contrary to the inception superheat, the criterion of ONB in term of the inception heat flux will be exhibited a great difference between normal and microgravity. Generally, the cooling of single phase natural convection before ONB in normal gravity can be so efficient that at high subcooling the inception superheat will not be attained. Thus, to attain the inception of pool boiling, a much higher heat flux will be needed in normal gravity than that in microgravity.

4 Bubble Dynamics During Nucleation Pool Boiling

After ONB, nucleate boiling ensues which is characterized, particularly in the partial nucleate boiling at low heat fluxes, by the appearance of vapor bubbles at discrete locations on the heater surface. A bubble continues to grow until forces causing it to detach from the surface exceed those pushing the bubble against the wall. After departure, liquid from the bulk fills the space vacated by the bubble, and the thermal layer at and around the nucleation site reforms. When the required critical superheat is attained, a new bubble starts to form at the same site, and the process repeats. It is called ebullition cycle, including the following four stages: bubble nucleation, growth period, bubble departure, and waiting period. It is believed that the quasi-periodic cycle of bubble growth, departure, and waiting processes plays a fundamental role in nucleate pool boiling.

The growth period is generally divided into inertia-controlled and thermally controlled regions. Inertia-controlled growth occurs in the first very short period of bubble growth, in which the bubble radius is proportional to the time, namely $R_B \sim t$. Superheated liquid provides the thermal energy for rapid evaporation at the interface in this region, but the bubble growth is limited by the inertia of the surrounding liquid. Subsequent to the initial rapid growth, evaporation is limited by transient conduction heat transfer. Thus, bubble growth is thermally controlled, which is commonly observed in pool boiling phenomenon.

Analytical solutions for transient heat conduction in the case of uniformly heated liquid over a spherical bubble are generally of the following form

$$R_B(t) = C(\alpha t)^{1/2} \tag{5}$$

where α is the thermal diffusivity of the liquid, and C is a function of the thermo-physical properties and the wall superheat, being mostly expressed with an empirical or analytical factor and the Jakob number

$$Ja = \frac{\rho_L c_{p,L} \Delta T_W}{\rho_V h_{LV}} \tag{6}$$

here c_p is the specific heat at constant pressure. Scriven [22], for example, proposed for small superheat $C = (2Ja)^{1/2}$ and for large superheat $C = (12/\pi)^{1/2} Ja$, which corresponds to the asymptotic solution of Plesset and Zwick [23]. Ground experimental observations, for example, Dergarabedian [24], among many others, of the growth of vapor bubbles in superheated water and the calculations using the Plesset-Zwick method agree quite well, although the analytical model can only be valid under the idealized assumptions of the boundary conditions which were realizable only in microgravity. Straub [1] studied the bubble growth experimentally and numerically in microgravity in drop towers and space shuttle experiments, where a bubble was initiated by a heat pulse and grew in an overall supersaturated liquid that was generated by a pressure drop. A mean values of $C = 2.03Ja$ and for the exponent 0.43 are obtained, confirming that the analytical model of Plesset and Zwick [23] and Scriven [22] is the best to describe bubble growth in a uniformly heated liquid, or bubble growth after homogeneous nucleation.

The above analytical model should not be applied to the growth of bubbles in boiling, because these bubbles grow in a thermal layer in which strong temperature gradient exists, and the bubbles, attached to the heater wall, are not spherical. By incorporating the shape of a bubble growing in the vicinity of a heated surface, Mikic et al. [25] derived the following equation, which covers the entire growth cycle, including the inertia-controlled region for small values of t and the thermally controlled region for large values of t.

$$R^+ = \frac{2}{3}\left[\left(t^+ + 1\right)^{3/2} - \left(t^+\right)^{3/2} - 1\right] \tag{7}$$

where

$$R^+ = \frac{A}{B^2} R_B(t), \ t^+ = \frac{A^2}{B^2} t, \ A = \left(\frac{2\rho_V h_{LV} \Delta T_W}{\rho_L T_{SAT}}\right)^{1/2}, \ B = \left(\frac{12\alpha_L}{\pi}\right)^{1/2}\left(\frac{2\rho_L c_{p,L} \Delta T_W}{\rho_V h_{LV}}\right).$$

Equation (7) indicates that the exponent will decrease quickly from 1 to 0.5 during bubble growth.

Straub and his colleagues observed the growth of a single bubble on a flat gold-coated heater in saturated R113 in microgravity [1]. A values of $C = 0.42Ja$ and for the exponent 0.555 are obtained. They found that the bubble grew to a radius of about 15 mm, considerably farther than the extension of the thermal layer reached, and therefore the growth rate became slower than the prediction by the analytical model of growth in a uniformly heated liquid. However, 1/2-power law of bubble growth may be available only for saturated boiling. A mean exponent of 1/3 was observed under subcooled conditions from TEXUS. The exponent depends on the bubble size and decreases with the growing bubble from 0.527 for small bubbles with $R < 1.5$ mm, up to 0.286 for $R \le 3.5$ mm, where R is the equivalent radius of a spherical bubble calculated from the measured volume. Wan et al. [26] also observed that the exponent changes from 1/3 at the beginning with $R < 1$ mm to 1/5 for $R > 1$ mm for a single bubble growing in subcooled R113 in microgravity. For larger bubble in subcooled liquid, its growth stops when the condensation mass flow at the top reaches the evaporating one at the base. Marangoni effect, caused by un-even distribution of temperature along the interface, may affect the bubble behaviors, particularly in large subcooled boiling in microgravity. Li et al. [27] studied the growth of a single bubble on a smooth surface of a self-heating silicon chip in gas-saturated FC-72 in short-term microgravity in the Drop Tower Beijing. It is found that the bubble growth during the early period before $t = 0.04$ s can be described by $R_B = k \cdot t^{1/2}$ with an empirical value $k = 5.6$, which is consistent with the model based on classical thermal-controlled mechanism, and the empirical parameter locates in the range reported in the literature. The growth rate decreases quickly after $t = 0.04$ s, to 0 at $t = 0.1$ s, and even exhibits a slightly negative rate later.

Recently, a single bubble pool boiling on a plain plate was experimentally studied in microgravity aboard the Chinese recoverable satellite SJ-10 [28, 29]. With the benefit of frame by frame playback of the recorded video images, an axisymmetric isolated bubble is observed at the center of the top surface immediately after the activation of the bubble trigger. The bubble grows quickly both in radial and axial directions within the first 2 s, and in approximately unchanged shape in the following 6 s. After that, the bubble is observed to slide continually on the surface of the heater and merge with small bubbles appeared on the edge of the heating area. The variation of the radius is obtained by image analyses of the recorded video images from the two CCDs. The stage of steady growth of the bubble adhering to the excitation point can be divided into two sub-stages, i.e. expansion and retreat of the bubble base. During the bubble base expansion sub-stage, its radius can be expressed as an exponential function of time. The exponent decreases from 0.42 for the smaller size to 0.28 for the medium one, and finally to 0. The bubble size slightly retraces at the beginning of the bubble base retreat, and then slowly increases again until the subcooled liquid penetrated the bottom of the bubble completely, causing the bubble to detach from the heating surface and then slid on the heating surface under the external disturbance.

Furthermore, the bubble growth problem is much more complex. Most of the evaporation occurs at the base of the bubble, in which the micro-layer between the vapor-liquid interface and the heater surface plays an important role. Snyder and Edwards [30] were the first to propose this mechanism for evaporation. Subsequently, Moore and Mesler [31] deduced the existence of a micro-layer under the bubble from the oscillations in the temperature measured at the bubble release site. Cooper and Lloyd [32] further confirmed the existence of the micro-layer. Following these pioneer works, the micro-layer model is often used in analyzing the bubble dynamics and the relevant heat transfer in boiling. Later, Stephan and Hammer [33] proposed a contact line model, which is often used in the literature. One major difference between the contact line model and the micro-layer model is the wall heat flux profile which has only a peak at the triple-phase-line in the contact line model whereas it exhibits a peak at the triple-phase-line followed by a plateau along the micro-layer in the micro-layer model. This difference is caused by the different assumptions on the size of the liquid film (thicker than molecular level) underneath the growing bubble. In the micro-layer model, the liquid film, or micro-layer, exists in the nearly entire region underneath the growing bubble, while in the contact line model just in a tiny space, or micro region, adjacent to the bubble base. Experimental evidences of the transition between contact line and micro-layer regime have been reported recently for a configuration involving a liquid meniscus moving on a heated wall [34]. Further progress has been obtained most recently by numerical simulations [35–37] comparing with the latest experimental data. There are, however, remaining gaps in building a generally applicable model for the formation and depletion of the micro-layer during boiling process.

As the bubble grows, it experiences forces causing to detach from the surface or pushing it against the wall. As a result of all these forces, a bubble departs from the heater surface after attaining a certain size. The diameter to which a bubble grows before departing is dictated by the balance of forces acting on the bubble.

Fritz [38] correlated the bubble departure diameter by balancing buoyancy, which acts to lift the bubble from the surface, with the surface tension force, which tends to hold the bubble to the wall, so that

$$D_d = 0.0208\theta \sqrt{\frac{\sigma}{g(\rho_L - \rho_V)}} \tag{8}$$

where, θ is the contact angle measured in degrees. It provides a correct length scale for the boiling process, though significant deviations of the bubble diameter at departure have been reported in the literature. Several other expressions that are obtained either empirically or analytically by involving various forces acting on a bubble have also been reported in the literature, which are not always consistent with each other. Generally, these models predict an n-power scaling behavior of the departure bubble diameter related to gravity with a wide range of the exponent n from 0 to $-1/2$.

A qualitative model for bubble departure from cylindrical heating surface was proposed by Zhao et al. [39], in which the Marangoni effect was taken into account (Fig. 4)

$$f(y) = C_4 y^4 + C_3 y^3 + C_1 y + C_0 \tag{9}$$

where,

$$y = \tau^{1/2}, \quad C_4 = \frac{4}{3}\pi E^3(\rho_L - \rho_V)g, \quad C_3 = -2K\pi|\sigma_T|E^2\nabla T, \quad C_1 = 4\sigma R_0 \sin^2\theta + \frac{\pi}{3}\rho_L E^4,$$

$$C_0 = R_0 E^3 \rho_L \sin^2\theta \left(\frac{1}{3} - \frac{3}{8}C_d\right), \quad E = \frac{1}{2\sqrt{\pi}}Ja\sqrt{\alpha_L}$$

where τ, σ, σ_T, R_0, C_d and denote the growing time of bubble, surface tension and its temperature coefficient, wire radius, and drag coefficient, respectively. K is an empirical parameter to count the departure from the linear theory for the case of finite Reynolds and Marangoni numbers, and the function $f(y)$ denotes the sign of the resultant force acting on a growing discrete vapour bubble. If $f(y) < 0$, the departure

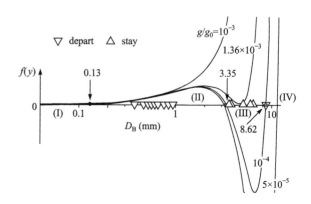

Fig. 4 The sign of the resultant force acting on a growing discrete vapour bubble in different gravity levels

force is larger than the resistant force, so the bubble will stay on the heater's surface; if $f(y) > 0$, the departure force is smaller than the resistant force, so the bubble will depart from the heater's surface.

The predictions by Eq. (9) are plotted in Fig. 4 at different gravity. In normal gravity, the function for the resultant force acting on the growing bubble, $f(y)$, has only one zero-value point, indicting only one critical diameter for bubble departure. When the residual gravity decreases to no more than $1.36 \times 10^{-4} g_0$, the second and third zero-value points will be predicted by the new model. It is consistent with the observation of special bubble behaviors in long-term microgravity during the TCPB (Temperature-Controlled Pool Boiling) experiment aboard the 22nd Chinese recoverable satellite RS-22 [40]. It is observed that there exist three critical bubble diameters in microgravity, which divided the observed vapor bubbles into four regions: Tiny bubbles were continually forming and growing on the surface before departing slowly from the wire when their sizes exceeded the first critical value. The bigger bubbles, however, were found staying on the surface again when their diameters were larger than the second critical value. If they grew further larger than the third critical value, departure would be observed once again. Furthermore, the first critical value exhibited no obvious difference between in normal gravity and in microgravity. Comparing the prediction of Eq. (9) at $g = 10^{-4} g_0$ (the level of residual gravity was estimated in the range of 10^{-3} to $10^{-5} g_0$) with the observation, the agreement is quite evident.

Knowing the growth rate and the diameter to which a bubble grows before departing, the growth time, t_g, can be calculated. After bubble departure, cooler bulk liquid fills the space vacated by the bubble. The thermal layer reforms over the area surrounding the nucleation site, and then a new bubble at this location will form and grow if the superheated liquid layer is re-established and the inception criterion is satisfied. The time taken by the thermal layer to develop prior to inception is termed the waiting period, t_w. Conceivably, a theoretical evaluation of the bubble release frequency can be made from expressions for the waiting time and the growth time. Such an approach, however, meets with little success comparing with experimental data because of the extreme complexity of ebullition cycle. Thus, correlations, including both the bubble diameter at departure and the bubble release frequency f, have been reported in the literature, usually in the following form

$$f D_d^n = const \qquad (10)$$

The constant parameter may be gravity-dependent, resulting in an m-power scaling behavior of the bubble release frequency with a range of the exponent m from 1/4 to 1.

One of the most comprehensive correlations of this type is given by Malenkov [41] with the value $n = 1$, while $n = 2$ is another alternative commonly used in the literature [42]. Besides, Di Marco and Grassi [43] suggested the following correlation for the bubble release frequency at different gravity

$$\frac{f_a}{f_0} = \left(\frac{a}{g_0}\right)^{3/4} \tag{11}$$

where f_0 is the bubble release frequency in normal gravity, and a is residual gravity.

Numerical simulation is a powerful alternative of experimental study for understanding bubble dynamics in pool boiling. Zhao et al. [44] studied numerically the growth processes of a single bubble on a fixed single site and the relative heat transfer in different gravity for saturated water at 0.1 MPa including the contact line model [33]. The Level Set Method and constant superficial contact angle are used to capture the interface between the vapor and liquid phases and the motion of contact line on the heating wall, respectively. The numerical results indicate that the equivalent diameter is proportional to about (1/3–1/2)-power of the growing time in spite of whatever gravity levels. But gravity has great influences on both the departure diameter and the growing time. The bubble departure diameter is proportional inversely to about 1/3-power of gravity, while the growing time is proportional inversely to about 4/5-power of gravity. The area-averaged heat fluxes are approximately proportional to the 3/2-power of the wall superheat when the number density of active nucleation sites fixes, which is consistent with the hypothesis proposed by Zuber [45] for small superheat. Furthermore, this trend has no change with the decrease of gravity, which is also confirmed later for the case including the influence of heater thermal capability [19–21].

By using the Ghost Fluid Method for sharp interface representation, Zhang et al. [19, 20] and Li et al. [21] extended the above numerical simulation to include the influence of heater thermal capability on nucleate pool boiling. Multi-cycle simulations are carried out to eliminate the influence of unreal initial conditions. Equation (3) was adopted for determining ONB. A constant and uniform temperature is fixed on to the bottom surface of the solid wall in the simulations, and thus, both the spatio-temporal averaged heat flux and superheat on the top surface, which contacts the working fluid directly, are dependent variables instead of controllable ones. They found that the surface temperature of solid wall can vary both temporally and spatially and solid wall thickness and material properties are observed to affect waiting time significantly, and the heater thickness will also affect the surface temperature recovery during nucleate boiling. Additionally, highly conductive materials are able to recover faster than poorly conductive materials. A slight dependence of the superheat related to gravity, i.e. $\Delta T_W \sim g^{-0.05}$ in the range of $(10^{-2}$ to $10^0)$ g_0, is observed in the case of single bubble boiling of saturated FC-72 on SiO_2 solid wall. The departure diameter of bubbles is proportional to $g^{-0.5}$, while the bubble release frequency is proportional to g^1. Thus, fD_d^2 can keep constant, which agrees with the experimental observations by Siegel and Keshock [42].

Zhang et al. [19, 20] and Li et al. [21] analyzed in detail the transient heat conduction inside the solid wall. A sharp drop of the wall temperature is evident in the vicinity of the contact line due to violent evaporation in this tiny region. The area of the temperature drop moves with the contact line, resulting in a pseudo-periodical process of heat storage and release inside the solid wall, which exhibits a coupling effect with bubble dynamics and heat transfer. The thermal penetration depth caused by the processes of bubble growth and departure in a single bubble cycle is about 0.5 mm in both steady and quasi-steady cases, which is much smaller than the heater thickness. Based on the analysis of the thermal penetration depths caused by the processes of bubble growth and departure, a suitable thickness of about 2 mm is proposed for the substrate of the integrated micro heater used in the SOBER-SJ10 experiment aboard the Chinese recoverable satellite SJ-10 [46, 47].

Based on the measured local temperature data underneath a single growing vapor bubble of FC-72 on a flat heater surface in microgravity obtained from the SOBER-SJ10 experiment, the spatio-temporal evolution of the temperature on the heated surface (Fig. 5) is reached under the axisymmetric hypothesis [29]. A narrow superheat region corresponding to the local superheat process can be observed. Several discrete cold points, instead of continuous low temperature line due to limited spatial resolution of local temperature measurement, are also observed, which indicates the location of the trajectory of the contact line on the heated surface. A dashed line is drawn in Fig. 5 according to these cold spots to show the possible trajectory of the contact line moving on the heated surface. Furthermore, the variation of the radius is also shown in Fig. 5 for comparison. These data provides a benchmark for the validation and verification of the bubble growth models.

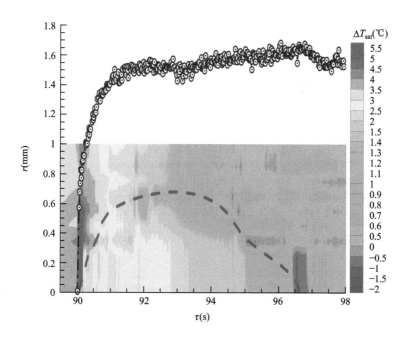

Fig. 5 The spatio-temporal evolution of local temperature on the heated surface

5   Gravity Scaling of Heat Transfer in Nucleation Pool Boiling

With high efficiency of boiling heat transfer due to the release of latent heat, nucleate pool boiling allows transferring high heat fluxes at moderate wall super-heats, and then is widely used in industry. To meet the demands of industrial applications, numerous heat transfer correlations have been developed theoretically, semi-empirically, or empirically in the past decades, and are still subject of many ongoing research activities all over the world. A successful correlation must be built on a reliable physical concept and mechanisms of the phenomenon. Particularly, for space applications, the correlation must correctly represent gravity as a real parameter, not just an irrelevant constant in classical boiling study which is based on experiments performed in normal gravity on the ground.

It has been recognized that the heat transfer during nucleate pool boiling is closely related to the bubble activity over a heater surface. In principle, the heat transfer rate can be predicted by knowing the bubble frequency, active nucleating cavity site density (per unit heater surface area), and heat transferred during each bubble ebullition cycle. However, empirical correlations are commonly used in fact. The well-known and often-used one is the semi-empirical model developed by Rohsenow [48], namely

$$q'' = \mu_L h_{LV} \left[\frac{g(\rho_L - \rho_V)}{\sigma} \right]^{1/2} \left(\frac{c_{p,L} \Delta T_w}{C_{sf} h_{LV} \, \mathrm{Pr}_L^n} \right)^3 \tag{12}$$

where μ is the dynamic viscosity, and the Prandtl number $Pr = \nu/\alpha = \mu c_p/k$, in which ν and k denote the kinematic viscosity and thermal conductivity, respectively. Rohsenow introduced an empirical constant C_{sf} to account for the fluid-surface effect on nucleate boiling, while the parameter n depends on the working fluid. Rohsenow model is based on assumptions for forced convection caused by the motion of the detached bubbles which depends on the buoyancy force exerted on the bubbles, resulting in the heat transfer intensity being a function of gravity. It implies a scaling exponent $m = 1/2$ related to gravity for the same superheat, namely

$$\frac{q''}{q_0''} = \left[\frac{a}{g_0} \right]^{1/2} \tag{13}$$

Consequently, at a given superheat, the heat flux and heat transfer coefficient would be reduced by a factor of 10^{-2} to 10^{-3} if the gravity were reduced to the level of 10^{-4} to 10^{-6} g_0. Fortunately, it is not confirmed in microgravity experiments.

Generally, it is believed that the pool boiling curve near the ONB is independent of the acceleration level, i.e., m approaches zero. On the other hand, at higher superheat approaching CHF, the power law coefficient approaches a value of 0.25, as what is predicted by the CHF correlations of Kutateladze [49] and Zuber [50]. There is, however, a wide range from -0.35 to 1 for the gravity scaling exponent according

Fig. 6 A schematic of gravity scaling exponent of heat transfer in nucleate pool boiling

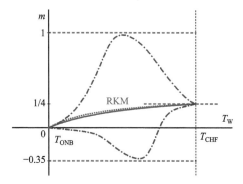

to the existing correlations of heat flux in nucleate pool boiling. There may exist numerous possible trajectories of gravity scaling exponent of heat transfer in nucleate pool boiling (Fig. 6). No common trend concerning the gravity scaling behavior of nucleate pool boiling could be achieved now.

Raj et al. [51, 52] observed a sharp transition in the heat transfer mechanism at a threshold gravity level a_{tran} ($= 4.41\sigma/[L_h^2(\rho_{L-}\rho_V)]$, here L_h is the characteristic length of the heater) based on quasi-steady pool boiling experiments over a continuous gravity range of $(0–1.8)\,g_0$. Below this threshold (surface tension dominated boiling, or SDB, regime), a non-departing primary bubble governed the heat transfer and the effect of residual gravity was small. Above this threshold (buoyancy dominated boiling, or BDB, regime), bubble growth and departure dominated the heat transfer and gravity effects became more important. A gravity scaling model in the BDB regime was developed, namely

$$\frac{q''_{\mathrm{BDB}}}{q''_0} = \left(\frac{a_{\mathrm{BDB}}}{g_0}\right)^m, \quad m = \frac{0.9T^*}{1 + 2.6T^*} \quad \text{for} \quad \frac{L_h}{L_0} \geq 2.1 \qquad (14)$$

where

$$T^* = \frac{T_W - T_{\mathrm{ONB}}}{T_{\mathrm{CHF}} - T_{\mathrm{ONB}}}, \quad L_0 = \sqrt{\frac{\sigma}{a(\rho_L - \rho_V)}} \qquad (15)$$

The dimensionless wall temperature T^* is defined based on the assumption that CHF for all gravity levels occurs at the same wall superheat and ONB is independent on gravity. In Eq. (14), q''_0 denotes the heat flux of nucleate pool boiling at the reference gravity level, usually at the normal gravity g_0.

Later, Raj et al. [53] slightly modified the expression of the exponent m in the BDB regime as

$$m_{BDB} = \frac{0.65T^*}{1 + 1.6T^*} \qquad (16)$$

and proposed $m_{SDB} = 0.025$ in the SDB regime. A general for the scaling behavior in the SDB regime including the jump in heat flux ($\Delta q''$) due to subcooling and dissolved gas is developed as follows

$$\frac{q''_{SDB}}{q''_0} = \left(\frac{a_{tran}}{g_0}\right)^{m_{BDB}} K_{jump} \left(\frac{a_{SDB}}{a_{tran}}\right)^{0.025}, \quad \text{for} \quad \frac{L_h}{L_0} < 2.1 \qquad (17)$$

where

$$K_{jump} = 1 - e^{-CMa}, \quad \text{and} \quad Ma = \frac{\sigma_T \Delta T_{SUB} L_h}{\mu_L \alpha_L} \qquad (18)$$

In the above equation, $\sigma_T = -d\sigma/dT$ denotes the temperature coefficient of surface tension. The value of the empirical parameter C for FC-72 was found to be 8.3×10^{-6}.

The heat transfer, however, measured in the SDB regime is likely to be artificially high as the bubble responds to the g-jitter in the aircraft data which is on the order of 10^{-2} g_0, and then it is observed that the power law coefficient for gravity obtained in the SDB regime ($m_{SDB} = 0.025$) had much scatter. Based on experimental data obtained in the true microgravity environment provided by the ISS, a modified power law coefficient of $m_{SDB} = 0$ in the SDB regime is proposed by Raj et al. [54]. A schematic of heat flux versus acceleration is shown in Fig. 7. It is physically reasonable. Once in the SDB regime where a non-departing, coalesced bubble covers the heater, a small change in the gravity level would only change the bubble shape without affecting the steady state value of heat transfer significantly.

The trajectories of the gravity scaling exponent m_{BDB} in the above model, namely Eqs. (14) and (16), are plotted in Fig. 6, labeled as "RKM". They show a monotonically increasing trend from 0 at ONB to 1/4 at CHF. Many unexplained trends in boiling literature can be explained and modeled using this scaling framework, which demonstrates its robustness in predicting low gravity heat transfer [54–56]. Further endeavors, however, are needed because of the following questions.

Fig. 7 A schematic of the gravity scaling model proposed by Raj et al. [54]

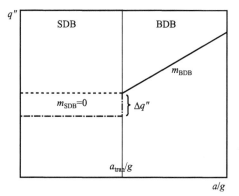

At first, the gravity scaling model of Raj et al. [54] is developed mainly based on the naturally transient data obtained from the pool boiling experiments aboard parabolic aircrafts, in which a continuous gravity change occurs from 0 (or 1.8 g_0) to 1.8 g_0 (or 0) in a very short period of about 5 s. As the gravity changes, time is required for the flow field and heat transfer profiles to develop and achieve steady state. Before the transition from high-g to low-g, the natural convection flow field was fully developed. During the transition from high-g to low-g, the flow field required more time to achieve steady state than was available, resulting in higher heat transfer than the expected quasi-steady value. Similarly, during the transition from low-g to high-g, the heat transfer was lower than the expected quasi-steady value. It is assumed that if there is no difference in the two curves of heat flux versus acceleration during the transition from hypergravity to low-g and vice versa at the same superheat, the flow field and heat transfer profiles may have sufficient time to achieve steady state at each acceleration level. However, a hysteresis in the heat flux curve is present at the lower superheat whenever the superheat was not sufficient to initiate nucleation, and heat transfer was by natural convection in high-g condition. If the superheat is large enough, and the heat transfer is independent of the direction of transition, the heat transfer during the transitions when boiling occurs will be considered as in quasi-steady state. At present, there are no real steady data of pool boiling in long-term partial gravity regime, even the data regarding pool boiling in the partial gravity regime are very scarce.

Secondly, the assumption that CHF for all gravity levels occurs at the same wall superheat is questionable. Recently, Ma et al. [57] simulated numerically pool boiling heat transfer from a horizontal hydrophilic surface under constant wall temperature in different gravity levels based on an improved liquid-vapor phase-change lattice Boltzmann method with the imposition of a conjugate thermal boundary condition at solid/liquid interface. It is shown that gravity has significant effects on pool boiling curves. In contrast to the assumption of constant CHF superheat, the critical heat flux occurs at a lower wall superheat and is lower in microgravity than in normal gravity. Similar experimental observations have also been reported in the literature. The same observations have also been obtained by Feng et al. [58].

Thirdly, the influence of heater geometry is not taken into account. Drastically different behaviors have been observed between pool boiling on flat plates and on cylinders in microgravity. For example, Zhao et al. [17] reported a slight enhancement of heat transfer of nucleate pool boiling on thin wires in short- and long-term microgravity comparing with that in normal gravity. On the contrary, boiling heat transfer on plates in microgravity was generally deteriorated comparing with that in normal gravity, particularly at high superheats or heat fluxes [18, 55, 59]. The deterioration of heat transfer of nucleate pool boiling on plates may be characterized by the gravity scaling model of Raj et al. [54], but it never was for the enhancement on wires.

Finally, the 1/4-power scaling law of CHF with the gravity is not always maintained. For CHF on cylinders, Zhao et al. [17] found that the Lienhard-Dhir-Zuber model [60], established on the mechanism of hydrodynamic instability, can provide a relative good prediction on the trend of CHF in different gravity conditions, though the value of dimensionless radius $R' = R[(\rho_{L-}\rho_V)g/\sigma]^{1/2} = Bo^{1/2}$ was far beyond the initial application range of the model. This observation was consistent with Straub [1]. Furthermore, it was inferred, as pointed out by Di Marco and Grassi [61], that the dimensionless radius R', or equivalently the Bond number, may not be able to scale adequately the effects and to separate groups containing gravity due to the competition of different mechanisms for small cylinder heaters. Zhao et al. [62] revisited the scaling behaviors of CHF with respect to R' at small value of the Bond number in normal gravity conditions. It has been found that interactions between the influences of the subcooling and size on CHF will be important for the small Bond number, and that there may exist some other parameters, which may be material-dependant, in addition to the Bond number that play important roles in the CHF phenomenon with small Bond number.

A parameter, named as the limited nucleate size d_{LN}, and a corresponding dimensionless coefficient $\Gamma = d_{LN}/d_{wire}$ were introduced to interpret this phenomenon [17]. It was assumed that the limited nucleate size is not dependent with gravity but with the other parameters of the boiling system, such as the material parameters of the working fluid and the heater, the heater surface condition, an so on. If Γ is small enough, the initial vapor bubbles will be much smaller than the heater surface and then the occurrence of the CHF will be caused by the mechanism of hydrodynamic instability. On the contrary, it will be caused by the mechanism of local dryout if Γ is so large that the initial bubble larger than the wire diameter d_{wire} may easily encircle the heater. Further researches, however, are needed for the delimitation of the two mechanisms.

6 Enhancement of Pool Boiling Heat Transfer in Microgravity

The loss of buoyancy in microgravity has been found to change the nature of terrestrial nucleate boiling, which is driven primarily by the gravity on Earth. It is found that the boiling process in microgravity appears to be strongly dependent on heat flux and subcooling levels. For the boiling heat transfer in microgravity, the diminished buoyancy effect results in a longer stay time for the bubble departure, which prevents the effective access of fresh bulk liquid to the heater surface in time, and then leads to a lower boiling heat transfer performance and a strong increase of the heater temperature as with film boiling at high heat flux. How to improve boiling heat transfer effectively in microgravity is an important issue. A straightforward solution is to impose another force on the boiling process in microgravity to replace the buoyancy force. Sitter et al. [63, 64] and Moehrle and Chung [65] examined boiling on a wire in the presence of an acoustic field in terrestrial and microgravity experiments and found that the acoustic actuation led to an increase in the heat transfer coefficient on the wire by directly coupling with the natural oscillations of vapor bubbles through the action of the primary Bjerknes force. Complete boiling curves are presented to show how the applied acoustic field enhanced boiling heat transfer and increased critical heat flux in the terrestrial environment, while in microgravity the acoustic field was found to be capable of filling the role of terrestrial gravity in maintaining nucleate boiling. Applying a static electric field [66–71] or a magnetic field [72] can also provide additional volume forces able to replace buoyancy, to reduce the size of detaching bubbles and to lead them away of the surface, restoring efficient heat transfer conditions in microgravity.

On the other hand, drastically different heat transfer performances were observed between pool boiling on flat plates and on cylinders in microgravity, which is caused by the possible fresh liquid supply. In nucleate pool boiling on cylinders (Fig. 8a), bubble grows on a side of the cylindrical surface, and the bubble, even very bigger than the cylinder diameter, may not enwrap the cylinder, resulting in a plenty supply of fresh liquid from other sides to the heater surface to maintain the same heat transfer efficiency of microgravity nucleate pool boiling as that in normal gravity, or even to obtain a higher efficiency due to the advantage of sustained phase change on the heated surface in microgravity comparing with convection driven by the rising of detached bubbles in normal gravity. On the contrary, vapor bubbles cannot depart easily from the smooth surface of a flat plate in microgravity (Fig. 8b), and then can grow attaching to the surface and coalesced with each other. As the increase of their sizes, the coalesced bubbles can cover the heater surface and prevent the fresh liquid from moving to the heater surface, thus local dryout may occur, resulting in deterioration of heat transfer. These observations inspire us to propose a new passive method for nucleate boiling heat transfer enhancement (Fig. 8c): A reasonable design of the micro-structure of the heater surface can form effective path of fresh liquid supply to ensure that even if bubbles staying on the top of the heater surface can not be detached in microgravity, the fresh bulk liquid may still access to the heater surface through interconnect tunnels formed by the micro-structure due to the capillary forces, which is independent of the gravity level.

A serial of experiments on boiling enhancement in microgravity by use of micro-pin-fins which were fabricated by dry etching have been performed in the drop tower Beijing [73–78]. Unlike much obvious deterioration of heat transfer of nucleate pool

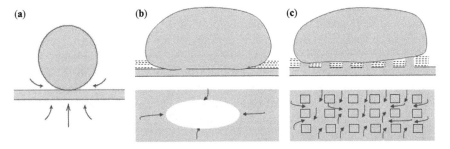

Fig. 8 Schematic path of fresh liquid supply to the heater surface underneath the growing bubble on different heater geometries: **a** cylinder, **b** smooth flat plate, and **c** flat plate with micro-structure surface

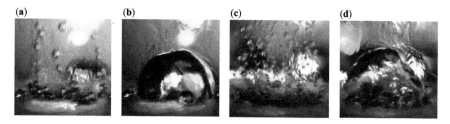

Fig. 9 Bubble behaviors in nucleate pool boiling on smooth and micro-pin-finned surfaces in normal and microgravity. **a** 1 g, smooth surface; **b** µg, smooth surface; **c** 1 g, micro-pin-finned surface; **d** µg, micro-pin-finned surface

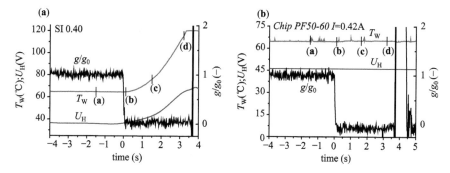

Fig. 10 Heat transfer performances of nucleate pool boiling on smooth and micro-pin-finned surfaces in normal and microgravity. **a** smooth surface; **b** micro-pin-finned surface

boiling on the smooth surface in microgravity, constant heater surface temperature of nucleate pool boiling for the micro-pin-finned surface was observed, even though a large coalesced bubble completely covered the surface under microgravity condition (Figs. 9 and 10). The critical heat flux on micro-pin-finned surface in microgravity can reach about two-thirds of that in normal gravity, but almost three times as large as that for the smooth surface in microgravity. It is also found that the fin pitch and configuration have significant effects on the boiling heat transfer coefficient as well as critical heat flux.

Therefore, it is confirmed that the micro-pin-finned surface can provide large capillary force and small flow resistance, driving a plenty of bulk liquid to access the heater surface for evaporation in high heat flux region, which results in large boiling heat transfer enhancement. Since the capillary force is no relevant to the gravity level, the micro-pin-finned surface appears to be one promising enhanced surface for efficient electronic components cooling schemes not only in normal gravity but also in microgravity conditions, which is very helpful to reduce the cooling system weight in space and in planetary neighbors.

Further study on the enhancement of the micro-pin-finned surfaces combined with forced convection has also been performed in the drop tower Beijing [79], and long-term microgravity experiments are also proposed, which will be conducted on board the Chinese Space Station (CSS) in future.

7 Conclusion Remarks

Pool boiling is a daily phenomenon transferring effectively high heat flux, and then is widely used in industrial processes. It is, however, a very complex and illusive process due to interrelation of numerous factors and effects. Among many sub-processes in boiling phenomenon, gravity can be involved and play much important roles, even enshroud the real mechanism underlying the phenomenon. The detailed knowledge of the mechanisms underlying this phenomenon is important and essential for industrial applications. Microgravity experiments offer a unique opportunity to study the complex interactions without external forces, such as buoyancy, which can affect the bubble dynamics and the related heat transfer, and provide an effective means to study the actual influence of gravity on the boiling phenomenon. Furthermore, many potential applications exist in space and in planetary neighbors due to its high efficiency in heat transfer. Thus, pool boiling in microgravity has become an increasing significant subject for investigation.

Up to now, great progress on understanding of pool boiling phenomenon has been achieved based on a substantial amount of solicited experiments both in normal and microgravity. Based on the outcome of the current trends in pool boiling research, some recommendations for future work, particularly on resolving many outstanding issues related to gravity effect on boiling and building a comprehensive database for the development of a pool boiling regime map, can be summarized as follows:

(1) Visualization of pool boiling in various gravity levels, particularly in long-term partial gravity regime, and successful development of a gravity scaling model reflecting the actual mechanism underlying this phenomenon.
(2) Verification of the heat transfer enhancement of nucleate pool boiling by micro-pin-finned surface in long-term microgravity environment, and optimization of the micro-structure parameters to obtain highly efficient heat transfer capability in various gravity levels.
(3) Research with high spatial resolution heaters at smaller scales of both length and time in normal gravity on Earth and in partial and microgravity in space to reveal micro-convection mechanism in the immediate vicinity of growing bubbles during single- and/or multiple-bubble pool boiling experiments.
(4) Extension of single bubble boiling models to actual experimental conditions with multiple bubbles using a fractal pattern approach, and corresponding numerical studies with a moving liquid-vapor interface including the transient thermal response of the solid wall.

Acknowledgements The studies presented here were supported financially by the National Natural Science Foundation of China (U1738105, 11802314, 11672311, 11372327, 11402273, 10972225, 10432060, 51636006, 51611130060), and the Chinese Academy of Sciences (QYZDY-SSW-JSC040, XDA04020404, XDA04020202-04).

References

1. Straub J (2001) Boiling heat transfer and bubble dynamics in microgravity. Adv Heat Transf 35:57–172
2. Di Marco P (2003) Review of reduced gravity boiling heat transfer, European research. J Jpn Microgravity Appl 20(4):252–263
3. Ohta H (2003) Review of reduced gravity boiling heat transfer: Japanese research. J Jpn Soc Microgravity Appl 20(4):272–285
4. Kim J (2003) Review of reduced gravity boiling heat transfer, US research. J Jpn Microgravity Appl 20(4):264–271
5. Kim J (2009) Review of nucleate pool boiling bubble heat transfer mechanisms. Int J Multiphase Flow 35:1067–1076
6. Zhao JF (2010) Two-phase flow and pool boiling heat transfer in microgravity. Int J Multiphase Flow 36(2):135–143
7. Nukiyama S (1934) Maximum and minimum values of heat transmitted from metal to boiling water under atmospheric pressure, JSME J 37:367. See also: Int J Heat Mass Transf 9(12):1419 (1966); 27(7):959 (1984)
8. Carey VP (2008) Liquid vapor phase change phenomena. Taylor & Francis Group, New York, USA
9. Bankoff SG (1958) Entrapment of gas in the spreading of liquid over a rough surface. AIChE J 4:24–26
10. Griffith P, Wallis JD (1960) The role of surface conditions in nucleate boiling. Chem Eng Prog Symp Ser 56(30):49–63
11. Hsu YY, Graham RW (1961) An analytical and experimental study of the thermal boundary layer and ebullition cycle in nucleate boiling, NASA TND-594. NASA Lewis Research Center, Cleveland, OH, USA
12. Hsu YY (1962) On the size range of active nucleation cavities in a heating surface. Trans ASME J Heat Transf 84:207–216
13. Davis EJ, Anderson GH (1966) The incipience of nucleate boiling in forced convection flow. AIChE J 12(4):774–780
14. Bergles AE, Rohsenow WM (1964) The determination forced-convection surface boiling heat transfer. J Heat Transf 86:365–372
15. Sato T, Matsumura H (1964) On the condition of incipient subcooled boiling with forced convection. Bull JSME 7(26):392–398
16. Wang CH, Dhir VK (1993) Effect of surface wettability on active nucleation site density during pool boiling of saturated water. J Heat Transf 115:659–669
17. Zhao JF, Wan SX, Liu G, Yan N, Hu WR (2009) Subcooling pool boiling on thin wire in microgravity. Acta Astronaut 64(2–3):188–194
18. Zhao JF, Li J, Yan N, Wang SF (2009) Bubble behavior and heat transfer in quasi-steady pool boiling in microgravity. Microgravity Sci Tech 21(S1):S175–S183
19. Zhang L, Li ZD, Li K, Li HX, Zhao JF (2014) Influence of heater thermal capacity on pool boiling heat transfer. J Comput Multiphase Flows 6(4):361–375
20. Zhang L, Li ZD, Li K, Li HX, Zhao JF (2015) Influence of heater thermal capacity on bubble dynamics and heat transfer in nucleate pool boiling. Appl Therm Eng 88:118–126
21. Li ZD, Zhang L, Zhao JF, Li HX, Li K, Wu K (2015) Numerical simulation of bubble dynamics and heat transfer with transient thermal response of solid wall during pool boiling of FC-72. Int J Heat Mass Transf 84:409–418

22. Scriven LE (1959) On the dynamics of bubble growth. Chem Eng Sci Genie Chim 10:1–13
23. Plesset MS, Zwick SA (1954) Growth of vapor bubbles in superheated liquids. J Appl Phys 25:493–500
24. Dergarabedian P (1953) The rate of growth of vapor bubbles in superheated water. ASME J Appl Mech 20:537–545
25. Mikic BB, Rohsenow WM, Griffith D (1970) On bubble growth rates. Int J Heat Mass Transf 13:657–666
26. Wan SX, Zhao JF, Liu G (2009) Dynamics of discrete bubble in nucleate pool boiling on thin wires in microgravity. J Therm Sci 18(1):13–19
27. Li J, Zhao JF, Xue YF, Wei JJ, Du WF, Guo D (2012) Experimental study on growth of an isolated bubble utilizing short-term microgravity drop tower. Chin J Space Sci 32(4):544–549
28. Liu P, Wu K, Du W, Zhao JF, Li HX, Li K (2018) Experimental study on bubble behaviors in microgravity pool boiling. Chin J Space Sci 38(2):221–226
29. Wu K, Liu P, Du WF, Zhao JF, Li HX, Li K (2018) Heat transfer and bubble dynamical behavior during single bubble pool boiling in microgravity. In: Proceedings of the 16th international heat transfer conference (IHTC-16), 10–15 August 2018, Beijing, China, Paper no. IHTC16-22294
30. Snyder NR, Edwards DK (1956) Summary of conference on bubble dynamics and boiling heat transfer. Memo 20–137, Jet Propulsion Laboratory, Pasadena, CA, USA, pp 14–15
31. Moore FD, Mesler RB (1961) The measurement of rapid surface temperature fluctuations during nucleate boiling of water. AIChE J 7:620–624
32. Cooper MG, Lloyd AJP (1969) The microlayer in nucleate pool boiling. Int J Heat Mass Transf 12:915–933
33. Stephan P, Hammer J (1994) A new model for nucleate boiling heat transfer. Heat Mass Transf 30:119–125
34. Fischer S, Gambaryan-Roisman T, Stephan P (2015) On the development of a thin evaporating liquid film at a receding liquid/vapour-interface. Int J Heat Mass Transf 88:346–356
35. Urbano A, Tanguy S, Huber G, Colin C (2018) Direct numerical simulation of nucleate boiling in micro-layer regime. Int J Heat Mass Transf 123:1128–1137
36. Guion A, Afkhami S, Zaleski S, Buongiorno J (2018) Simulations of microlayer formation in nucleate boiling. Int J Heat Mass Transf 127:1271–1284
37. Hänsch S, Walker S (2019) Microlayer formation and depletion beneath growing steam bubbles. Int J Multiphase Flow 111:241–263
38. Fritz W (1935) Maximum volume of vapor bubbles. Physik Zeitschr 36:379–384
39. Zhao JF, Liu G, Wan SX, Yan N (2008) Bubble dynamics in nucleate pool boiling on thin wires in microgravity. Microgravity Sci Technol 20(2):81–89
40. Zhao JF, Liu G, Li ZD, Wan SX (2007) Bubble behaviors in nucleate pool boiling on thin wires in microgravity. In: 6th international conference multiphase flow, 9–13 July 2007, Leipzig, Germany
41. Malenkov IG (1971) Detachment frequency as a function of size of vapor bubbles. Translated Inzh Fiz Zhur 20:99
42. Siegel R, Keshock EG (1964) Effects of reduced gravity on nucleate boiling bubble dynamics in saturated water. AIChE J 10:507–517
43. Di Marco P, Grassi W (2000) Pool boiling in microgravity: assessed results and open issues. In: Proceedings of the 3rd European thermal sciences conference
44. Zhao JF, Li ZD, Zhang L (2012) Numerical simulation on single bubble pool boiling in different gravity conditions. Chin J Space Sci 32(4):537–543
45. Zuber N (1963) Nucleate boiling: the region of isolated bubbles and the similarity with natural convection. Int J Heat Mass Transf 6(1):53–79
46. Hu WR, Zhao JF, Long M, Zhang XW, Liu QS, Hou MY, Kang Q, Wang YR, Xu SH, Kong WJ, Zhang H, Wang SF, Sun YQ, Hang HY, Huang YP, Cai WM, Zhao Y, Dai JW, Zheng HQ, Duan EK, Wang JF (2014) Space program SJ-10 of microgravity research. Microgravity Sci Technol 26:159–169
47. Wu K, Li ZL, Zhao JF, Li HX, Li K (2016) Partial nucleate pool boiling at low heat flux: preliminary ground test for SOBER-SJ10. Microgravity Sci Technol 28:165–178

48. Rohsenow WM (1952) A method of correlating heat transfer data for surface boiling of liquids. Trans ASME 74:969–976
49. Kutateladze SS (1948) On the transition to film boiling under natural convection. Kotloturbostroenie 3:10–12
50. Zuber N (1959) Hydrodynamic aspects of boiling heat transfer. PhD thesis, University of California, Los Angeles, CA, USA
51. Raj R, Kim J, McQuillen J (2009) Subcooled pool boiling in variable gravity environments. J Heat Transf 131(9):09152
52. Raj R, Kim J, McQuillen J (2010) Gravity scaling parameter for pool boiling heat transfer. ASME Trans J Heat Transf 132(9):091502
53. Raj R, Kim J, McQuillen J (2012) On the scaling of pool boiling heat flux with gravity and heater size. ASME Trans J Heat Transf 134(1):0115021
54. Raj R, Kim J, McQuillen J (2012) Pool boiling heat transfer on the international space station: experimental results and model verification. J Heat Transf 134:10154
55. Wang XL, Zhang YH, Qi BJ, Zhao JF, Wei JJ (2016) Experimental study of the heater size effect on subcooled pool boiling heat transfer of FC-72 in microgravity. Exp Therm Fluid Sci 76:275–286
56. Zhao JF, Wei JJ, Li HX (2017) Influences of gravity on bubble dynamics and heat transfer in nucleate pool boiling. In: Keynote lecture. 2nd international conference of interfacial phenomena & heat transfer (IPHT 2017), 7–10 July 2017, Xi'an, China
57. Ma X, Cheng P, Gong S, Quan X (2017) Mesoscale simulations of saturated pool boiling heat transfer under microgravity conditions. Int J Heat Mass Transf 114:453–457
58. Feng Y, Li HX, Guo KK, Zhao JF, Wang T (2018) Numerical study of single bubble growth on and departure from a horizontal superheated wall by three-dimensional lattice Boltzmann method. Microgravity Sci Technol 30(6):761–773
59. Xue YF, Zhao JF, Wei JJ, Li J, Guo D, Wan SX (2011) Experimental study of nucleate pool boiling of FC-72 on smooth surface under microgravity. Microgravity Sci Technol 23(S1):S75–S85
60. Lienhard JH, Dhir VK (1973) Hydrodynamic prediction of peak pool boiling heat fluxes from finite bodies. J Heat Transf 95:152–158
61. Di Marco P, Grassi W (1999) About the scaling of critical heat flux with gravity acceleration in pool boiling. In: Proceedings of XVII UIT national heat transfer conference, Ferrara, pp 139–149
62. Zhao JF, Lu YH, Du WF, Li ZD (2015) Revisit on the scaling of the critical heat flux on cylinders. Interfacial Phenomena Heat Transf 3(1):69–83
63. Sitter JS, Snyder TJ, Chung JN, Marston PL (1998) Acoustic field interaction with a boiling system under terrestrial gravity and microgravity. J Acoust Soc Am 104:2561–2569
64. Sitter JS, Snyder TJ, Chung JN, Marston PL (1998) Terrestrial and microgravity pool boiling heat transfer from a wire in an acoustic field. Int J Heat Mass Transf 41:2143–2155
65. Moehrle RE, Chung JN (2016) Pool boiling heat transfer driven by an acoustic standing wave in terrestrial gravity and microgravity. Int J Heat Mass Transf 93:322–336
66. Snyder TJ, Chung JN (2000) Terrestrial and microgravity boiling heat transfer in a dielectrophoretic force field. Int J Heat Mass Transf 43(9):1547–1562
67. Di Marco P, Grassi W (2002) Motivation and results of a long-term research on pool boiling heat transfer in low gravity. Int J Therm Sci 41(7):567–585
68. Di Marco P, Grassi W (2009) Effect of force fields on pool boiling flow patterns in normal and reduced gravity. Heat Mass Transf 45(7):959–966
69. Di Marco P, Grassi W (2011) Effects of external electric field on pool boiling: comparison of terrestrial and microgravity data in the ARIEL experiment. Exp Therm Fluid Sci 35(5):780–787
70. Iacona E, Herman C, Chang SN, Liu Z (2006) Electric field effect on bubble detachment in reduced gravity environment. Exp Therm Fluid Sci 31(2):121–126
71. Schweizer N, Di Marco P, Stephan P (2013) Investigation of wall temperature and heat flux distribution during nucleate boiling in the presence of an electric field and in variable gravity. Exp Therm Fluid Sci 44:419–430

72. Munasinghe T (2009) Studying the characteristics of bubble motion in pool boiling in microgravity conditions under the influence of a magnetic field. In: Proceedings of the 4th international conference on recent advances in space technologies, 11–13 June 2009, Istanbul, Turkey, pp 700–703
73. Wei JJ, Zhao JF, Yuan MZ, Xue YF (2009) Boiling heat transfer enhancement by using micro-pin-finned surface for electronics cooling. Microgravity Sci Technol 21(S1):S159–S173
74. Wei JJ, Xue YF, Zhao JF, Li J (2011) Bubble behavior and heat transfer of nucleate pool boiling on micro-pin-finned surface in microgravity. Chin Phy Lett 28(1):016401
75. Xue YF, Zhao JF, Wei JJ, Zhang YH, Qi BJ (2013) Experimental study of nucleate pool boiling of FC-72 on micro-pin-finned surface under microgravity. Int J Heat Mass Transf 63:425–433
76. Zhang YH, Wei JJ, Xue YF, Kong X, Zhao JF (2014) Bubble dynamics in nucleate pool boiling on micro-pin-finned surfaces in microgravity. Appl Therm Eng 70:172–182
77. Zhang YH, Zhao JF, Wei JJ, Xue YF (2017) Nucleate pool boiling heat transfer on a micro-pin-finned surface in short-term microgravity. Heat Transf Eng 38(6):594–610
78. Qi BJ, Wei JJ, Wang XL, Zhao JF (2017) Influences of wake-effects on bubble dynamics by utilizing micro-pin-finned surfaces under microgravity. Appl Therm Eng 113:1332–1344
79. Zhang YH, Liu B, Zhao JF, Deng YP, Wei JJ (2018) Experimental study of subcooled flow boiling heat transfer on micro-pin-finned surfaces in short-term microgravity. Exp Therm Fluid Sci 97:417–430

Study on Thermocapillary Convection in an Annular Liquid Pool

Li Duan, Qi Kang, Di Wu, Li Zhang, Di Zhang, Huan Jiang, Chu Zhang, Pu Zhang, Yongli Yin and Wenrui Hu

Abstract Thermocapillary convection is an important content in the study of microgravity fluid physics. It is not only the problem of fluid physics mechanism such as convection stability, but also closely related to spacecraft flow control, efficient heat transfer, etc., and has direct guiding significance for material growth processes such as floating zone method and Czochralski method. This chapter mainly introduces the research status of thermocapillary convection about an annular liquid pool, the scientific experiment and partial analysis results carried out by the project group in the early stage, and the development of space experimental payload, space experiment and preliminary experimental results about the annular liquid pool from SJ-10 mission.

Keywords SJ-10 mission · Space experiment · Microgravity fluid physics · Thermocapillary convection · Annular liquid pool · Convection instability · Chaos · Bifurcation · Transition routes · Volume ratio effect

1 Introduction

Among many new energy sources, solar energy is the best energy choice known to mankind for some advantages such as renewability, cleanliness, wide applicability, and economical efficiency. Solar photovoltaic industry is based on silicon materials. The monocrystalline silicon is the first discovered, studied and applied. It is still

L. Duan · Q. Kang (✉) · D. Wu · L. Zhang · D. Zhang · H. Jiang · C. Zhang · W. R. Hu
Key Laboratory of Microgravity (National Microgravity Laboratory) Institute of Mechanics, Chinese Academy of Sciences, No.15 Beisihuanxi Road Haidian District, Beijing 100190, China
e-mail: kq@imech.ac.cn

L. Duan · Q. Kang · W. R. Hu
School of Engineering Science, University of Chinese Academy of Sciences, No. 19(A) Yuquan Road, Shijingshan District, Beijing 100049, China

Y. L. Yin
Astronaut Research and Training Center, Youyi Road, Haidian District, Beijing 100094, China

© Science Press and Springer Nature Singapore Pte Ltd. 2019
W. R. Hu and Q. Kang (eds.), *Physical Science Under Microgravity: Experiments on Board the SJ-10 Recoverable Satellite*, Research for Development,
https://doi.org/10.1007/978-981-13-1340-0_5

one of the most important materials for solar cells. Monocrystalline silicon, a single crystal of silicon, is a kind of semiconductor material with good performance. There are different crystal growing methods, including Czochralski (Cz) method, float zone method, and epitaxial method. The monocrystalline silicon by Cz method is the most widely used material. That is the most important method for the growth of semiconductor crystals. It is divided into seven stages including filling material, melting material, contacting the seed crystal with the melt, pulling the thin neck, pulling the shoulder, pulling the body and pulling the end zone. The technology of pulling single crystals by seeded crucible melts was firstly used in 1918 by J. Czochralski to grow metallic crystals. It is called Cz method. Cz method uses the principle of solidifying crystallization of the melt to establish a temperature field, form a certain degree of super-cooling on the solid-liquid interface, and ensure the other regions not overheated at the same time; the melt is crystallized on the seeded crucible, and the rod like crystal is formed with the rising of the pulling rod. Therefore, in Cz method, the control of heat transfer is the key for the successful growth of crystals. As a result, to establish an appropriate temperature field is a very important condition to grow high quality crystals. The flow in the melt is closely related to the temperature field, and the change in the flow state will influence the distribution of molten materials and impurities as well as the performance of the crystal.

During the crystal growing process by Cz method, momentum transport (distribution of fluid velocity), energy transport (distribution of temperature) and mass transport (distribution of melting concentration) are involved. When the crystal grows, it melts and cools at the same time, and there is a temperature gradient on the crystal surface. This kind of temperature distribution, on the one hand, causes the difference in density of silicon melt at the wall and the core of the crucible, and causes buoyancy convection, thus, the silicon melt with high density and low temperature near the core is pressed to the bottom of the crucible, and then it is heated up at the bottom and rises up along the wall of the crucible, forming a convection cycle; on the other hand, thermocapillary convection driven by the gradient in surface tension on the free surface, which is also called Marangoni flow, is also an important factor leading to the decrease in uniformity and unidirectional property of the temperature field of melt as well as the crystal. In addition, with the forced convection caused by the crystal rotation, during the practical process of crystal growth, variable factors act on the melt and make the melt flow very complicated, leading to the un-uniformity in properties of the crystal, producing mechanical stress and dislocation, and as a result, affecting the quality of the crystal (Fig. 1).

The instability of flow field has attracted more and more attention since it was firstly discovered. In the last century, many scientists have simulated the flow in the Cz furnace, and analyzed the change rule of the configuration of crystallization interface. From the whole view of the work done, natural convection in the melting flow takes the dominating role, and forced convection takes the secondary role [2–5].

Convection is a common physical phenomenon in nature. In the gravitational field on the ground, in a fluid region with warmer fluid on the bottom and cooler fluid on the top, buoyancy will drive the warmer fluid up to form a convection, and this is called buoyancy convection. On the liquid free surface in an open container, if there

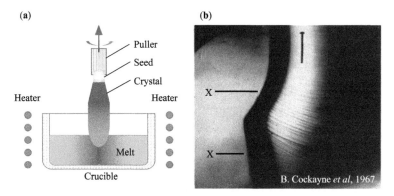

Fig. 1 The physical model of Cz method (left) and the slice map of crystal with striations defects (right). Reprinted from Ref. [1], copyright 1967, with permission from Springer Nature

is a temperature gradient on the liquid surface, there will be a gradient of surface tension. Because the surface tension is inversely proportional to the temperature, under the effect of surface tension gradient, the fluid on the free surface will flow from the hot end to the cold end. In addition, because of viscosity and conservation of mass of the fluid, the free surface flow will drive the flow in the whole flow field. As early as in 1686, Heyde has observed the movement of camphor fragments on the free surface of olive oil. Due to various factors such as temperature and concentration, there is always difference in the distribution of surface tension on the liquid free surface, thus there exists gradient of surface tension. The fluid on the liquid surface usually moves to the area with higher surface tension, thus the convection is formed. The camphor fragments used as tracer particles showed the phenomenon of convection. Afterwards, the convection driven by the gradient of surface tension is called Marangoni convection, among which, the convection caused by un-uniformity in surface tension due to un-uniformity in surface temperature is called thermocapillary convection. Buoyant effect cannot be neglected in the study of thermocapillary convection in the gravity environment on the ground. The fluid flow is the coupled flow of thermocapillary convection and buoyant convection, so it is usually defined as buoyant-thermocapillary convection. People gradually realized that, under some extreme conditions, such as in the microgravity environment in space or in a small scale fluid system, the influence of gravity will be greatly weakened, and thermocapillary convection is the dominant factor in the process of natural convection [6–8].

More and more frequent aerospace activities and opening of the space station provide people many opportunities and resources to carry out research on pure thermocapillary convection. From the formation mechanism of thermocapillary convection, it is a nonlinear result of coupling effects of flow field and temperature field as well as mechanical movement and thermal movement. With the change of control parameters, viewed from the change of state of fluid, it may transit from original two-dimensional basic flow or axisymmetric flow to three-dimensional full field flow, from steady convection to unsteady convection or even oscillatory convection, from periodic oscillatory convection to chaos, and at last, under a certain condition, it may transit from laminar flow to turbulent flow as many other classic flow models do. Many problems in this process need to solve, so thermocapillary convection and its instability property are still open problems.

At present, the study on chaos and turbulence of fluid is a long-term and challenging project. For an unknown nonlinear system, the chaos phenomenon is ubiquitous, and the chaos theory is possible to explain the transition process of fluid from simple laminar flow to complex turbulent flow. However, there are few related reports on chaos study and analysis about experimental measurements of a nonlinear system. Therefore, numerical analysis on chaotic dynamics to the nonlinear model of annular liquid pool with Cz method is a very good try out.

SJ-10 recoverable satellite provides us a very good experimental condition under microgravity environment in space, and gives us the opportunity to carry out deep study on the phenomenon and mechanism of pure thermocapillary convection.

2 Research Status on Thermocapillary Convection

The instability of thermocapillary convection was first brought up by Smith and Davis [9]. They performed linear stability analysis on thermocapillary convection in the infinite horizontal thin liquid layer with a free surface driven by the horizontal temperature gradient, and discovered two kinds of instability, stable axial rolling cells and hydrothermal waves. They suggest that the critical Marangoni number (Ma_c) and the critical wave number are determined by Prandtl number (Pr) of the fluid. Generally, Marangoni number Ma characterizes the strength of thermocapillary convection; and Prandtl number characterizes the momentum exchange and the heat exchange in the fluid flow. Their definition is as Eqs. (1) and (2).

$$Ma = \frac{\sigma_T \Delta T L}{\rho \nu \kappa} \tag{1}$$

$$Pr = \frac{\nu}{\kappa} \tag{2}$$

where, ν is the kinematic viscosity, ρ is the density of fluid, κ is the thermal diffusivity, σ is the surface tension, σ_T is the temperature coefficient of surface tension, and ΔT is temperature difference.

Later researchers studied stability of thermocapillary convection and buoyant-thermocapillary convection under various conditions with this method [10–19], obtained critical conditions of flow transition process, analyzed influences of various factors on this process, and determined various possible flow models after destabilization. Hydrothermal waves are widely used in explaining the transition mechanism of thermocapillary convection from quasi-steady flow to oscillatory flow. With the perturbation theory, we can get perturbation solutions by solving eigenvalue equations and boundary condition equations, and for the basic state flow, we can get travelling wave solutions for perturbations of hydrothermal waves. This indicates the transition from steady single cell convection to oscillatory convection of hydrothermal waves. Generally, the relationship between the critical Ma number and Pr number can be obtained through calculation.

For thermocapillary convection, as a fluid mechanics system, its oscillation characteristics and the whole transition process from laminar flow to turbulent flow must be of great interest for theoretical research. On the other hand, many applications of thermocapillary convection should avoid the occurrence of oscillation. With the development of manned space activities, people have a great interest in the space material processing, especially the container free process for growing crystal in space, and try to realize the material production by pure distribution process by using the property of almost disappeared buoyancy effect in the microgravity environment, which is the aspirational material producing environment for mankind. However, in fact, in the microgravity environment, some secondary effects such as surface tension take the dominant role, and thus thermocapillary convection is prominent. Therefore, the study on thermocapillary convection has significant meanings in both theory and applications. Kang and Duan et al. [20–25] studied transition problems of thermocapillary convection in detail. They discovered a few transition routes including quasi-periodic transition, period-doubling bifurcation, quasi-periodic transition with tangent bifurcation, and period-doubling bifurcation with tangent bifurcation, and also found the laws for the occurrence of various transition routes. Tang, Li, and Hu et al. [26, 27] discussed transition routes of thermocapillary convection in the liquid bridge model of half-floating zone through numerical simulations, found the period-doubling bifurcation, and proved that the period-doubling bifurcation process of thermocapillary convection satisfies Feigenbaum general theorem. The quasi-periodic transition with tangent bifurcation and the period-doubling bifurcation with tangent bifurcation are relatively uncommon. There are abundant contents in the research on transition routes, and more models need to be further studied in this area.

Because the annular liquid pool is geometrically similar to the model for crystal growth by Cz method, the research on it has been widely concerned. The experimental study on thermocapillary convection in an annular pool was firstly carried out by Kamotani et al. [28]. 2cSt silicone oil with Pr number of 27 was used in their experiments, and the diameter of the liquid pool was 3.98 cm, 4.78 cm and 5.87 cm respectively. The oscillation phenomenon of thermocapillary convection was observed. After the onset of oscillation, the flow field turns into a non-axisymmetric three-dimensional structure; the velocity of surface flow is greatly higher than the velocity of inverse flow in the lower part of the fluid; the curve of temperature oscillation is approximately sinusoidal, and if Ar is fixed, the critical temperature difference for the onset of thermocapillary convection is not related to the radius of the liquid pool, so oscillation cannot be simply characterized by Ma number alone; surface thermal images show the periodic change of temperature oscillation, indicating that the intensity of convection on the radial plane is unstable and fluctuating.

In 1992, Kamotani and Ostrach et al. [29–31] finished the first experiment under microgravity condition, observed stable thermocapillary convection, and numerically studied the axisymmetric flow in steady state. The experimental fluid was 10cSt silicone oil, the diameter of the liquid pool was 10 cm, and the depth was 5 cm. Two sets of heating systems were used. One was the laser heating system with the laser diameter of 0.5–3.0 cm, and the other was the electric film heating system with the film diameter of 1.11 cm. By using the flow visualization technology (PTV), they observed flow fields with different heating methods and different shapes of free surface, and obtained 18 sets of experimental results, which were in good agreement with corresponding numerical results. The obtained $Ma = 3.1 \times 10^5$, was 5 times as high as that obtained in the ground experiment, but the oscillation had not been observed yet.

In 1995, Kamotani and S. Ostrach [32–37] finished the second experiment under microgravity condition in USML-2 space laboratory. The experimental fluid was 2cs silicone oil with $Pr = 27.33$, the diameter of the liquid pool was 1.2 cm, 2.0 cm and 3.0 cm respectively $Ar = 1$, $Ma < 6 \times 10^5$, and the temperature of the wall was 14 °C. When Ma number is high enough, the driven force of thermocapillary convection is mainly concentrated at the hot zone and the cold zone, and the oscillation phenomenon is closely related to the flow in the hot zone. The flow fields and the temperature fields of steady flow as well as oscillatory flow were observed, and the critical oscillation conditions were determined in this space experiment. The flow field was observed qualitatively, and the experimental result of thermocapillary convection in the liquid pool with 1.2 cm in diameter was in agreement with the result of ground experiments under 1-g condition. This indicates that the influence of gravity can be neglected in experiments with equipment equal to or smaller than this size. The surface deformation in the oscillation was also observed in the experiment, and a surface deformation parameter was defined. This parameter can well indicate the occurrence of oscillation, and when the free surface is very curved or the liquid layers very shallow, the oscillation will be delayed. The influence of heating rate on critical conditions was studied too. It turns out that the critical temperature difference is not related to the heating rate. Isotherms obtained by the thermal infrared imager discovered the change in temperature field on the surface during the oscillation process; the 2-earlobe oscillation mode was found in all experiments, and as the temperature difference was further increased, the pulsating pattern of 2-earlobe and 3-earlobe appeared. This indicates that, in an oscillation period, convection with fluctuating intensity will occur in different areas of the free surface. Later on, more experimental studies and numerical simulations about oscillatory flow in an annular pool have been completed by Ostrach et al. The silicone oil with Pr number of 27 was still used in their experiments, and the flowing states in an annular liquid pool were studied with the consideration of buoyant effect.

In 2001, Renaud et al. [38] experimentally and numerically studied nonlinear thermocapillary convection of the fluid with large Pr number in an annular pool, elaborated the oscillation characteristics of thermocapillary convection, and obtained distributions of the flow field and the temperature field. The results show that, oscillating thermocapillary convection is related to the capillary Re number; when the oscillation begins, it is in the standing wave form, and with the increase of acting force, it transforms into the travelling wave form; the number of hydrothermal waves depends on geometric parameters, and the mechanism of oscillation starting is related to fluid inertia, surface tension and buoyancy.

In the past decades, theoretical analyses and experimental studies on thermocapillary convection in various geometric structures have been completed, and transition processes of thermocapillary convection and various flow structures have been discovered. From 1990s to 2004, Sim and Schwabe et al. reported results of a series of three-dimensional numerical simulations on thermocapillary convection of fluids with medium Pr number in annuli [39, 40] as well as that of fluids with low Pr number in the annular shallow pool and the Cz structure shallow pool [41, 42]. They verified the existence of flow transitions and various oscillation modes. In recent years, people carried out further in-depth study on thermocapillary convection in the annular pool. In 2004, Sim et al. [43] studied axisymmetric thermocapillary convection with interface deformations in the open cylindrical annular pool. In this two-dimensional model, there is only steady convection no matter whether the interface deforms and how large Re number is. Therefore, even with large Re number and capillary number Ca, free surface deformations will not lead to oscillatory axisymmetric convection, so the deformed free surface is not the reason why the flow transits into oscillatory axisymmetric convection. When the free surface is convex at the cold wall and concave at the hot wall, there are two peaks if Re number is small and more waves if Re number is large. The heat dissipation on the free surface leads to more intensified convection when the Bi number is high, so obviously, the convection depends on the curvature of free surface. In addition, they compared thermocapillary convection in the cylindrical liquid bridge with that in the annular liquid pool [44]. These two models are based on the physical models for two methods of crystal growth technology, Cz method and floating-zone method, respectively. They also discussed dynamic free surface deformations in the axisymmetric model, and obtained results that were in agreement with the documentary [43]. That is, within a certain range of parameters, there is only steady convection in the axisymmetric model no matter how large Re number is, well, when Re number exceeds a certain critical value, the generated three-dimensional oscillatory flow depends on the aspect ratio (depth to diameter), Pr number and the melt volume V.

The earlier studies on restraining thermocapillary convection include: applying a magnetic field to the conductive melt, vibrating the wall to generate a flow field to restrain the surface flow, or injecting air parallel to the surface directly. These techniques are all attempted to reduce Ma number in order to weaken the fluctuation. The defect of these methods is that, due to the damping effect of ground state convection, the weakening of global mixing capacity results in macro segregation of chemical components. A way of improvement is to focus on instability of thermocapillary convection. Since the temperature of free surface determines the instability of flow, it is possible to change oscillation by changing temperature. Thus, the method of weakening fluctuations by changing the instability without changing the flow state is useful to improve the uniformity of single crystal growth no matter macroscopically or microscopically.

Schwabe et al. [45, 46] experimentally studied buoyant-thermocapillary convection in the annular liquid pool with inner and outer radius of 20 mm and 40 mm respectively and depth 2.5–20 mm by heating the outer wall and cooling the inner wall. It is found in their experiments that, the flow is steady multicellular flow under a small horizontal temperature difference. With the increase of temperature difference, the flow will lose its stability and show hydrothermal waves at first, and then, as the temperature difference is further increased, more complicated oscillatory flow will occur. In 2006, Schwabe and Benz et al. reported results of a set of successful experiments on thermocapillary convection in an annulus on the satellite FOTON-12 of Russia [47, 48]. In the experiments, the outer wall was heated and its radius was 40 mm, the inner wall was cooled and its radius was 20 mm, and the depth of the annulus was $d = 2.5$–20 mm. The experimental results show that, the flow is steady multicellular flow under a small horizontal temperature difference; as the temperature difference is increased, the flow will lose its stability and transits into hydrothermal waves first, then, as the temperature difference is further increased, more complicated oscillatory flow will occur. The results also show that, as the thickness of liquid layer increases, the number of flow cells in the multicellular structure decreases. The results of later numerical simulations are in agreement with the results of experiments.

Li et al. and Shi et al. carried out in-depth studies on thermocapillary convection of silicon melt in the annular shallow pool. In 2003, Li et al. [49] carried out a series of unsteady two-dimensional numerical simulations on thermocapillary convection in the annular liquid pool with the horizontal temperature gradient under the microgravity condition by the finite volume method. The outer wall of the annular liquid pool was heated and its radius was 40 mm, the inner wall was cooled and its radius was 20 mm, and the depth of the liquid pool was 3–14 mm. 0.65cSt silicone oil ($Pr = 6.7$) was used as the experimental fluid. The numerical results are depicted as a stability chart by the aspect ratio Ar (the ratio of depth to groove width) and thermocapillary Reynolds number Re, and show that time-dependent movements will only occur when the aspect ratio is larger than a certain value (about 2.29). In 2004, Li et al. [50] carried out unsteady three-dimensional numerical simulations on thermocapillary convection of the melt with low Pr number in the shallow pool of Cz structure under the action of horizontal temperature gradient by the finite difference method, and verified the existence of hydrothermal waves and the second transition in the flow.

In 2006, Shi et al. [51] calculated the critical conditions for the generation of hydrothermal waves in the annular pool under the normal gravity condition and the microgravity condition respectively. They analyzed characteristics of thermocapillary convection of the silicon melt and hydrothermal waves when the outer wall was heated and the inner wall was cooled. Research shows that the hydrothermal wave is one of the forms of unsteady flow in the annular pool, and its wave number and angular velocity change with the change of Ma number. When Ma number is small, a single group of hydrothermal waves spread to the entire region gradually with the increase of computation time, but when Ma number is increased, multiple groups of hydrothermal waves propagating in different azimuthal directions will exist at the same time, in addition, the curvature of temperature curve of the hydrothermal wave is large near the hot wall. In 2007, Peng et al. [52] studied three-dimensional buoyant-thermocapillary convection of silicone oil in the annular pool. Their results show that, there are three types of flow patterns when Ma number is large: when the liquid pool is shallow ($d = 1$ mm), there exist hydrothermal waves in the pattern of curved spoke, and with the increase of Ma number, a single group of hydrothermal waves transform into two coexisting groups of hydrothermal waves that have different wave numbers and propagate in different directions; when the liquid pool is deep ($d \geq 5$ mm), Rayleigh-Bénard instability appears due to the influence of buoyancy, and the flow pattern of straight spoke appears in the entire surface region; when $2 \leq d \leq 4$ mm, hydrothermal waves and three-dimensional oscillatory flow exist simultaneously, in addition, hydrothermal waves are at the inner wall of the liquid pool, well, in the region near the hot wall, under the influence of hydrothermal waves, the generated pairs of longitudinal flow cells rotating counterclockwise propagate circumferentially at the same angular velocity of hydrothermal waves.

The study on thermocapillary convection of silicon melt in the rotating annular pool was completed by Shi et al. [53]. In the rotating annular pool, the propagating direction of the hydrothermal wave is opposite to the rotating direction of the liquid pool, which is because the flow field in the azimuthal direction generated by Corilois force supplied extra energy. In the rotating pool, under a certain Ma number, the wave number of azimuthal wave increases, and a branch comes out on the spiral wave at the inner wall; as Ma number is increased to a certain value, two groups of hydrothermal wave are generated, and the group with a smaller wave number propagates in the same direction as the rotating direction of the pool. Both numerical calculation and linear stability analysis show that, within a certain range of rotating velocity, the rotation of liquid pool will have influence on the steady axisymmetric thermocapillary convection, and even when the rotating velocity is very low, the influence on the instability of convection is significant. Li et al. [54] carried out in-depth study on the transition process of thermocapillary convection in the liquid pool rotating slowly. It is found from the results that, when the radial temperature gradient on the free surface is increased, two types of flow transition will occur. Under a certain rotating velocity, two-dimensional steady flow transits into the first type of hydrothermal wave, with further increase of temperature difference, the flow will transit into the second type of hydrothermal wave with fewer wave numbers. The critical values for the generation of hydrothermal waves and the critical region where these two types of hydrothermal wave transform to each other all depend on the rotating velocity. The critical temperature difference at which the second type transits into the first type and the critical temperature difference at which the first type transits into the second type have hysteresis, and in the critical region, there exists the phenomenon of having both types of hydrothermal wave. With the increase of rotating velocity of the annular pool, the propagating direction of the second type of hydrothermal wave transforms gradually from the opposite of the rotating direction of the liquid pool to the same of that. Afterwards, the higher the rotating velocity, the faster the propagating speed of the hydrothermal wave, therefore, the rotating velocity of the liquid pool decides the propagating direction of the hydrothermal wave.

3 Ground Experimental Study on Thermocapillary Convection

Through years of research, people have got fundamental understanding about the scientific laws of microgravity. The equation used to describe the fluid movement in microgravity environment is Navier-Stokes equation with the term of gravity neglected. The work has been focused on critical conditions, especially the influences of surface tension and the gradient of surface tension on the fluid interface as well as the contact angle effect on the liquid-solid interface. Perturbation analysis is usually adopted for some simple processes, that is, by applying a little linear perturbation to a steady basic state, to solve the perturbed linear system of equations, and analyze its linear instability. Some people also solve the process of instability by the energy method. At present, three-dimensional unsteady computational fluid dynamics method is widely used in the research on nonlinear dynamics problems.

Conducting small-scale ground experiments with small Bond number Bo is an effective method of simulation study on thermocapillary convection. Bo number is used to analyze the relative importance between the actions of gravity and surface tension. We can see from the expression of Bo number that, a small Bo number can be realized by the small fluid density, the small acceleration of gravity and the small size. The simulation experiment emphasizes the overall effect. The gravity still exists during the process, and it will cause deformations of free surface and introduce new characteristic scales. Thus, the scale of the ground experiment should be as small as possible, and the diagnostic technique of the experiment is required to be very high. Since the opportunity for the microgravity experiment in space is very rare, a large number of ground experiments with small Bo number are carried out as a preliminary study of the space experiment.

In recent years, the National Microgravity Laboratory carried out a lot of ground experimental studies on thermocapillary convection in the annular liquid pool. It is found that, under the action of horizontal radial temperature gradient, oscillations of internal temperature and free surface will occur. The critical conditions have been obtained. For the same silicone oil, the critical conditions of two kinds of oscillation have the same trend of change with the change of the thickness of liquid layer [55–57].

3.1 Experimental Model

An experimental system of buoyant-thermocapillary convection in the annular liquid pool is established to visualize flow behavior, as shown in Fig. 2. The annular pool is made of red copper with good thermal conductivity, and its inner radius is $Ri = 4$ mm, outer radius is $Ro = 20$ mm, and depth is $d = 12$ mm. There is a resistance wire heating film in the central column for heating the fluid medium in the annulus. Six Peltier elements are attached to the external side of the liquid pool's outer wall for transferring the heat on the outer wall to the external environment, and keep the

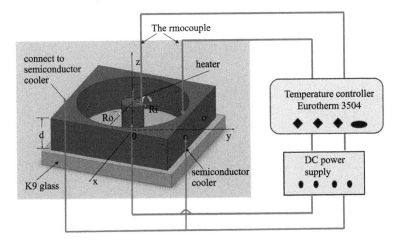

Fig. 2 Experimental system of Buoyant-thermocapillary convection

temperature of the cold end constant at a relative low level. The bottom of the pool is made of adiabatic material. The experimental media are KF96 1.5 and 2cSt silicone oil.

The Electric heating film and the Peltier elements are controlled to work according to the predetermined temperature control program through a DC power supply and a temperature controller, thus a horizontal temperature difference is generated in the annulus, and two thermocouples are used to monitor the temperature on the central column and the outer wall in real time. As the temperature difference between the two ends is increasing, the flow in the layer of silicone oil will transit from the steady state to the unsteady state, and the phenomenon of oscillation will appear.

3.2 Measurement System of Thermocapillary Convection

The temperature measurement system consists of T-type thermocouples, the nano-voltmeter, and the computer. In the experiment, T-type thermocouples are connected with the 2812A nanovoltmeter produced by Keithly to make up the temperature measurement system. The diameter of T-type thermocouple is 60 μm, and the soldered joint is 0.2 mm, one end is the measuring point and the other end is the cold end dipped in the ice water; the sensitivity of nanovoltmeter is as high as 1pV. The resolution of the whole temperature measurement system can reach 0.001 °C. The T-type thermocouples are put in the silicone oil directly, and the measured temperature signals are converted into voltage signals that can be recognized by the nanovoltmeter. Then the corresponding temperature values are calculated according to the relation between temperature and EMF (electromotive force), and they are recorded in the computer through the data acquisition instrument in real time.

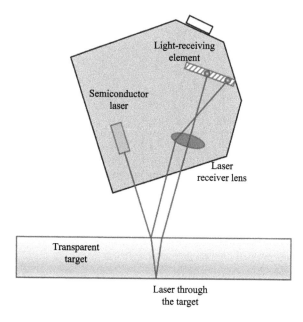

Fig. 3 Triangular measurement principle of the displacement sensor

The free surface oscillation measurement system consists of the LK-H080 high-precision laser displacement sensor produced by KEYENCE of Japan, the signal controller and the computer. The laser displacement sensor is based on triangular measurement principle, that is, the sensor calculates the distance to the object by focusing the light beam reflected from the object onto the light receiving element, as shown in Fig. 3.

The measurement of temperature distribution on the liquid surface uses the E60 thermal infrared imager produced by FLIR. It is set up right above the annular liquid pool and is focused on the fluid surface. The resolution of image is 320×240 pixels, the sensitivity of measurement is 0.05 °C, the precision of temperature measurement is ± 2 °C, and the maximum acquisition rate can reach 30 frames/s. The imager can set parameters of radiance rate and working distance and convert thermal infrared graph into temperature automatically. Then the temperature data is output in the form of CVS table. We can also extract temperature data series with time of a few points in the radial or circumferential direction on the liquid surface for the analyses of frequency and amplitude.

We also used the optical shadow method to measure the temperature field inside the liquid layer. This method is a kind of flow visualization technique based on the principle that light beams are deflected in different directions in different flow states and they are focused or diverged to form an optical image. The light path diagram of the principle of shadow method is shown in the left part of Fig. 4. The laser beam passes through the beam expander and collimating lens to form parallel light; the

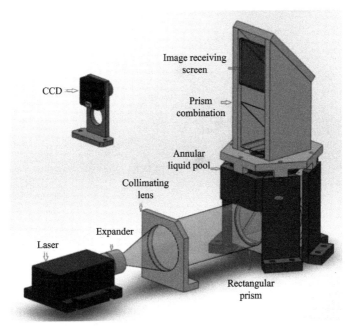

Fig. 4 The measurement system of shadow method

parallel lights totally reflected by the right angle prism that is installed right below the liquid pool, then it is projected onto the screen through the transparent bottom of the pool and the experimental medium, so the integral changing information of the fluid refractivity field (the corresponding temperature field) will be displayed on the screen in real time; videos are recorded by a CCD camera. Though the shadow method is usually used for qualitative visualization, since its pattern of light intensity has some regularity, in our analysis, we also extract a gray scale sequence at a certain point in the images to explain the change of temperature field.

3.3 Experimental Process and Results Analysis

In the experiment, the fluid in the annular liquid pool is 1.5 or 2cSt silicone oil, and the thickness of liquid layer increases from 0.8 to 3 mm with 0.25 mm per step. Various working conditions are experimentally studied. A T-type thermocouple is used to measure temperature oscillation signals at a single point in the liquid layer. With the increase of temperature difference, the internal temperature of fluid increases as well, and when the temperature difference reaches a critical value, the internal temperature of fluid starts to oscillate regularly, and the corresponding fluid flow transits from steady state to oscillatory convection. Figure 5 shows the original

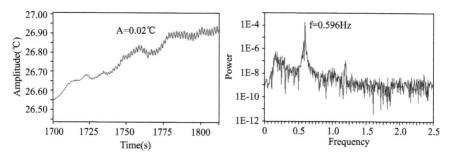

Fig. 5 Original temperature signals (left) and frequency spectrum analysis (right)

temperature signals (left) and frequency spectrum analysis (right) of 1.5cSt silicone oil. When the experiment time is at 1724 s, the temperature starts to oscillate, and the temperature difference is 16 °C at this time, then temperature oscillates regularly. The fundamental frequency in this stage is analyzed through fast Fourier transform (FFT). The fundamental frequency is 0.6 Hz, and the oscillation amplitude in this stage is 0.02 °C.

The free surface oscillation is an important characteristic in the transition process of thermocapillary convection. In the experiment, the free surface oscillation of thermocapillary convection in the annular liquid pool is measured with a displacement sensor simultaneously with the temperature measurement. Figure 6 shows the original displacement signals (left) and frequency spectrum analysis (right) of the measured point on the free surface in the working condition with 2 mm thick liquid layer of 1.5cSt silicone oil. At about 1710 s, when the displacement of the measured point on the free surface exceeds the threshold value, the displacement starts to oscillate regularly, and the temperature difference at this time is 16 °C. The fundamental frequency in this stage is analyzed through fast Fourier transform (FFT). The fundamental frequency is 0.6 Hz, and the oscillation amplitude is 1.09 μm.

The oscillation of free surface is the basic characteristic in the transition process of a dynamics system, and it is a physical quantity that is more sensitive than the oscillation of internal temperature of the fluid. With the increase of the temperature

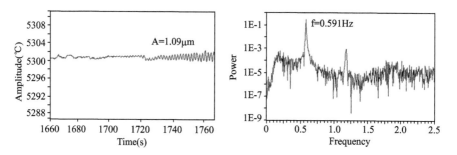

Fig. 6 Original displacement signals (left) and frequency spectrum analysis (right)

Fig. 7 Critical conditions

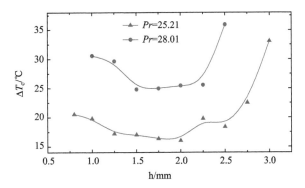

difference between the cold end and hot end, the fluid temperature changes as well, and when the critical temperature difference is reached, the free surface starts to oscillate regularly.

The critical values of temperature difference for the oscillation of buoyant-thermocapillary convection in 1.5cSt ($Pr = 25.21$) and 2cSt ($Pr = 28.01$) silicone oil with various thickness of liquid layer have been obtained. The critical curves are given as shown in Fig. 7. The critical temperature difference of buoyant-thermocapillary convection in 2cSt silicone oil is about 10°C higher than that in 1.5cSt silicone oil.

In the experiment of this study, a thermal infrared imager is used to capture the evolutionary process of the surface temperature field with the gradual increase of temperature difference between the inner and outer walls. The fluid media is 1cSt silicone oil with different thicknesses of liquid layer, and the purpose is to quantitatively observe the circumferential oscillation laws of temperature. It is found that a surface circumferential standing-wave mode exists in the transition of convection system in the annular liquid pool. With the increase of the radial temperature difference in the experiment, 7 standing-wave modes with different wave numbers ($m = 0, 1, 5, 6, 7, 8,$ and 9), and a hydrothermal traveling-wave mode are roughly found, as shown in Fig. 8. The radial oscillation exists in the seven types of surface standing-wave modes, and the energy is transferred from the wave abdomen to the side wall. Meanwhile, the standing-wave mode is a two-dimensional oscillation flow; the traveling wave mode of hydrothermal wave is characterized by circumferential spiral-wave in one-way or two-way rotation; the wave number changes and temperature turning points are also different in the same heating process, under the different thicknesses of liquid layers. The first critical temperature difference is between 9 and 12°C. Direct transition to the hydrothermal wave only occurs in thin liquid layers ($h < 1.2$ mm, $Bo < 0.25$). When the liquid layer is larger than 1.2 mm, radial concentric circles oscillation ($m = 0$) will occur, and then circumferential standing waves will occur. The patterns observed in 1.5 mm thick liquid layer are the most abundant ($m = 0,9,6,8,7$), which may be because of the suited aspect ratio.

During the heating process of the model of buoyant-thermocapillary convection, the projected images by the flow visualization of shadow method are screened in full. In the experiment, the shadow images at different stages of the convection have been

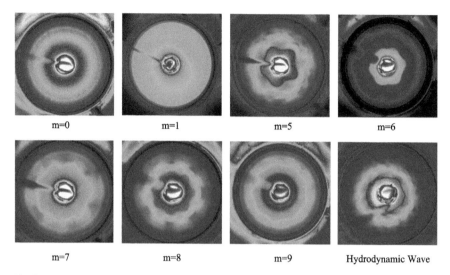

m=0   m=1   m=5   m=6

m=7   m=8   m=9   Hydrodynamic Wave

Fig. 8 Flow pattern transitions of Buoyant-thermocapillary convection, reprinted from Ref. [57], copyright 2016, with permission from Springer Nature

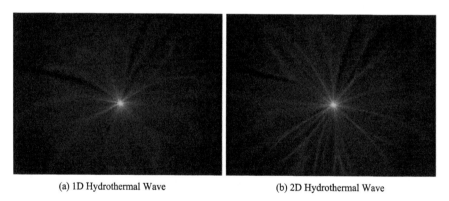

(a) 1D Hydrothermal Wave   (b) 2D Hydrothermal Wave

Fig. 9 Shadow images of hydrothermal waves

captured. When the temperature difference is small, the distribution of light intensity of the shadow image is uniform. With the increase of temperature difference, the distribution of light intensity of the shadow image gradually transforms from a single uniform mode to a rotating pattern of two modes. One is a kind of spoke-shaped helicon waves that start from the source at the top left of the center propagating circumferentially in the opposite directions and at the same angular velocity to a centrosymmetric sink, as shown in Fig. 9a; the other is a kind of spoke-shaped helicon waves that start from the source propagating circumferentially to the sink and move forward in the radial direction, as shown in Fig. 9b.

By analysis in comparison with the temperature measurement method, it is found that, when the flow field is in static state or steady basic flow state, the radial temperature distribution is mostly linear with a small gradient because the distribution of temperature field is uniform. The temperature at each point in the space can be considered constant for a certain period of time, thus light can pass through evenly, and there is no local converging or diffusion; with the increase of temperature difference, the steady basic flow firstly transforms into the hydrothermal wave propagating circumferentially in one-dimensional rotation, i.e., 1D Hydrothermal Wave (HW1). The temperature on the concentric circumference is not the same any more, but in the distribution of helicon wave with alternate high and low temperature and rotating at a certain angular velocity. At this time, the temperature on the circumferential points change regularly with time, and there exists phase difference spatially, as a result, the change in refractivity field caused by the change in thermal field will change the path of light passing through, which will reflect the shadow image that is consistent with the propagation of hydrothermal wave. When the temperature difference is increased to a certain value, the radial oscillatory flow appears that has been in one-dimensional circumferential unsteady state, that is, the radial flow field shows an oscillatory flow propagating from inside to outside with the change of convective state, and this is called 2D Hydrothermal Wave (HW2). It can also be considered as a 2D oscillatory flow. The ununiformity and periodic change of temperature field can also be reflected in the shadow image through the deflection effect of radial light.

4   Experimental Study in Space on Thermocapillary Convection

Because the simulation experiment on the ground can hardly avoid the influence of gravity totally, it is necessary to carry out microgravity experiment in space. SJ-10 recoverable satellite for scientific experiments has been launched successfully on April 6th at 1:38 am. 19 experimental projects of microgravity science and space life science are carried out on SJ-10 this time, and the space experimental study on surface waves of thermocapillary convection is one of them.

4.1   Scientific Objectives

The scientific objectives of this project is to establish thermocapillary convection system in an annular (cylindrical) liquid pool in space, study destabilization laws and transition routes of thermocapillary convection, study volume ratio effect for the first time, and understand the instability and oscillation mechanism of thermocapillary convection.

4.2 Development of Payload

The space experimental payload has 8 systems, including the liquid pool system, the liquid storage and injection system, the temperature control system, the temperature measurement system with thermocouples, the thermal infrared imager, the displacement sensor, the CCD image acquisition system and the electrical control system, as shown in Fig. 10.

Thermocapillary flow system in the annular liquid pool is established by heating and cooling the fluid system in space (see Fig. 11). High-sensitivity thermocouples are used to measure the fluid temperature, the displacement sensor is used to measure deformations of liquid surface. By these methods together with the infrared imaging, the oscillation characteristics, transition conditions, transition processes and flow pattern transitions of thermocapillary convection are obtained. In the meantime, the problem of volume ratio effect is also studied, which cannot be carried out on the ground.

The technical parameters of the space model of annular liquid pool are same as that of the ground experiment model. During the space experiment, the liquid is injected into the pool from the liquid injection hole on the bottom of the pool. Because of the loss of gravity, the fluid will climb along the walls of the pool, so in order to maintain the fluid interface, the wedge-shaped edge is designed on the central column and the outer wall of the annulus at the same height. This realized the key technique for maintenance and control of the liquid surface in space.

The liquid storage and injection system consists of the electric motor, liquid cylinder, transmission rod, pipeline and solenoid valve as shown in Fig. 12. Before the space experiment, the fluid is stored in the liquid cylinder, and at the beginning of the experiment, the fluid is injected into the annular liquid pool by the stepper motor to form the experimental fluid system. M-227 motor produced by PI of Germany is selected in this experiment, and it is driven by C-863 controller. The transmission rod is designed to connect the stepper motor and the piston of liquid cylinder.

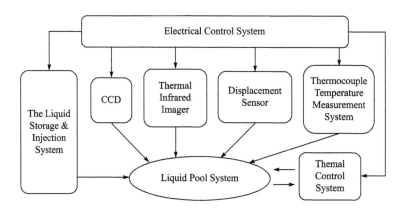

Fig. 10 The composition of payload

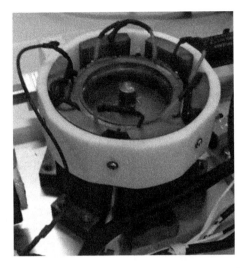

Fig. 11 The annular liquid pool, reprinted from Ref. [57], copyright 2016, with permission from Springer Nature

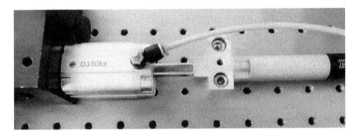

Fig. 12 The liquid storage and injection system, reprinted from Ref. [57], copyright 2016, with permission from Springer Nature

The liquid cylinder, pipeline and the solenoid valve are all products of FESTO of Germany. The model of liquid cylinder is ADVU-40-25-P-A. Its internal diameter is 40 mm, and its effective movement region is 25 mm. The model of solenoid valve is MSFG-12-0D. The transmission of experimental working fluid between the liquid pool and the liquid cylinder is controlled through opening and breaking of the solenoid valve. In the process of liquid loading, the negative pressure injection method is used, which ensures that there is no air bubble in the liquid model in the space experiment. This is also one of the key techniques in our project.

The temperature control system mainly includes high temperature sensor, low temperature sensor and ambient temperature sensor, the heating column and Peltier elements. Each part of the temperature control system is under the unified control and management of the electric control system. The temperature on the inner wall and the outer wall of the liquid pool is adjusted by PID control to ensure the temperature difference between the two ends of fluid within the required range of the experiment.

The image acquisition system includes a CCD camera and LED light source. The CCD camera is the model WAT-230VIVID produced by WATTEC of Japan, with effective pixels of 752×582, and analog video outputs. The LED light source selects hemispherical high-brightness LEDs, and the model is 1025MW7C. In the experiment, two LEDs are connected in series and parallel with another two in series, with the protecting resistor connected in series, to make sure the light source can work reliably and safely.

Two methods of temperature measurement are adopted in this experiment. One is using thermocouple sensors that measure the temperature directly and transform temperature signals into EMF signals. This thermocouple has a measuring range of -40 to $+125\,°C$, and a sensitivity of $0.05\,°C$. The other technique of temperature measurement is using the thermal infrared imager to measure the temperature distribution on the liquid surface of thermocapillary convection. The thermal infrared imaging technology uses various detective sensors to receive infrared radiation from the object, then processes the photoelectric information, and finally displays the temperature distribution of fluid surface by digits, signals and images. HL-C1 ultra-high-speed laser displacement sensor produced by Panasonic of Japan is selected to measure deformations of fluid surface. It can realize the ultra-high-precision measurement with a high sampling speed of $100\,\mu s$, resolution of $1\,\mu m$, and linearity of $\pm0.1\%$ F.S.

The computer control system implements the ON/OFF operations for the whole experimental device, receives the 28 V power supply from the payload manager, and uniformly provides and distributes power supplies to each subunit in the whole device; it injects instructions and uploads engineering parameters and scientific data through the communication ports; it manages equipment such as stepper motor controller, CCD camera, LED light source, thermal infrared imager, displacement sensor, solenoid valve, and so on; according to the ambient temperature and the temperature measurement conditions of high temperature and low temperature sensors, it controls the heating column and Peltier elements to reach the temperature difference requirement and collects and downloads temperature signals in real time.

This project is for the study on volume ratio effect and transition problems of pure thermocapillary convection in the microgravity environment in space. For thermocapillary convection system in an annular liquid pool, the study on volume ratio effect has to be carried out in the microgravity environment in space; about the problem of oscillation transition, a new phenomenon has been found in our ground experiment, that is, after the onset of oscillation, the oscillation disappears with the increase of temperature difference, and this has not been discovered in any other's experimental studies, and whether it is an inherent characteristic of thermocapillary convective oscillation in the annular liquid pool or due to the influence of gravity needs to be verified in the space experiment; in addition, it is found in the ground experiment that multiple oscillation frequencies exist simultaneously, and they increase with the increase of temperature difference, and this phenomenon of increasing fundamental frequencies also needs to be further studied in the space experiment. Therefore, the study on volume ratio effect and transition problems of pure thermocapillary flow system reflects the innovative idea of space experiments. The internal structure and the external appearance of the experiment case are shown in Fig. 13.

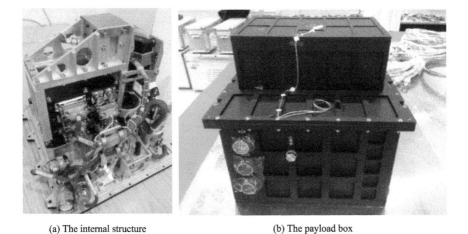

(a) The internal structure   (b) The payload box

Fig. 13 The internal structure and external appearance of the payload

4.3 Preliminary Results of the Space Experiment

In this space mission, 23 space experiments on surface waves of thermocapillary convection have been completed, including 13 times and 10 times respectively before and after the separation of orbit capsule and reenter capsule; 17 experiments used the linear heating mode and 6 experiments used the step heating mode; the volume ratio is 0.363–1.220; the max temperature difference is 25 °C, 29 °C, 31 °C, 32 °C, 34 °C, 35 °C, and 40 °C respectively; the heating rate includes 0.5 °C/min and 1.0 °C/min. Figure 14 shows the picture of fluid model captured by CCD during the experimental process in space and the temperature control curve, which indicates that the liquid has been injected successfully in the microgravity environment and the temperature has been controlled successfully.

 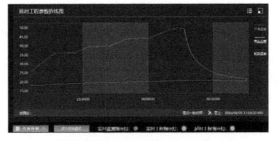

(a) The space experimental model   (b) The temperature control curve

Fig. 14 The fluid model of space experiment and the temperature control curve

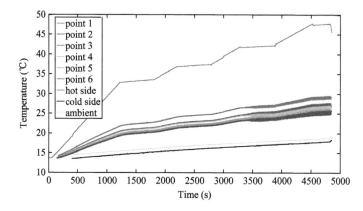

Fig. 15 The original temperature data in the space experiments

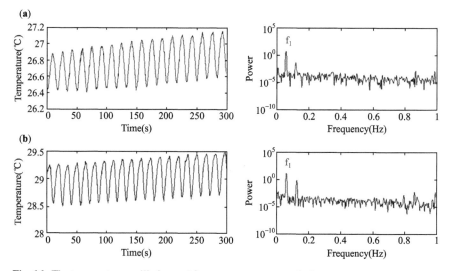

Fig. 16 The temperature oscillation and frequency spectrum analysis

The phenomenon of temperature oscillation is found according to the measuring results of thermocouples, as shown in Fig. 15. The oscillation frequencies are significantly lower than that measured in the ground experiments, with a difference of almost an order of magnitude, as shown in Fig. 16. The analysis shows that, the critical curves are separated into two branches. When the volume ratio is less than about 0.65, the critical temperature difference increases with the increase of volume ratio, and when the volume ratio is larger than about 0.65, the critical temperature difference decreases with the increase of volume ratio, as shown in Fig. 17. The quasi-periodic bifurcation and the period-doubling bifurcation have been discovered.

From the comparison between deformation measurements of liquid surface by the displacement sensor and the fluid temperature measurements, it is found that the

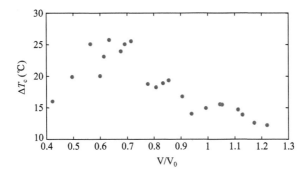

Fig. 17 The critical conditions

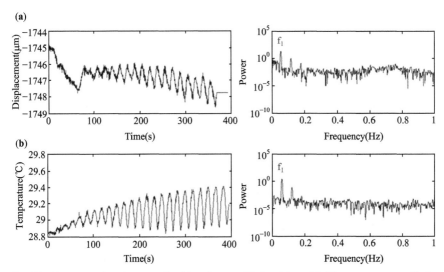

Fig. 18 The comparison between liquid surface oscillation (**a**) and fluid temperature oscillation (**b**)

liquid surface and the temperature start to oscillate simultaneously with the same oscillation frequency, as shown in Fig. 18.

The infrared measurement results of temperature distribution on the fluid surface at 23 working conditions with various volume ratios have been obtained in the space experiments. It is found that, the temperature on the fluid surface starts the axial oscillation first and transits to circumferential rotation, and then transits to the coexisting mode of axial oscillation and circumferential rotation together. This means that, thermocapillary convection transits from steady state to standing wave mode and then to travelling wave mode, and finally to the coupled mode of standing wave and travelling wave. Two oscillation modes with wave numbers of 3 and 4 have been found, as shown in Fig. 19. A large number of space experimental results have been processed, and more research results have been reported [58–60].

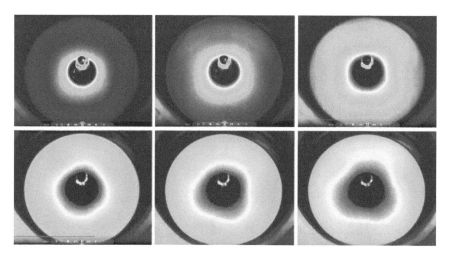

Fig. 19 The transition from wave number $m = 4$ to $m = 3$

5 Summary

This paper introduces the ground experiments and the space experiments on the thermocapillary convection in an annular liquid pool. In the ground experiments, thermocapillary convections with different depths (0.8 to 3 mm) are studied. The critical temperature difference of 2cSt silicone oil is about 10°C higher than that in 1.5cSt silicone oil. The transition from steady state to oscillatory convection is observed by the thermal couples, displacement sensor, infrared image and shadow graph. In the space experiments, thermocapillary convection with different volume ratio are studied. The critical condition is sensitive to the variation of volume ratio, which the critical curve can be divided into two branches. Due to the effect of buoyancy, the oscillatory convection only presents in a very thin layer in the ground experiments. However, in the space experiments, the oscillatory convection occurs in the deep layer (12 mm) with different surface configurations. It is found that thermocapillary convection transits from steady state to standing wave mode and then to travelling wave mode, and finally to the coupled mode of standing wave and travelling wave. In the future, we will give an in-depth analysis on the space experiments to give a better understand the characteristic, transition and nonlinear dynamic of the oscillatory thermocapillary convection.

Acknowledgements This project was funded by the National Natural Science Foundation of China (U1738116), and the Strategic Priority Research Program on Space Science of Chinese Academy of Sciences—SJ-10 Recoverable Scientific Experiment Satellite (XDA04020405 and XDA04020202-05).

The development of this space experimental payload got assistance from Astronaut Research and Training Center and Shenyang Zhixing Science and Technology Company Limited etc. The related experimental results and processing results of space experimental data obtained by the staffs and students of our project group are presented in this paper. Appreciations to all of them.

References

1. Cockayne B, Gates MP (1967) Growth striations in vertically pulled oxide and fluoride single crystals. J Mater Sci 2(2):118–123
2. Kobayashi N (1978) Computational simulation of the melt flow during Czochralski growth. J Cryst Growth 43(3):357–363
3. Kobayashi N, Wilcox WR (1982) Computational studies of convection due to rotation in a cylindrical floating zone. J Cryst Growth 59(3):616–624
4. Langlois WE, Shir CC (1977) Digital simulation of flow patterns in the Czochralski crystal-pulling process. Comput Methods Appl Mech Eng 12(2):145–152
5. Langlois WE (1985) Buoyancy-driven flows in crystal-growth melts. Annu Rev Fluid Mech 17(1):191–215
6. Ostrach S (1982) Low-gravity fluid flows. Annu Rev Fluid Mech 14(1):313–345
7. Napolitano LG, Monti R, Russo G (1984) Some results of the Marangoni free convection experiment. In: Materials science under microgravity. Results of Spacelab 1, vol 1, 15–22
8. Schmidt RJ, Milverton SW (1935) On the instability of a fluid when heated from below. Proc R Soc London Ser A Math Phys Sci 152(877):586–594
9. Smith MK, Davis SH (1983) Instabilities of dynamic thermocapillary liquid layers. Part 1. Convective instabilities. J Fluid Mech 132:119–144
10. Hu W et al (2010) Introduction to Microgravity Science. Science Press, Beijing (in Chinese)
11. Hashim I, Sarma W (2007) On oscillatory Marangoni convection in a rotating fluid layer subject to a uniform heat flux from below. Int Commun Heat Mass Transfer 34:225–230
12. Nepomnyashchy AA, Simanovskii IB, Braverman LM (2001) Stability of thermocapillary flows with inclined temperature gradient. J Fluid Mech 442:141–155
13. Kats-Demyanets V, Oron A, Nepomnyashchy AA (1997) Linear stability of a tri-layer fluid system driven by the thermocapillary effect. Actaastronautica 40(9):655–661
14. Ermakov MK, Ermakova MS (2004) Linear-stability analysis of thermocapillary convection in liquid bridges with highly deformed free surface. J Cryst Growth 266(1):160–166
15. Shiratori S, Kuhlmann HC, Hibiya T (2007) Linear stability of thermocapillary flow in partially confined half-zones. Phys Fluids (1994–present) 19(4):044103
16. Xu B, Ai X, Li BQ (2007) Stabilities of combined radiation and Rayleigh–Bénard–Marangoni convection in an open vertical cylinder. Int J Heat Mass Transf 50(15):3035–3046
17. Li MW, Zeng DL, Zhu TX (2002) Instability of the Marangoni convection in a liquid bridge with liquid encapsulation under microgravity condition. Int J Heat Mass Transf 45(1):157–164
18. Guo WD, Narayanan R (2007) Onset of Rayleigh–Marangoni convection in a cylindrical annulus heated from below. J Colloid Interface Sci 314(2):727–732
19. Hoyas S, Herrero H, Mancho AM (2004) Thermocapillar and thermogravitatory waves in a convection problem. Theoret Comput Fluid Dyn 18(2–4):309–321
20. Zhu P, Duan L, Kang Q (2013) Transition to chaos in thermocapillary convection. Int J Heat Mass Transf 57:457–464
21. Zhu P, Zhou B, Duan L, Kang Q (2011) Characteristics of surface oscillation in thermocapillary convection, Exp Thermal Fluid Sci 35:1444–1450
22. Jiang H, Duan L, Kang Q (2017) A peculiar bifurcation transition route of thermocapillary convection in rectangular liquid layers. Exp Thermal Fluid Sci 88:8–15
23. Jiang H, Duan L, Kang Q (2017) Instabilities of thermocapillary–buoyancy convection in open rectangular liquid layers, Chin Phys B 26(11):114703
24. Wang J, Di W, Duan L, Kang Q (2017) Ground experiment on the instability of buoyant-thermocapillary convection in large-scale liquid bridge with large Prandtl number. Int J Heat Mass Transf 108:2107–2119

25. Wang J, Duan L, Kang Q (2017) Oscillatory and chaotic buoyant-thermocapillary convection in the large-scale liquid bridge, Chin Phys Lett 34(7):074703
26. Tang ZM, Hu WR (1995) Fractal feature of oscillatory convection in the half-floating zone. Int J Heat Mass Transfer 38(17):3295–3303
27. Li K, Xun B, Hu WR (2016) Some bifurcation routes to chaos of thermocapillary convection in two-dimensional liquid layers of finite extent. Phys Fluids 28:054106
28. Kamotani Y, Lee J, Ostrach S (1992) An experimental study of oscillatory thermocapillary convection in cylindrical containers. Phys Fluids A 4:955–962
29. Kamotani Y, Chang A, Ostrach S (1994) Effects of heating mode on steady antisymmetric thermocapillary flows in microgravity. Heat Transf Microgravity Systems Trans ASME 290:53–59
30. Kamotani Y, Ostrach S, Pline A (1995) A thermocapillary convection experiment in microgravity. J Heat Trans-T ASME 117:611–618
31. Kamotani Y, Ostrach S, Pline A (1994) Analysis of velocity data taken in surface tension driven convection experiment in microgravity. Phys Fluids A 6:3601–3609
32. Kamotani Y, Ostrach S, Masud J (2000) Microgravity experiments and analysis of oscillatory thermocapillary flows in cylindrical containers. J Fluid Mech 410:211–233
33. Kamotani Y, Ostrach S, Pline A (1998) Some temperature field results from the thermocapillary flow experiment aboard USML-2 spacelab. Adv Space Res 22:1189–1195
34. Kamotani Y (1997) Surface tension driven convection in microgravity. Adv Astronaut Sci 96:487–499
35. Kamotani Y, Masud J, Lee JH, Pline A (1995) Oscillatory convection due to combined buoyancy and thermocapillarity. AIAA-paper 95-0817
36. Kamotani Y, Masud J, Pline A (1996) Oscillatory convection due to combined buoyancy and thermocapillarity. J Thermophys Heat Tr 10:102–108
37. Lin J, Kamotani Y, Ostrach S (1995) An experimental study of free surface deformation in oscillatory thermocapillary flow. Acta Astronaut 34:525–536
38. Renaud L, Gustav A, Henrik A (2001) Experimental and numerical investigation of nonlinear thermocapillary oscillations in an annular geometry. J Mech B Fluids 20:771–797
39. van der Vorst HA (1992) Bi-CGSTAB: a fast and smoothly converging variant of Bi-CG for the solution of nonsymmetric linear systems. SIAM J Sci Stat Comput 13:631–644
40. Li YR, Peng L, Akiyama Y, Imaishi N (2003) Three-dimensional numerical simulation of thermocapillary flow of moderate Prandtl number fluid in an annular pool. J Crystal Growth 259:374–387
41. Li YR, Imaishi N, Peng L, Wu SY, Hibiya T (2004) Thermocapillary flow in a shallow molten Silicon pool with Czconfiguration. J Crystal Growth 266:88–95
42. Li YR, Imaishi N, Azami T, Hibiya T (2004) Three-dimensional oscillatory flow in a thin annular pool of Silicon melt. J. Crystal Growth 260:28–42
43. Sim BC, Kim WS, Zebib A (2004) Axisymmetric thermocapillary convection in open cylindrical annuli with deforming interfaces. J Heat Mass Transfer 47:5365–5373
44. Sim BC, Zebib A (2004) Thermocapillary convection in cylindrical liquid bridges and annuli. CR Mecanique 332:473–486
45. Schwabe D, Benz S, Cramer A (1999) Experiment on the multi-roll-structure of thermocapillary flow in side-heated thin liquid layers. Adv Space Res 24(10):1367–1373
46. Benz S, Schwabe D (2001) The three-dimensional stationary instability in dynamic thermocapillary shallow cavities. Exp Fluids 31:409–416
47. Schwabe D, Benz S (2002) Thermocapillary flow instabilities in an annulus under microgravity results of the experiment MAGIA. Adv Space Res 29:629–638
48. Schwabe D (2002) Buoyant-thermocapillary and pure thermocapillary convective instabilities in Czochralskisystems. J Crystal Growth 237–239:1849–1853
49. Li YR, Peng L, Wu SY (2004) Thermocapillary convection in a differentially heated annular pool for moderate Prandtl number fluid. Int J Therm Sci 43:587–593
50. Li YR, Imaishi N, Azami T et al (2004) Three-dimensional oscillatory flow in a thin annular pool of silicon melt. J Crystal Growth 260:28–42

51. Shi WY, Imaishi N (2006) Hydrothermal waves in differentially heated shallow annular pools of silicone oil. J Crystal Growth 290:280–291
52. Peng L, Li YR, Shi WY (2007) Three-demensional thermocapillary-buoyancy flow of silicone oil in a differentially heated annular pool. J Heat Mass Transfer 50:872–880
53. Shi WY, Ermakov MK, Imaishi N (2006) Effect of pool rotation on thermocapillary convection in shallow annular pool of silicone oil. J Crystal Growth 294:474–485
54. Li YR, Xiao L, Wu SY (2007) Effect of pool rotation on flow pattern transition of silicon melt thermocapillary flow in a slowly rotating shallow annular pool. Int J Heat Mass Transfer
55. Zhang L, Duan L, Kang Q (2014) An experimental research on surface oscillation of buoyant-thermocapillary convection in an open cylindrical annuli. Acta Mechanica Sinica 30(5):681–686
56. Zhang D, Duan L, Kang Q (2016) Critical condition of buoyancy—thermal capillary convection oscillations in annular liquid pool. Mechanics and practice (in Chinese)
57. Kang Q, Duan L, Zhang L, Yin Y, Yang J, Hu W (2016) Thermocapillary convection experiment facility of an open cylindrical annuli for SJ-10 satellite. Microgravity Sci Technol 28:123–132
58. Kang Q, Jiang H, Duan L, Zhang C, Hu W (2019) The critical condition and oscillation - transition characteristics of thermocapillary convection in the space experiment on SJ-10 satellite. Int J Heat Mass Transf 135:479–490
59. Kang Q, Wang J, Duan L, Su YY, He JW, Wu D, Hu WR (2019) The volume ratio effect on flow patterns and transition processes of thermocapillary convection. J Fluid Mech 868:560–583
60. Kang Q, Wu D, Duan L, He J, Hu L, Duan L, Hu W (2019) Surface configurations and wave patterns of thermocapillary convection onboard the SJ10 satellite. Phys Fluids 31:044105

Droplet Manipulation and Colloidal Particle Self-assembling in Space

Yuren Wang, Ding Lan and Weibin Li

Abstract The manipulation of liquid drops could be very promising in space applications, such as fluid management, heat exchangers, life support systems, etc. Colloidal material box is one of a subsystem and payload of SJ-10 satellite, which was designed for studying the aqueous drop control and colloidal assembling in space. Based on space experiment platform (colloid material box), we have carried out the drop manipulation experiment in space successfully by using the hydrophilic/hydrophobic patterned surface technique, and the mechanism of the drop wetting and control on the patterned substrate during drop evaporation have been further studied. In addition, the drying process of the constrained colloidal drops on the patterned surface has been investigated. Through analyzing deposition patterns of both sessile and pendant droplets, we found that the deposition morphology is the result of competition and cooperation interactions of the free setting, the interface shrinkage and the outward capillary flow. Two different regimes for the relative motion between the particles and the interface was proposed: the pursuit regime (sessile droplet) and the meeting regime (pendant droplet), which furthered the theory of particles deposition in drying droplet.

Keywords Complex fluids · Drop manipulation · Colloidal assembling · Deposition patterns · Evaporation

Y. Wang (✉) · D. Lan · W. Li
Key Laboratory of Microgravity (National Microgravity Laboratory), Institute of Mechanics, Chinese Academy of Sciences, No. 15 Beisihuanxi Road, Haidian District, Beijing 100190, China
e-mail: yurenwang@imech.ac.cn

School of Engineering Science, University of Chinese Academy of Sciences, No. 19(A) Yuquan Road, Shijingshan District, Beijing 100049, China

© Science Press and Springer Nature Singapore Pte Ltd. 2019
W. R. Hu and Q. Kang (eds.), *Physical Science Under Microgravity: Experiments on Board the SJ-10 Recoverable Satellite*, Research for Development,
https://doi.org/10.1007/978-981-13-1340-0_6

1 Introduction

In this chapter, we focused on the droplet manipulation and colloidal particle self-assembling in space. Gravity driven flow is naturally occurred on earth and is utilized to manipulate fluid flow in abundant applications in industry. However, this privilege is no longer possessed in space. In space, absence of gravity induced surface tension gradient and capillary force dominating the fluid flow. Under this circumstance, a lot of work has been done aimed at solving fluid transport in microgravity. In the last two decades, most of studies in this field concentrated on controlling large volume fluid in space, such as fuel transport in the fuel tank. Recently, the studies turn to concern controlling small volume liquid transport inside experimental and life support system, such as droplet movements in an open cabin and liquid–liquid separation in spacecraft. It is of vital significance for human beings long-term staying in space. In 2017, several space experiments on the study of surface tension measurement and control were done in ISS, including "Surface Tension Containment Experiment-1", "Capillary Structures for Exploration Life Support" and "Continuous Liquid–Liquid Separation in Microgravity".

This chapter was divided into five parts. In Sect. 2, we introduced the Colloidal Material Box (CMB) facility boarding SJ-10 recoverable satellite. Section 3 focused on the water and colloidal droplets manipulation technique and the evaporation processes in an open space cabin. Subsequently, the colloidal self-assembling physics in the evaporation process of colloidal droplets was described in Sect. 4. Conclusion and outlook were summarized in the final part.

2 The First Attempt for the Colloidal Assembly in the Space

Colloidal system is an ideal model to study phase transitions and defect formation [1], because in comparison to atoms, micron-sized colloidal particles are big enough and move sufficiently slowly to allow direct observation with optical microscopy, thus the phase change and defect formation processes on the "atomic" scale could be visualized [2]. However, due to the gravity, the colloidal particles will sediment and result in uneven concentrations of colloidal system. It will be difficult to accurately distinguish the relationship between the phase transition and the precipitate [3].

Convective assembly is a widely adopted method to form ordered structure in the colloidal suspension [4]. A template assisted self-assembly (TASA) method was developed by Xia and Yin, which was used to produce a variety of complex aggregates with well-controlled sizes, shapes and structures [5]. More recently, D.J. Norris explored the properties of photonic crystals by self-assembly from colloidal microspheres, and found that solvent flow played a critical role in controlling the formation process [6]. Convective assembly relies on the accumulation of particles near three phase contact lines driven by solvent evaporation, which was attributed to capillary flow resulting from non-uniform evaporation flux and contact line pinning [7]. However, when the convective process induced by evaporation occurs at ground-level, the self-assembling behavior is obscured by gravitational effects including sedimentation and buoyancy convection, and particles are assembled in every desired orientation to the gravitational vector [8, 9]. Under microgravity conditions, the self-assembly process induced by evaporation occurs in a uniform environment that is devoid of gravity-driven, deleterious convective mass transport or sedimentation effects [10]. Therefore, the change to a gravity-free environment is expected to have a remarkable effect on the assembly dynamics and final deposition patterns.

As a colloidal suspension, the phase transition process of liquid crystal is affected by parameters such as gravity, diffusion force and particle dispersion, but the details of these effects remain unclear [11]. A theory was formulated to describe the mechanism of phase transition in liquid crystal [12], hypothesizing that the liquid crystal phase transition is an entropy-driven process based on the results of competition, interaction of the excluded volume, and orientational entropy [13]. Onsager theory assumed that the phase transition was an ideal process without consideration of effects of gravity. However, due to the presence of gravity, the diffusion force, polydispersity of particles and long-range interactions between molecules on the ground, the actual phase transition likely differs from the ideal process. Thus, it is important to verify and correct Onsager theory in microgravity conditions.

Most research demonstrated that gravity plays a prominent role in affecting the structure of colloidal crystals and the phase transition process of colloidal suspensions [14, 15]. The phase behavior of colloidal particles under microgravity conditions has been studied, and experimental results have also been obtained from space research, including nucleation and growth processes in colloidal crystallization [16, 17], kinetics of colloidal alloy crystallization of binary mixtures [18, 19], and heterogeneous nucleation induced by seed or wall [20]. The positive influence of gravity on the self-assembly and crystal-growth processes was detected [10]. Further study of these processes should be performed as space experiments. The recoverable satellite is an effective tool for space experiments in the microgravity environment [21, 22], and many space microgravity experiments in China have been completed aboard recoverable satellites since the late 1980s. In the mid-2000s, the Chinese National Space Administration (CNSA) developed SJ-10 recoverable satellites, which were mainly used to perform experiments of microgravity science and space life science. The engineering phase of program SJ-10 was started by Chinese Academy of Sciences (CAS) since the beginning of 2013, and the satellite was launched in early 2016 [23], and stayed for 15 days in Earth orbit. The colloidal material box (CMB) is one of several experimental boxes aboard the SJ-10 satellite, which was used to study the self-assembly dynamics of the colloidal spheres (with or without Au-coated nanoparticles) under microgravity, and to test the liquid crystal phase transition model, namely the Onsager model in space.

2.1 The Structure and Functional Units of the Colloidal Material Box (CMB)

The structure and function of the CMB which consists of three modules: (i) Colloidal evaporation experimental module, made up of a sample management unit, an injection management unit, and an optical observation unit, using to observe the self-assembly process of colloidal spheres at the solid/liquid interface and the change of the droplet profile under microgravity. (ii) Liquid crystal phase transition experiments module, including a sample management unit and an optical observation unit, to observe the process of liquid crystal phase transition and verify the mechanism of phase formation driven by entropy alone. (iii) Electronic control module, to manage the experimental procedure in Earth orbit and respond to interface of the integrated electrical control box on the satellite.

Two experimental projects to be implemented in space are listed in Table 2 of Chap. 1 (i) Self-assembly of colloidal spheres induced by evaporation. As shown in Fig. 1, five colloidal drops with different volumes will be injected in sequence onto the five sample positions installed on the linear displacement platform. The change of colloidal droplet profile with evaporation will be recorded by CCD1 camera, capturing side-view images at 0.5 fps. These images will be used to calculate the apparent contact angle, volume, radius, and height of the droplet. The time-dependent contact angle of the droplet will be obtained. In addition, the dynamic deposition process of colloidal microspheres during evaporation will be captured by CCD2 camera at 5 fps, in which continual microscopic images will reveal the movement direction and deposition rate of the particles at the solid/liquid interface during droplet evaporation. We will also obtain images of ordered and disorder phases of the final deposition by moving the linear displacement platform. (ii) Formation and evolution process of ordered phase driven by entropy alone. Six liquid crystal samples of different concentrations will be put in the sample positions within the analyzer and polarizer as shown in Fig. 1. We will observe the phase transition process using CCD3 camera, and obtain color macroscopic images of the liquid crystal phase transition at 30-min intervals. From the images, we will observe the morphology of ordered phase with different concentrations, analyze the liquid crystal orientation, and obtain the phase transformation time.

2.2 Colloidal Evaporation Experimental Module of the CMB

The CMB includes three parts as shown in Fig. 2. (i) Colloidal evaporation experimental module, composed of a sample management unit, an injection management unit, and an optical observation unit. (ii) Liquid crystal phase transition experimental module, which consists of a sample management unit and an optical observation unit. (iii) Electronic control module, to manage the experimental procedure of the first two modules and respond to the interface of the integrated electrical control box located

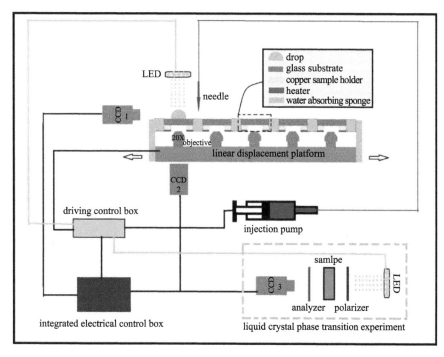

Fig. 1 Experimental principle diagram of the CMB, reprinted from Ref. [24], copyright 2016, with permission from Springer

in the instrument compartment on the SJ-10 satellite, allowing image acquisition and transmission.

The sample management unit was designed to provide sample positions necessary fora drop evaporation experiment, heating the samples, and implementing conversion of the sample positions. The optical observation unit will be used to observe the outer droplet profile and the motion of colloidal particles within the droplet during evaporation. These two units were integrated in the colloidal experimental table as shown in Fig. 3a.

There are five colloidal sample positions with sample substrates, one motor, a linear displacement platform, sets of parts for sample heating and a thermal sensor (DS18B20), five microscope objectives, one deflective optical path, and two CCDs. The motor can move the linear displacement platform back and forth linearly, and the five sample positions can be switched sequentially. Figure 3b shows that each sample position has a heating plate. The heating range can be controlled between 40 and 55 degrees Celsius, and the stability of temperature control is less than 1 degrees Celsius. The thermal sensor (DS18B20) contacting the sample substrate is used for temperature acquisition.

The five microscopic objectives were fixed on the linear displacement platform, and each of them was placed just below the sample position, so they could move

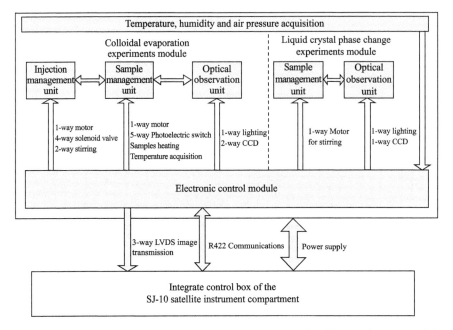

Fig. 2 Flow chart of composition and application of the CMB, reprinted from Ref. [24], copyright 2016, with permission from Springer

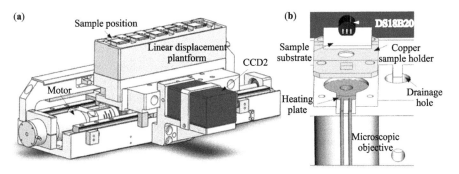

Fig. 3 **a** Schematic of colloidal experiments table; **b** The schematic of the structure of a colloidal sample position on the experimental table. Reprinted from Ref. [24], copyright 2016, with permission from Springer

synchronously with displacement of the platform (Fig. 3b). CCD2 was fixed on the experimental table, which allows observation of microscopic images in the corresponding sample position by deflective optical path, the hole for positioning the optical path, and the axis of CCD2. The side-view CCD (CCD1) was fixed for macroscopic observation of the droplet profile, as shown in Fig. 1.

Since drainage holes (Fig. 3b) and sample positions are arranged alternately, space experiments in all five sample positions can be completed by moving the platform in one direction. The engineering states of the five positions is exactly the same, so the use of samples, image observing, and heating process can be mutual backups, which increases the chance of success for the colloidal evaporation experiments in space.

3   How to Control the Droplet in the Space

The manipulation of liquid drops under microgravity environment could be very promising in space applications, such as life support systems, waste water treatment, heat exchangers, machining biologics and pharmacy [25, 26]. Over the past two decades, a variety of microgravity experiments concerning growth and manipulation of liquid drops had been performed. Antar et al. performed drop coalescence experiments through two hypodermic needles which were contacted using manual force, allowing the drops to coalesce [27]. Savino et al. examined wetting and coalescence of silicone oil sessile drops formed over copper substrates using a cylindrical copper needle with a coaxial hole [28]. To investigate gravity effect on the contact angle and the drop interface shape, Brutin et al. created drops by injection through a hole of the substrate [29]. All these works had already led to major advances for the drop manipulation in space. However, these solutions were too complicated; furthermore, the drops couldn't be fully constrained, which was likely to be unstable under vibration or attitude adjustment of the spacecraft. A more simple and reliable technique was necessary for the control of liquid drops under the conditions of weightlessness.

3.1   The Confined Substrate for Droplet Pinning

Patterned substrates on the basis of wetting enhancement or wetting barrier were always designed to manipulate the drop in terrestrial condition [30–34]. For examples, Tenjimbayashi et al. introduced a liquid manipulation strategy to design dynamically hydrophobic and statically hydrophobic/hydrophilic patterned surfaces [35]. Dong and McCarthy described a method for preparing super-hydrophobic surfaces containing guiding lines that control water motion [36]. Draper et al. designed the patterned surface with a super-hydrophobic polymer coating for the manipulation of droplets within microfluidic channels [37]. Zhang et al. designed super-hydrophobic TiO_2 nanostructures for condensate micro-drop self-propelling [38]. Wu et al. designed Laplace pressure pattern based on conical morphology and wetting heterogeneity for micro-droplet manipulation [39]. These works illustrated that the patterning technology had excellent performance in drop manipulation under terrestrial condition. We might wonder whether it was feasible to expand this technology to space as well. In fact, the drops could be easily kept stable on a substrate under the gravity force on earth that has nothing to do with the surface modification of the substrate. However, in microgravity environment, gravitational effects were negligibly small and surface tension effects dominated the liquid behavior [40], thus the liquid can climb along the wetting surface [41, 42], which might be easily to result in catastrophic effects in space. Therefore, the applicability of drop manipulation through modifying the surface wettability in microgravity environment was still a problem that await for proof.

The rough and the non-uniform wettability of the contact line anchored the border of the droplet, which is referred to as canthotaxis effect [43]. In addition, the particles within the droplet will accumulate at the contact line during evaporation, which could change the roughness and the wettability of the contact line and result in the self-pinning effect [44]. The sample substrate will be placed on the copper sample holder, composed of hydrophilic and hydrophobic areas. The round hydrophilic area is a quartz surface with a diameter of 5 mm, which is surrounded by a super-hydrophobic coating (SHOS150; Shunytech). As can be seen in Fig. 4a, the white part is the super-hydrophobic coating and the center part is the confined hydrophilic area. Figure 4b shows that the super-hydrophobic contact angle of the coating is about 150°, and its thickness is nearly 60 μm, as shown in Fig. 4c. The sample substrate could confine a droplet with extremely volume of 100 μL within the hydrophilic area, and prevent it from floating away in the microgravity of space. The polystyrene colloidal microspheres, dispersed in super-pure water with a mean diameter of 2.2 μm were purchased from Duke (5200A). The polystyrene colloidal suspension will be prepared by diluting the original solution to 0.1% (w/w) with demonized water.

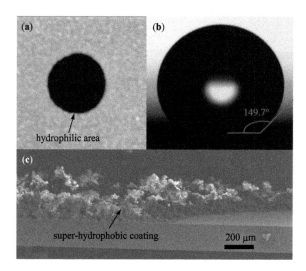

Fig. 4 Substrate with confined hydrophilic area. **a** A quartz crystal wafer with confined area treated by super-hydrophobic coating, the white part is the coating while the center part is the confined hydrophilic area; **b** The super-hydrophobic contact angle is about 150°; **c** SEM image of the cross section of substrate shows that the thickness of the coating is nearly 60 μm. Reprinted from Ref. [24], copyright 2016, with permission from Springer

3.2   The Profiles of the Confined Colloidal Droplet on the Ground or in the Space

The drop shape was mainly dominated by two forces: the surface tension force, which tended to minimize the area of the surface to decrease the surface energy (producing typically a spherical shape), while the gravitational force which tended to flatten the drop. The balance of these two forces was described by the Bond number Bo, $Bo = (\rho_L - \rho_V)gL^2/\sigma$, where, ρ_L was the density of the liquid, ρ_V was the density of the vapor, L was the characteristic length, g was the gravitational acceleration, and σ was the liquid–vapor surface tension [45]. It was critical to choose the characteristic length.

The patterned substrate could be used for capturing water or water-based drops in space. The capture ability of the patterned substrate was attributed to its surface wettability. Evaporation studies were useful in characterizing wetting behavior because drops with various sizes can be created to evaluate the transition criterion [46, 47]. For a large drop resting on an ideal patterned substrate, with the decrease of the drop volume (V) as evaporation, three different regimes could be distinguished, as shown in Fig. 5a [48]. The intrinsic contact angles, which was determined by the chemical compositions, were θ_α and θ_β for α phase (hydrophobic region) and β phase (hydrophilic region), respectively. In regime 1 and 3, the apparent contact angle (ACA) was equal to the intrinsic contact angles. Different chemical compositions determined the wetting barrier at the interphase boundary between the hydrophilic region and the hydrophobic region. At this wetting barrier, the liquid surface turned from the contact angle with the better-wetting face to the contact angle with the worse-wetting face [49]. Therefore, in regime 2, the contact radius kept constant as R and the apparent contact angle (ACA) was in a closed interval $[\theta_\beta,\theta_\alpha]$. Thus, the aqueous drop of the specific volume in between $[V(R,\theta_\beta),V(R,\theta_\alpha)]$ could be pinned and constrained by the wetting barrier in this regime, which was responsible for the drop capturing. The super-wettable surfaces (α phase was super-hydrophobic, and the β phase was super-hydrophilic) could be chosen to achieve large drop capturing.

Figure 5a showed ideal homogeneous surfaces (α phase and β phase) that contact angle hysteresis (CAH) was negligible. In fact, the hysteresis could arise from any surface roughness or heterogeneity. We could speculate that the range of ACA in regime 2 would be wider for real surfaces, i.e., the maximum and minimum ACA were equal to the advancing contact angle of α phase and the receding contact angle of β phase. The advancing contact angle and receding contact angle for α phase and β phase could be marked as $\theta_{\alpha a}$, $\theta_{\alpha r}$, $\theta_{\beta a}$ and $\theta_{\beta r}$ in sequence. Thus, the ACA was in a closed interval $[\theta_{\beta r}, \theta_{\alpha a}]$ in regime 2. The volume of aqueous drop that could be constrained was in between $[V(R,\theta_{\beta r}),V(R,\theta_{\alpha a})]$, which was larger than the ideal patterned surface. In our experiment, the quartz glass and SHOS150 coating could be considered as α phase and βphase, respectively. We had measured the contact angles and hysteresis for water and water-based suspension on these two surfaces, as shown in Table 1. The aqueous drop with ACA between 26° and 156° could be constrained. For water-based suspension drop, the ACA was in between 17° and 137°. It was

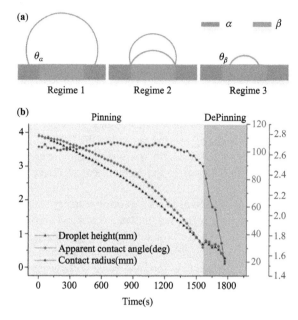

Fig. 5 **a** The three wetting regimes of a drop on the ideal patterned substrate as with evaporation, where the intrinsic contact angles are θ_α and θ_β in regimes 1 and 3, respectively. The drop could be confined in regime 2. **b** Evaporation of an aqueous drop on a patterned substrate with initial volume of 60 μL under microgravity environment. The black, red, and blue line represent the drop height, the apparent contact angle, and the contact radius variation versus time t, respectively. Reprinted from Ref. [50], copyright 2018, with permission from American Chemical Society

Table 1 Contact angles and hysteresis for water and water-based suspension on homogeneous surfaces, reprinted from Ref. [50], copyright 2018, with permission from American Chemical Society

Surfaces	Contact liquids					
	Water			Water-based suspension		
	θ_a (deg)	θ_r (deg)	$\Delta\theta$ (deg)	θ_a (deg)	θ_r (deg)	$\Delta\theta$ (deg)
Quartz glass	62° ± 2°	26° ± 2°	36° ± 4°	58° ± 3°	17° ± 1°	41° ± 4°
SHOS150 coating	156° ± 2°	134° ± 3°	22° ± 5°	137° ± 3°	131° ± 3°	6° ± 6°

Notes θ_a is the advancing contact angle, θ_r is the receding contact angle, and $\Delta\theta = \theta_a - \theta_r$ is the contact angle hysteresis of homogeneous surface

realized that the contact angles for water-based suspension were less than the purity water, which could be attributed to the decrease the surface energy of the aqueous solution because of impurities (PS microspheres).

The aqueous drops in regime 2 came in different shapes in space and on the ground, which meant that the drop volumes that could be captured by the patterned substrate might be unequal under the normal gravity and microgravity. We had further studied

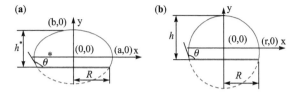

Fig. 6 **a** Geometry of an ellipsoidal cap-shaped drop model; **b** A spherical cap-shaped drop model. Reprinted from Ref. [50], copyright 2018, with permission from American Chemical Society

the gravity effect on the capacity of the substrate. The drop resting on the substrate on earth could be easily distorted by gravity, its shape could be considered as an oblate ellipsoidal cap geometry [51], where h^* was the drop height, θ^* was the ACA, R was the contact radius, as shown in Fig. 6a. The eccentricity value, e, was held constant for all times. The volume of the oblate ellipsoidal cap geometry was given by

$$V^* = \frac{\pi}{6(1-e^2)}[3(1-e^2)R^2 + h^{*2}]h^* \tag{1}$$

The ACA of the drop could be expressed as

$$\theta^* = \arctan[\frac{2h^*R(1-e^2)}{(1-e^2)R^2 - h^{*2}}](1 > e \geq 0) \tag{2}$$

For a drop in microgravity condition, it could be considered having the shape of a spherical cap, as shown in Fig. 6b. The drop volume V could be expressed as a function of the contact radius R, and the ACA,

$$V(R,\theta) = \frac{\pi R^3}{3}\frac{(1-\cos\theta)^2(2+\cos\theta)}{\sin^3\theta} \tag{3}$$

The ACA could be expressed as a function of the drop height h and the contact radius R,

$$\theta = 2\arctan\frac{h}{R} \tag{4}$$

3.3 The Evaporation of the Droplet in the Space

To reveal the confinement mechanism of the substrate, evaporation experiments of water drops with different initial volumes (30, 40, 50, 60, 70 μL) had been conducted under microgravity conditions. A typical drop with initial volume of 60 μL was chosen to describe the wetting transition. The time evolution of the drop height, the ACA, and the contact radius, were shown in Fig. 5b. The evaporation process could be divided into two stages. In the first stage, the height and the ACA of the drop nonlinearly decreased with time, but in the meanwhile, the contact radius kept unchanged, and the drop evaporated in constant contact radius (CCR) mode. The drop evaporation spends almost all of its time in this stage, which could be characterized by the quasi-steady diffusion-driven evaporation model. In the last minute of evaporation, the contact line began to slip suddenly and the evaporation entered into the second stage. We noted that the ACA at the transition point (30.8°) was nearly equal to the receding contact angle of the quartz glass (26°), which fitted basically with the theoretical prediction. In the second stage, the contact radius decreased rapidly, the height and the ACA showed a broken line variation, and the drop evaporated as a skip-slip mode. What we cared about was the first stage of evaporation, which corresponded to regime 2. For water-based suspension drop, the ACA at the transition point (~17°) was less than the water drop's (26°). Actually, there was no transition point for an evaporating colloidal drop, because a ring-like stain formed near the contact line, which enhanced the pinning effect and made the drop evaporating as CCR mode. The surface irregularities and unevenness of the patterned substrate was inevitable in our experiment, which would enhance the pinning effect and make the drop being confined more firmly.

The drop was considered to be equivalent to the spherical drop with the same volume, and we chose the diameter of this spherical drop as the characteristic length, thus the Bond number (Bo) of the drops with different equivalent volumes (V_e) in normal gravity could be calculated. The differences of the ACA ($\Delta\theta = \theta^* - \theta$) and the height ($\Delta h = h - h^*$) of the drop between normal gravity and microgravity were plotted versus the Bond number, as shown in Fig. 7d. $\Delta\theta$ and Δh could be used to judge the shape deviation of the drop in normal gravity from an ideal spherical cap drop in microgravity. When the $Bo < 1$ ($V_e < 10.3$ μL), $\Delta\theta$ and Δh was close to zero, which meant that the drop shape could be hardly affected by gravity. When the $Bo = 1$ ($V_e = 10.3$ μL), the characteristic length $L = 2.7$ mm (the capillary length), the equivalent volume of the drop was 10.3 μL, which coincided with the experimental results (10 μL), as shown in Fig. 7c. When the $Bo > 1$ ($V_e = 10.3$ μL), $\Delta\theta$ and Δh increased as with increasing of the Bond number, so the flattening effect of the drop resulting from gravity was more pronounced for higher values of the Bond number. It indicated that the ACA could be easier to reach to the advancing contact angle of the hydrophobic region for drops in normal gravity, so this patterned substrate could confine drops with greater volume in space than on earth. Therefore, the gravity effect significantly influenced the confinement capacity of the patterned substrate, the bigger the drop volume, the larger the gravity effect on its confinement capacity.

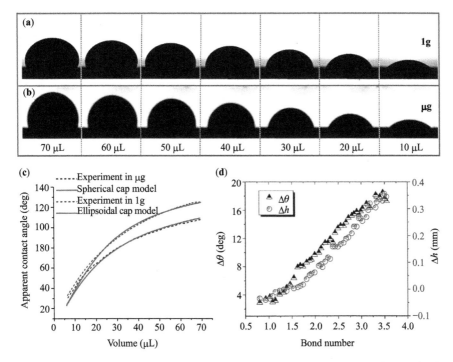

Fig. 7 Profile images of drop with different volume at normal gravity (**a**) and microgravity (**b**) conditions during the evaporation process; **c** The theoretical and experimental apparent contact angles were plotted versus volume for evaporating drops in normal gravity and microgravity ($e = 0.56$); **d** The differences of the apparent contact angle ($\Delta\theta$) and the height (Δh) of the drop between normal gravity and microgravity were plotted versus the Bond number (Bo). Reprinted from Ref. [50], copyright 2018, with permission from American Chemical Society

4 The Self-assembly of the Colloidal Particles

4.1 Multi-physical Effects During the Droplet Evaporation

The evaporation of a sessile colloidal droplet occurs commonly in nature, and the evaporation process involves multiple physical phenomenon including pinning effects [52, 53], convection flow [52, 53], dewetting [54], and capillary forces [55, 56]. These physical effects influence a wide range of applications, such as photonic crystal formation [57], nano-material assembly [58], injection printing [59], and biotechnology [60, 61]. When the contact angle is small, the evaporation will induce the outward capillary flow mainly which bring the particles to the edge, eventually forming the familiar "coffee-ring" pattern [62]. As far as applications are concerned, the most important thing is to know how the deposition morphologies form and then how to control the deposition pattern. During droplet evaporation, surfactant [63] or a temperature gradient [64] along the droplet surface induced Marangoni flow, which could affect the flow field and the distribution of particles on the substrate, but it was the only element to determine the ultimate patterns. It was found that minimizing contact angle hysteresis could lead to the formation of a central stain, rather than a ring [65, 66]. Particles that spontaneously adsorb at the substrate [67] or migrate at the liquid–vapor interface [68] also oppose the coffee ring effect. Even in the presence of pinning, the ring formation can be impeded by the presence of a reverse (i.e. inward) flow during evaporation [44, 69], which can be caused by a temperature-driven or surfactant-induced Marangoni effect [64, 70]. Even the shape of the particles can lead a different deposition process [71].

The works above might have created an impression that the final pattern morphologies were determined by various flows inside the droplet. However, in the final stage of evaporation, the pinned droplet evolves into a thin liquid film, which spontaneously undergoes dewetting and promotes particles redistribution to form the resulting patterns inside the coffee ring.

Recently, significant progress had been made in understanding the dewetting process of liquid film [72]. In the absence of particles, two different film rupture mechanisms have been identified that occur in film of nanoscale thickness, nucleation dewetting and spinodal dewetting [73–75]. The formation of characteristic patterns such as annular ring-like structures [76] and dendritic structure [77] by dewetting of thin film have been analyzed in liquid film that contains particles. In previous studies, the focus is on the dewetting of ultrathin films with or without nanoparticles, which provide a rapid, bottom-up approach to create textures on a surface [78, 79]. Compared with nanoscale liquid film, micro-sized film is more stable, and may more easily allow the assembly of particles into ordered structures, especially a close-packed hexagonal structure. However, the dewetting process of micro-sized film containing microspheres still remains unclear, though it is involved in the final stages of evaporation of most colloidal drops. The film stability–instability (rupture) is the subject of thermal–mechanical and capillary surface waves caused by spatial variations of the surface tension due to the temperature or surfactant concentration.

4.2   The Effect of the Gravity for the Deposition Pattern

The most common phenomenon is the coffee-ring effect, which was attributed to the outward capillary flow caused by the uneven evaporation flux, on the conditions of the droplet evaporating in constant contact radius (CCR) mode at a small contact angle. The capillary flow will bring the suspended particles from the center to the edge of the droplet, and these particles will accumulate near the contact line and finally form the ring-like stain. In contrast, the presence of Marangoni flow, induced by temperature gradients or concentration gradient on the droplet surface, could reverse the capillary flow and change the deposition patterns [64, 69, 70, 80]. In addition, particles will be inevitably affected by the gravitational effect in the process of droplet evaporation on the ground. It was demonstrated that gravity has a significant influence on deposition patterns [81, 82]. Further studies show that the deposition profiles could be determined by the capillary forces [83], liquid film dewetting [84], and interface capture [85].

In fact, the above-mentioned phenomena are frequently involved in the droplet evaporation process at the same time, which will cooperate or compete together. The final deposition morphology could be determined by these two or more effects. In the past, the deposition morphology which were determined by the interaction between the coffee ring effect and other phenomenon, such as Marangoni effect, shape-dependent capillary interactions, capillary force, contact angle hysteresis, were widely reported [68, 86–88]. Despite great progress, to our best knowledge, there are still no studies that examine the combined effects of the coffee ring and gravity sedimentation on deposition patterns.

For the differences of deposition patterns between sessile and pendant droplet is the gravity effect, which can be described by the gravitational Peclet number. The Peclet number is defined as:

$$Pe_G = \frac{\pi d^4 g \Delta\rho}{12 k_B T} \tag{5}$$

where d is the particle diameter, $\Delta\rho$ is the difference in density between the particles and the surrounding fluid, g is the gravitational acceleration, k_B is the Boltzmann constant, and T is the temperature. The smaller particles have a lower Peclet number, and thus follow the streamlines closely, but the larger particles can be easily pulled downward by gravity.

For both sessile and pendant droplet with a lower Peclet number, the gravity sedimentation is weak, thereby most particles still exist inside the droplet when the contact angle is small in the final stage of evaporation, which will be taken by the capillary flow towards the edge and finally formed the ring-like stain.

4.3 The Mechanism for Encounter and Pursuit

The interface shrinkage, the gravitational sedimentation and the capillary flow in different stages of evaporation. For an evaporating droplet, the terminal velocity of sphere falling in a fluid can be described by the Stokes law [89]:

$$u_p = \frac{d^2 \Delta\rho g}{18\eta} \tag{6}$$

where u_p is the terminal velocity, d the particles diameter, η is the viscosity, g is the gravitational constant, $\Delta\rho$ is the difference in density of the particle and the dispersed phase. As the particle size increases, the sedimentation rate increases and the particles will settle faster.

The average interface shrinkage rate could be expressed as follows:

$$u_i = \frac{h}{t_f} \tag{7}$$

where h is the initial height of the droplet, t_f is the final evaporation time, which can be calculated if the droplet evaporation is considered as the quasi-steady, diffusion-driven evaporation model [90]. The average interface shrinkage rate is determined by the droplet shape and the evaporation rate.

The capillary flow, which is caused by the uneven evaporation flux of the droplet surface, could take particles from the center to the edge. Close to the contact line, the height-averaged radial velocity u_c can be expressed by equation [91]:

$$u_c = \frac{D^*}{\theta} \frac{1}{\sqrt{R(R-r)}} \tag{8}$$

where D^* is the diffusion coefficient, R is the contact radius of the drop, r is the distance from the drop center, θ is the contact angle. There is an inverse correlation between the velocity of the capillary flow and the contact angle. As mentioned earlier, the droplet evaporates as CCR mode, thereby the contact angle decreases continuously. Therefore, the capillary flow enhanced in the evaporation process.

The competitive effect between the radial velocity and the terminal velocity can be described as the dimensionless group number from Eqs. (6) and (8):

$$u_p/u_c = \frac{\Delta\rho g \sqrt{R(R-r)}}{18\eta D^*} d^2\theta \tag{9}$$

The velocity ratio of the interface shrinkage rate and the radial velocity can be expressed through Eqs. (7) and (8)

$$u_i/u_c = \frac{\sqrt{R(R-r)}}{D^*} \frac{h\theta}{t_f} \tag{10}$$

The particle deposition in an evaporating droplet can be divided into two stages. In the first stage, the contact angle is large enough and the capillary flow is weak to be negligible. There are two different regimes for sessile and pendant droplets in this stage: the pursuit regime and the meeting regime, as shown in Fig. 8 In the pursuit regime, both the direction of the interface shrinkage and the particle sedimentation are downward, it seems like that the descending interface is pursuing the particles. If $u_p > u_i$, the particles fall fast and never meet with the liquid–air interface. Thus, parts of particles fall directly on the substrate and the rest particles remain inside the droplet. If $u_p < u_i$, the interface is able to catch up with particles and capture them. Then one part of these particles accumulates at the interface, the other part remains inside the droplet. In the meeting regime, the direction of the interface shrinkage and the particle sedimentation is opposite, it seems like that the ascending interface tends to meet with the particles. If the value $|u_p u_i|$ is large, the interface will meet with particles and capture them, nearly all of particles will assemble monolayer islands at

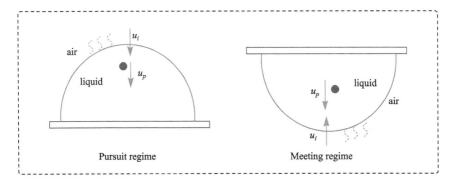

Fig. 8 Schematic of the pursuit regime and the meeting regime for sessile droplet and pendant droplet, respectively. Reprinted from Ref. [92], copyright 2019, with permission from American Chemical Society

the interface. If the value $|u_p u_i|$ is small, the interface will hardly encounter particles, thus most of particles remain inside the droplet. It is worth mentioning that these particles close to the liquid–air interface may be accelerately captured under the effect of recirculation flow. In the second stage, the contact angle is small and the outward capillary flow dominates the particles motion. If there are enough particles inside the droplet, these particles will be taken to the edge and form ring-like stain. Otherwise, for the particles being trapped at the liquid–air interface, they will transfer directly to the substrate and form centered stain. These theories mentioned above can explain the deposition patterns for both sessile and pendant droplet containing different particles well.

5 Conclusion and Outlook

In this chapter, we have introduced a method for capturing colloidal drops in space based on patterned substrate, and further studied the deposition behavior of colloidal particles inside the drop. The space experiments demonstrate how capillary forces work in space, how the patterned surface change the wetting behavior and achieve the manipulation of an aqueous drop, and how a constrained drop evaporates in space. The patterns formation of evaporating constrained drops containing particles has been investigated. It was found a network pattern formed inside the coffee ring. Results show that there is a complex and strong interplay between liquid film dewetting and particle aggregation: the non-uniform distribution of particles promotes the rupture of the liquid film and the liquid film in turn dominates the particle assembly process. We reported the interactions of different physical effects through drying sessile and pendant droplets and found that the deposition morphology is the result of competition and cooperation interactions of the free setting, the interface shrinkage and the outward capillary flow.

As we know, interest in wetting has increased enormously over the last two decades, particularly since 2010. This growth can be attributed to increasing interest of scientists worldwide to capillarity and wetting phenomena, which is caused by their wider application to various new areas. Understanding how microgravity amplifies the wetting behaviors could improve the reliability of such key processes as liquid fuel storage and supply, and general liquid transport aboard spacecraft. Some fluid handling processes in space are prone to involve complex fluids, such as colloidal suspension, emulsions and biological fluids. The microgravity environment may greatly influence the physical behavior of these complex fluids. A good knowledge of the mechanisms involved can help engineers to predict the behavior of complex fluids, and factor that into the design of space systems.

The wetting and evaporation of a colloidal drop attracts more many researchers' interests after the "coffee ring" effect found by Deegan since 1997. However, for the most part, asystematic, well-established, body of knowledge on deposition from drying droplets is not yet available. Future works are expected to study on the relationship between the macroscopic droplet shape and microscopic particle deposition, the underlying mechanisms of evaporation-induced changes of the local flow field, as well as the control of the "coffee ring" effect and the specific deposition patterns.

It would be nice to see the patterns from drying drops of complex fluids used for the early diagnosis of patients since it is a cheap and simple method. However, the understanding of mechanisms behind patterns left from complex fluids is still lacking. In addition, biological fluids are incredibly complex, containing numerous different materials that interact with each other to create very complicated patterns and make them very difficult to model. Finally, the gravity effect on the patterns formation of human body fluid is still unclear, which need to clarify under microgravity conditions. The patterns left from drops of human body fluid of similar condition should be regularly recorded under space and terrestrial conditions. With the increase in repeatable patterns, it can be hoped that this technique will be used to diagnose health status of astronaut.

References

1. Cao H, Lan D, Wang Y, Volinsky AA, Duan L, Jiang H (2010) Fracture of colloidal single-crystal films fabricated by controlled vertical drying deposition. Phys Rev E 82(3):031602
2. Schall P, Cohen I, Weitz DA, Spaepen F (2004) Visualization of dislocation dynamics in colloidal crystals. Science 305(5692):1944–1948
3. Pusey P, Van Megen W (1986) Phase behaviour of concentrated suspensions of nearly hard colloidal spheres. Nature 320(6060):340–342
4. Lin Z (2012) Evaporative self-assembly of ordered complex structures. World Scientific
5. Yin Y, Lu Y, Gates B, Xia Y (2001) Template-assisted self-assembly: a practical route to complex aggregates of monodispersed colloids with well-defined sizes, shapes, and structures. J Am Chem Soc 123(36):8718–8729
6. Norris DJ, Arlinghaus EG, Meng L, Heiny R, Scriven L (2004) Opaline photonic crystals: how does self-assembly work? Adv Mater 16(16):1393–1399

7. Deegan RD, Bakajin O, Dupont TF, Huber G, Nagel SR, Witten TA (1997) Capillary flow as the cause of ring stains from dried liquid drops. Nature 389(6653):827–829
8. Hampton MA, Nguyen TA, Nguyen AV, Xu ZP, Huang L, Rudolph V (2012) Influence of surface orientation on the organization of nanoparticles in drying nanofluid droplets. J Colloid Interface Sci 377(1):456–462
9. Van Blaaderen A, Ruel R, Wiltzius P (1997) Template-directed colloidal crystallization. Nature 385(6614):321–324
10. Dag Ö, Ahari H, Coombs N, Jiang T, Aroca-Ouellette PP, Petrov S, Sokolov I, Verma A, Vovk G, Young D (1997) Does microgravity influence self-assembly? Adv Mater 9(15):1133–1149
11. Priestly E (2012) Introduction to liquid crystals. Springer Science & Business Media
12. Onsager L (1949) The effects of shape on the interaction of colloidal particles. Ann N Y Acad Sci 51(4):627–659
13. Frenkel D (1987) Onsager's spherocylinders revisited. J Phys Chem 91(19):4912–4916
14. Baulin VA (2003) Self-assembled aggregates in the gravitational field: growth and nematic order. J Chem Phys 119(5):2874–2885
15. Murai M, Okuzono T, Yamamoto M, Toyotama A, Yamanaka J (2012) Gravitational compression dynamics of charged colloidal crystals. J Colloid Interface Sci 370(1):39–45
16. Okubo T, Tsuchida A, Okuda T, Fujitsuna K, Ishikawa M, Morita T, Tada T (1999) Kinetic analyses of colloidal crystallization in microgravity-aircraft experiments. Colloids Surf, A 153(1):515–524
17. Cheng Z, Zhu J, Russel WB, Meyer WV, Chaikin PM (2001) Colloidal hard-sphere crystallization kinetics in microgravity and normal gravity. Appl Opt 40(24):4146–4151
18. Ansari RR, Hovenac EA, Sankaran S, Koudelka JM, Weitz DA, Cipelletti L, Segre PN (1999) Physics of colloids in space experiment. In: Space technology and applications international forum-1999, vol 1. AIP Publishing, 108–113
19. Okubo T, Tsuchida A, Takahashi S, Taguchi K, Ishikawa M (2000) Kinetics of colloidal alloy crystallization of binary mixtures of monodispersed polystyrene and/or colloidal silica spheres having different sizes and densities in microgravity using aircraft. Colloid Polym Sci 278(3):202–210
20. Schöpe HJ, Wette P (2011) Seed-and wall-induced heterogeneous nucleation in charged colloidal model systems under microgravity. Phys Rev E 83(5):051405
21. Hu W (2008) Special issue: microgravity experiments on board the Chinese recoverable satellites preface. In: Springer 233 Spring St, New York, NY, USA
22. Li C, Zhao H, Ni R (2008) China's recoverable satellites and their onboard experiments. Microgravity Sci Technol 20(2):61–65
23. Hu W, Zhao J, Long M, Zhang X, Liu Q, Hou M, Kang Q, Wang Y, Xu S, Kong W (2014) Space program SJ-10 of microgravity research. Microgravity Sci Technol 26(3):159–169
24. Li W, Lan D, Sun Z, Geng B, Wang X, Tian W, Zhai G, Wang Y (2016) Colloidal material box: in-situ observations on colloidal self-assembling and liquid phase transition in space. Microgravity Sci Technol 28(2):179–188
25. Conrath M, Canfield P, Bronowicki P, Dreyer ME, Weislogel MM, Grah A (2013) Capillary channel flow experiments aboard the International Space Station. Phys Rev E 88(6):063009
26. Bostwick J, Steen P (2015) Stability of constrained capillary surfaces. Annu Rev Fluid Mech 47:539–568
27. Antar BN, Ethridge EC, Maxwell D (2003) Viscosity measurement using drop coalescence in microgravity. Microgravity Sci Technol 14(1):9–19
28. Savino R, Nota F, Fico S (2003) Wetting and coalescence prevention of drops in a liquid matrix. Ground and parabolic flight results. Microgravity Sci Technol 14(3):3–12
29. Brutin D, Zhu Z, Rahli O, Xie J, Liu Q, Tadrist L (2009) Sessile drop in microgravity: creation, contact angle and interface. Microgravity Sci Technol 21(1):67–76
30. Nakajima A (2011) Design of hydrophobic surfaces for liquid droplet control. NPG Asia Materials 3:49
31. Bormashenko E (2015) Progress in understanding wetting transitions on rough surfaces. Adv Coll Interface Sci 222:92–103

32. Herminghaus S, Brinkmann M, Seemann R (2008) Wetting and dewetting of complex surface geometries. Annu Rev Mater Res 38:101–121
33. Zhao XD, Fan HM, Liu XY, Pan H, Xu HY (2011) Pattern-dependent tunable adhesion of superhydrophobic MnO_2 nanostructured film. Langmuir 27(7):3224–3228
34. Lai Y, Pan F, Xu C, Fuchs H, Chi L (2013) In situ surface-modification-induced superhydrophobic patterns with reversible wettability and adhesion. Adv Mater 25(12):1682–1686
35. Tenjimbayashi M, Higashi M, Yamazaki T, Takenaka I, Matsubayashi T, Moriya T, Komine M, Yoshikawa R, Manabe K, Shiratori S (2017) Droplet motion control on dynamically hydrophobic patterned surfaces as multifunctional liquid manipulators. ACS Appl Mater Interfaces 9(12):10371–10377
36. Dong T, McCarthy TJ (2017) Superhydrophobic, low-hysteresis patterning chemistry for water-drop manipulation. ACS Appl Mater Interfaces 9(47):41126–41130
37. Draper MC, Crick CR, Orlickaite V, Turek VA, Parkin IP, Edel JB (2013) Superhydrophobic surfaces as an on-chip microfluidic toolkit for total droplet control. Anal Chem 85(11):5405–5410
38. Zhang S, Huang J, Tang Y, Li S, Ge M, Chen Z, Zhang K, Lai Y (2017) Understanding the role of dynamic wettability for condensate microdrop self-propelling based on designed superhydrophobic TiO_2 nanostructures. Small 13(4)
39. Wu L, Dong Z, Li F, Song Y (2018) Designing laplace pressure pattern for microdroplet manipulation. Langmuir 34(2):639–645
40. Stange M, Dreyer ME, Rath HJ (2003) Capillary driven flow in circular cylindrical tubes. Phys Fluids 15(9):2587–2601
41. Dreyer M, Delgado A, Path H-J (1994) Capillary rise of liquid between parallel plates under microgravity. J Colloid Interface Sci 163(1):158–168
42. Wang C-X, Xu S-H, Sun Z-W, Hu W-R (2010) A study of the influence of initial liquid volume on the capillary flow in an interior corner under microgravity. Int J Heat Mass Transf 53(9):1801–1807
43. Ondarçuhu T, Veyssié M (1991) J Phys II 1:75
44. Weon BM, Je JH (2013) Phys Rev Lett 110:028303
45. Diana A, Castillo M, Brutin D, Steinberg T (2012) Sessile drop wettability in normal and reduced gravity. Microgravity Sci Technol 24(3):195–202
46. Tsai P, Lammertink RG, Wessling M, Lohse D (2010) Evaporation-triggered wetting transition for water droplets upon hydrophobic microstructures. Phys Rev Lett 104(11):116102
47. Jung YC, Bhushan B (2007) Wetting transition of water droplets on superhydrophobic patterned surfaces. Scr Mater 57(12):1057–1060
48. Lenz P, Lipowsky R (1998) Morphological transitions of wetting layers on structured surfaces. Phys Rev Lett 80(9):1920
49. Langbein D (2002) Canthotaxis/wetting barriers/pinning lines. Capill Surf 149–177
50. Li W, Lan D, Sun H, Wang Y (2018) Drop capturing based on patterned substrate in space. Langmuir 34(16):4715–4721
51. Erbil HY, Meric RA (1997) Evaporation of sessile drops on polymer surfaces: ellipsoidal cap geometry. J Phys Chem B 101(35):6867–6873
52. Han W, Lin Z (2012) Angew Chem Int Ed 51:1534
53. Larson RG (2014) AIChE J 60:1538
54. Stannard A (2011) J Phys: Condens Matter 23:083001
55. Kralchevsky PA, Denkov ND (2001) Curr Opin Colloid Interface Sci 6:383
56. Xu J, Xia J, Hong SW, Lin Z, Qiu F, Yang Y (2006) Phys Rev Lett 96:066104
57. Norris DJ, Arlinghaus EG, Meng L, Heiny R, Scriven L (2004) Adv Mater 16:1393
58. Xie Y et al (2013) Langmuir 29:6232
59. Shimoni A, Azoubel S, Magdassi S (2014) Nanoscale 6:11084
60. Wong T-S, Chen T-H, Shen X, Ho C-M (2011) Anal Chem 83:1871
61. Askounis A, Takata Y, Sefiane K, Koutsos V, Shanahan ME (2016) Langmuir 32:4361
62. Deegan RD, Bakajin O, Dupont TF, Huber G, Nagel SR, Witten TA (1997) Nature 389:827
63. Marin A, Liepelt R, Rossi M, Kähler CJ (2016) Soft Matter 12:1593
64. Hu H, Larson RG (2006) J Phys Chem B 110:7090

65. Eral HB, Augustine DM, Duits MHG, Mugele F (2011) Soft Matter 7:4954–4958
66. Lafuma A, Quéré D (2011) EPL 96:56001
67. Crivoi A, Duan F (2013) Langmuir 29:12067–12074
68. Yunker PJ, Still T, Lohr MA, Yodh AG (2011) Nature 476:308–311
69. Hu H, Larson RG (2005) Langmuir 21:3972–3980
70. Still T, Yunker PJ, Yodh AG (2012) Langmuir 28:4984–4988
71. Yunker P, Still T, Lohr M, Yodh A (2011) Nature 476(7360):308–311
72. Nikolov A, Wasan D (2014) Adv Coll Interface Sci 206:207
73. Reiter G (1992) Phys Rev Lett 68:75
74. Elbaum M, Lipson S (1994) Phys Rev Lett 72:3562
75. Thiele U, Mertig M, Pompe W (1998) Phys Rev Lett 80:2869
76. Ohara PC, Gelbart WM (1998) Langmuir 14:3418
77. Harrington GF, Campbell JM, Christenson HK (2013) Cryst Growth Des 13:5062
78. Rezende CA, Lee L-T, Galembeck F (2007) Langmuir 23:2824
79. Gentili D, Foschi G, Valle F, Cavallini M, Biscarini F (2012) Chem Soc Rev 41:4430
80. Girard F, Antoni M, Sefiane K (2008) Langmuir 24:9207–9210
81. Sandu I, Fleaca CT (2011) J Colloid Interface Sci 358:621–625
82. Devlin NR, Loehr K, Harris MT (2016) AIChE J 62:947–955
83. Weon BM, Je JH (2010) Phys Rev E 82:015305
84. Li W, Lan D, Wang Y (2017) Phys Rev E 95:042607
85. Li Y, Yang Q, Li M, Song Y (2016) Sci Rep 6
86. Majumder M, Rendall CS, Eukel JA, Wang JY, Behabtu N, Pint CL, Liu T-Y, Orbaek AW, Mirri F, Nam J (2012) J Phys Chem B 116:6536–6542
87. Weon BM, Je JH (2013) Phys Rev E 87:013003
88. Li Y-F, Sheng Y-J, Tsao H-K (2013) Langmuir 29:7802–7811
89. Rhodes MJ (2008) Introduction to particle technology. Wiley
90. Sobac B, Brutin D (2012) Thermal effects of the substrate on water droplet evaporation. Phys Rev E 86(2):021602
91. Marín ÁG, Gelderblom H, Lohse D, Snoeijer JH (2011) Order-to-disorder transition in ring-shaped colloidal stains. Phys Rev Lett 107(8):085502
92. Li W, Ji W, Sun H, Lan D, Wang Y (2019) Patterns formation in drying sessile and pendant droplet: interactions of gravity settling, interface shrinkage and capillary flow. Langmuir 35(1):113–119

Influence of Gravity on Inorganic Liquid Crystal

Zengzi Wang, Yun Chen, Dejun Sun, Shenghua Xu, Zhiwei Sun, Ding Lan
and Yuren Wang

Abstract This chapter introduced the history of inorganic liquid crystals, and impact of gravity on phase transition was also discussed in detail. Inorganic liquid crystals are found to have different shapes including thread-like, rod-like or plate-like. They combine the good flowability and electrical properties of inorganic compounds, along with the superior thermal stability with low cost. However, for large size and weak particle–particle interactions, gravity can significantly influence the phase behavior of inorganic colloidal particles and liquid crystal phase transition. To rule out the impact of gravity, inorganic liquid crystal transition under microgravity will be a promising research aspect in future.

Keywords Microgravity · Inorganic liquid crystal · Polydispersity · Non-spherical particles · Dispersion

1 Introduction of Typical Inorganic Liquid Crystals

The liquid crystals have been investigated for over one hundred years, most of them are organic liquid crystals. In recent years, some researches on organic metal liquid crystals were reported [1]. However, real inorganic liquid crystals were still rare for the reasons listed below [2]: first, the building blocks must be highly anisotropic to form liquid crystal, while few inorganic compound can meet this standard; second, high melting point of inorganic compounds make it impossible to maintain anisotropic during the heating, and thus the thermotropic liquid crystals cannot be obtained. Another method to prepare inorganic liquid crystal is dissolving

Z. Wang · Y. Chen · D. Sun (✉)
Key Laboratory of Colloid and Interface Chemistry, Ministry of Education, Shandong University, Jinan 250100, Shandong, People's Republic of China
e-mail: djsun@sdu.edu.cn

S. Xu · Z. Sun · D. Lan · Y. Wang
Key Laboratory of Microgravity (National Microgravity Laboratory), Institute of Mechanics, Chinese Academy of Sciences, No. 15 Beisihuanxi Road, Haidian District, Beijing 100190, China

© Science Press and Springer Nature Singapore Pte Ltd. 2019
W. R. Hu and Q. Kang (eds.), *Physical Science Under Microgravity: Experiments on Board the SJ-10 Recoverable Satellite*, Research for Development,
https://doi.org/10.1007/978-981-13-1340-0_7

or dispersing anisotropic inorganic compounds in the solvents to prepare lyotropic liquid crystal. Nevertheless, low dimensional (1D or 2D) inorganic compounds with solubility in water or other solvents were rarely reported. It is of high necessity to investigate the solubility of these inorganic compounds, the interaction between compounds and solvents, and the properties of dispersions. In addition, the preparation of organic liquid crystal is much easier than inorganic liquid crystal.

Despite of the difficulties in finding and investigating inorganic liquid crystals, some progresses have been achieved in decades. Freundlich [3] firstly found that V_2O_5 sols can exhibit orientation order under electric field, magnetic field or flow. The orientation of long axes was in the same direction of electric field, magnetic field or flow. Once the orientation-inducing action was removed, the sols returned to their initial optically isotropic state. Zocher [4] further investigated the inorganic sols with induced optical anisotropy, especially for V_2O_5. They proved that the newly prepared V_2O_5 sol will not exhibit induced birefringence. The birefringence will be enhanced with increasing ageing time. However, Zocher and Török [5] believe that V_2O_5 sols should be regarded as "phase of a higher order" or "super phases" rather than lyotropic liquid crystals. Now it is clear that "super phases" are the same as thermotropic or lyotropic liquid crystals. The difference between V_2O_5 sol and conventional liquid crystal can be attributed to the nematic phase formed by rod like particles. Similar anisotropic bentonite sols were also reported by Langmuir [6].

The size of inorganic liquid crystal building blocks varies from 10 to 1000 nm, which can be regarded as colloids. They are found to have different shapes including thread-like, rod-like or plate-like. The inorganic lyotropic liquid crystals combine the flowability and anisotropic of liquid crystals with the electrical properties (including high conductivity or magnetism etc.) of inorganic compounds, and they have thermal stability superior to that of organic ones. In addition, many of these inorganic liquid crystals are prepared by natural minerals with low cost. These advantages of inorganic liquid crystals promise their potential application in future works, and in recent years they have been focused by researchers again.

1.1 Rod-like Inorganic Liquid Crystals

Two kinds of liquid crystal phases can be formed by rod-like inorganic particles: nematic phase and smectic phase. The mechanism of liquid crystals formed by rod-like particle suspensions can be explained by Onsager's theory.

Vanadium pentoxide (V_2O_5) suspension (Fig. 1) was the firstly reported inorganic lyotropic liquid crystals, nematic phases could be observed in this system. The real morphology of V_2O_5 particles is ribbon-like (Fig. 1b). In general, V_2O_5 particle have a thickness of 1 nm, a width of 25 nm, and a length up to thousand nanometers, which can be adjusted by preparation method and time. Researchers often regard them as rod-like to avoid more calculations. The phase behavior of suspension depends on the volume fraction (Φ) rather than temperature [7–9]. With the increasing concentration (critical $\Phi = 0.7$ vol%), transition from isotropic phase (I) to nematic phase (N) can

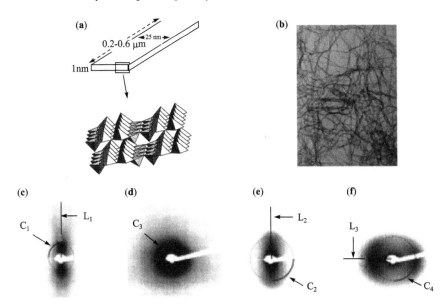

Fig. 1 **a** Structure of crystalline V_2O_5. **b** SEM micrograph of V_2O_5. **c** and **d** 2-D scattering patterns obtained at the 2 vol% sample, can be regarded as N_u; **e** and **f** at the 5 vol% sample, can be regarded as N_b. Reprinted with permission from (Langmuir **2000**, 16, (12), 5295–5303). Copyright (2000) American Chemical Society

be observed in the system. The primary phase transition can be explained by Onsager model; when Φ comes to 1.2 vol%, the sol–gel transition is observed in the system; with higher concentration ($\Phi = 5.0$ vol%), the system turns from uniaxial nematic (N_u) to biaxial nematic (N_b). The thermo-dynamic and fluid-dynamic properties can be explained by hard-core models. The V_2O_5 sol and gel show long range order and typical nematic textures. The sol–gel transition concentration depends on ionic strength of the system.

Magnetic field and electric field can influence the structure of tactoids. Commeinhes et al. [10] investigated the impact of magnetic field on V_2O_5 nematic sols. The application of magnetic field removes the topological defects and produces completely aligned nematic single domain. The orientation time of particles depends on the magnetic intensity and suspension concentration. For example, orientation time for 0.7 vol% suspension in 0.3 T and 1 T magnetic field is 2 h and 5 min, respectively. A sudden change in magnetic field orientation leads to transient hydrodynamic instability of samples.

One of main application of organic liquid crystal is panel display. However, compared to thermotropic liquid crystals, electro-optic effect is hard to be observed in the inorganic lyotropic liquid crystals, for the conductivity of water and the electrochemical reaction in aqueous systems. According to Lamarque-Forget et al. [11], V_2O_5 suspension also shows electro-optic effect under the AC field. Under effective voltages of 10 V, the response times are in the range of one second, making them suitable for slow display applications.

Researchers have made much progress in the application of V_2O_5. Desvaux et al. [12] applied the V_2O_5 suspension as anisotropic medium in the NMR of biomacromolecules. Camerel et al. [13] used V_2O_5 liquid crystals as templates to synthesize single-domain mesostructured inorganic composites. Several advantages including the low cost of V_2O_5 suspensions, the low magnetic field needed, and the simplicity of the synthesis facilitates many industrial applications.

The liquid crystal properties of boehmite (γ-AlOOH) needle-like crystallite were firstly reported by Zocher, and were investigated thoroughly by Buining et al. [14]. They prepared boehmite rod-like particles with diameter of 8 nm, length of 130 nm and 280 nm. I–N phase separation can be observed in long rods (280 nm) suspension ($\Phi = 0.1$ wt%) at low ionic strength after standing for one month. At boehmite concentration above 0.67 wt% or higher, permanently-birefringent monophasic dispersion will be obtained. For short rods (130 nm) suspension, liquid crystal phase can be observed with increasing concentration, while no phase separation appeared. Untreated boehmite particles are positively charged. To discuss their phase behavior, it is necessary to combine Onsager's model and DLVO theory together. Increasing ionic strength lead to the formation of gel network in the system and the phase separation will be influenced. The boehmite particles were also grafted by polyisobutene and dispersed in cyclohexane to obtain sterically stable suspensions [15]. The polymers prevent the contact of particles and weaken electrostatic repulsions and Van del Waals interaction. Only excluded volume effect and steric repulsion exist in the system, which can be explained by Onsager's theory.

Goethites (α-FeOOH) are widespread iron oxide and mainly served as industrial pigments. Davidson et al. [16] researched the liquid crystals in polydisperse goethite particles suspension. Unexpected physical properties were observed in this system. Without magnetic field, I/N phase separation occurs with the concentration increased to critical value, and the border of two phase is clear. Nematic phase is clearly observed and further proved by SAXS. In magnetic field, the particles orient along the field direction at magnetic intensities smaller than 350 mT, but they reorient perpendicular to the field beyond 350 mT. This outstanding behavior was also observed in isotropic phase, which has very strong magnetic-field induced birefringence. For bulk goethite, it is a typical antiferromagnetic material. While nanorod goethite is an unusual magnetic behavior and carries a small magnetic moment, which can be explained by uncompensated surface spins.

Rod-like akaganeite (β-FeOOH) nanoparticles can form smectic phase in aqueous dispersions. Zocher and Birstein [17] found a stable and ordered structure in the sol sediments of β-FeOOH, called "Schiller layers", which exhibits brilliant interference colors. The thickness of ordered layer structure is similar to wavelength of visible light. The interference colors can be explained by Bragg reflection. Using atomic force microscopy to observe the akaganeite liquid crystal, Maeda and Maeda [18] found there are several kinds of smectic phase. The particles show different orientation in different areas . In the iridescent regions, the akaganeite crystals are standing upright at a tilt (with respect to the plane of the smectic layer) and form approximately a square lattice. In contrast, in noniridescent regions, the crystals lie parallel or randomly oriented to each other.

1.2   Plate-like Inorganic Liquid Crystals

Langmuir [6] observed isotropic phase and birefringence phase, in the separated clay suspension after standing for hundreds of hours. With the help of crossed polarizer, Emerson [19] found band formation in the clay systems similar to that of Tobacco Mosaic Virus systems, and thus indicating clay suspensions are also able to form liquid crystals. The clay systems have been widely investigated, especially for mont-morillonite and laponite. Montmorillonite is one kind of expanding 2:1 layered clay (two silica-oxygen octahedron and one aluminum-oxygen tetrahedron) with 1 nm thickness after being exfoliated successfully. The diameter of montmorillonite ranges from 50 to 500 nm with high aspect ratios. Laponite is one kind of montmorillonite. The diameter of monodisperse laponite particles is 25 nm and the thickness is 1 nm. Montmorillonite systems are typical charged plate-like liquid crystal systems, their phase behavior can be explained by Onsager and DLVO theories. Using polarization microscopy to observe montmorillonite and laponite suspensions, threaded textures of nematic liquid crystal are obvious. The properties of clay suspensions depend on volume fraction of clay and ionic strength [20]. For clay suspensions, the sol/gel transition concentration is lower than I/N transition concentration. Under low ionic strength, with increasing clay concentration, suspensions turn from isotropic sol to isotropic gel, and finally form nematic gel. The network structure of gel hinders the macroscopic phase separation of birefringence phase and isotropic phase, and thus nematic phase will not form. Furthermore, the aspect ratio of clay particles is low. The competition of excluded volume effect and electrostatic interaction make the I/N phase unable to co-exist in clay suspension. Temperature, electric field and magnetic field do not influence the orientation of nematic phase gel [21, 22]. For clay suspensions, increasing ionic strength and suppressing double layer make it easier to gel and I/N phase separation will take place, thus make the nematic phase stable (Fig. 2e, f). However, further increasing ionic strength will totally screen the electrostatic repulsion and lead to flocculation of suspensions.

Onsager's theory and computer simulations illustrated the presence of I–N phase separation in the plate-like particle suspension. van der Kooij et al. [23] found a new model system and proved this theory. The $Al(OH)_3$ plate-like hexagonal particles were grafted by polyisobutene ($M_n \sim 1000$) and dispersed in cyclohexane or toluene. Only short range repulsion exists in this sterically stable system. At low concen-tration, I–N phase separation take place; with increasing concentration, the ratio of nematic phase increase gradually to 100%, along with increasing thermal stability of liquid crystal. The polydispersity of particles broaden the I/N biphasic region. and the width of biphasic region is proportional to the polydispersity of particles.

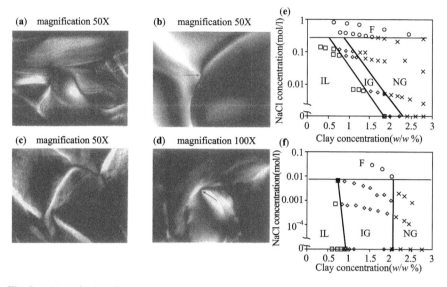

Fig. 2 a Nematic threaded texture of a bentonite suspension. **b** Detail of a 1/2 disclination line (arrow) in a bentonite suspension. **c** Nematic threaded texture of a laponite suspension. **d** Detail of a 1/2 disclination line (arrow) in a laponite suspension. **e** Phase diagram of the bentonite suspensions versus clay and NaCl concentrations. (○, F) Flocculated samples; (□, IL) isotropic liquid samples (◇, IG) isotropic gel samples; (×, NG) nematic gel. **f** Phase diagram of the laponite suspensions versus clay and NaCl concentrations. Reprinted with permission from (The Journal of Physical Chemistry **1996**, 100, (26), 11139–11143). Copyright (1996) American Chemical Society

Al_{13}-ions $\left(Al_{13}O_4(OH)_{24}(H_2O)_{12}^{7+}\right)$ were adsorbed to γ-Al(OH)$_3$ to prepare charged particles. Van der Beek and Lekkerkerker [24] used this system to investigate the phase behavior of charged plate-like particles. Al_{13}-ions increase the surface charge of Al(OH)$_3$ and thus enhancing the colloidal stability. The critical concentration at gel become higher than that of liquid phase transition, the gel structure will no more influence the liquid crystal phase in the system. According to the literature, with increasing particle concentration, I/N phase separation will be observed prior to the gel. The volume fraction of nematic phase decreases with increasing ionic strength and decreasing particle concentration.

Layered double hydroxides (abbreviate for LDHs) [25] are hybrid metal hydroxides composed by divalent and trivalent metal ions, which have hydrotalcite layered crystalline structure. LDHs particles are always positively charged for isomorphous replacement, the anions will adsorb between the layers to equilibrate charges. The interlayer spacing varies with different anions, and LDHs are regarded as anion clays; In contrast, montmorillonite clays are always negatively charged, which can be called cation clays. The anion clays rarely exist in nature but are easy to be synthesized in laboratory. The unique structure and electrical properties of LDHs endows their wide potential applications in catalyst, catalyst carrier, ion exchanger and pharmaceuticals.

The component of LDHs can be illustrate by following formula:

$$\left[M_{(1-X)}^{(II)}M_X^{(III)}(OH)_2\right]^{x+}\left(A_{-x/m}^m\right)\cdot mH_2O$$

$M^{(II)}$ is divalent metal cations (e.g. Mg^{2+}, Mn^{2+}, Fe^{2+}, Ni^{2+}, Cu^{2+}, Zn^{2+}, Ca^{2+}), $M^{(III)}$ is trivalent metal cation (e.g. Al^{3+}, Cr^{3+}, Mn^{3+}, Fe^{3+}, Co^{3+}, Ni^{3+}, La^{3+}); A is anion with n valence number (e.g. Cl^-, OH^-, NO_3^-, CO_3^{2-}, SO_4^{2-}), x is the number of trivalent metal ions, m is the number of bonded water.

LDHs can be structurally characterized as containing brucite ($Mg(OH)_2$) like layers [26], in which some divalent metal cations have been substituted by trivalent ions to form positively charged sheets. The metal cations occupy the centers of octahedra whose vertices contain hydroxide ion. These octahedra are connected to each other by edge sharing to form an infinite sheet. The cationic charge created in the layers is compensated by the presence of hydrated anions between the stacked sheets. These anions can be replaced by other anions.

Stupp and Braun [27] reported the effects of Ca–Al–LDHs on molecular manipulation of microstructures. Using organic molecules modified LDHs can synthesize monodisperse rosettes organic-inorganic composite. The molar ratio of Ca–Al in the composite is the same as in the LDHs, the mass proportion of organic molecules is 20%. The amount of organic molecules intercalated into the layers depend on the structures of LDHs. Removing the organic molecules by heating will not change the rosettes structure of composite. The modification of organic molecules significantly enhanced the mechanical properties of LDHs.

Sun et al. [28] investigated the LDHs lyotropic liquid crystals (Fig. 3). With the LDHs dispersion concentration increasing (>16 wt%), birefringent nematic phase initially forms; when the concentration came to 34 wt%, stable nematic phase formed and no phase separation is observed; under the polarization microscope, the dispersion at 18 wt% initially shows a band-type texture caused by shear or flow, then nematic droplets appear and finally develop into a threaded texture, typical of nematic phase. LDHs can serve as better model systems to investigate phase transitions in electrostatically stabilized platelike colloids.

Graphene oxide (GO) is one typical kind of 2D particles, which can be prepared by oxidation of graphite. Plenty of oxygenated functional groups (hydroxyl, epoxy, ketone and carboxyl) are connected both to the basal plane and edge, and thus render the GO particles to hydrophilic. Intrinsic anisotropy properties and high aspect ratio of GO makes it a good 2D particle to form liquid crystals in water, DMF or NMP dispersions.

Xu and Gao [29], Kim et al. [30] and other groups [31] reported the liquid crystallinity of graphene oxide aqueous dispersions almost at the same time (Fig. 4). With the increasing concentration (0.05–0.5 vol%), the volume fraction of nematic phase increases, and reaches maximum value at 0.4 vol% (Fig. 4b). Under the crossed polarizer, typical nematic schlieren texture can be observed at the bottom phase. Polarization microscopy observation proves the existence of birefringent texture of the graphene oxide liquid crystals, which exhibit a high density of $\pm 1/2$ disclinations (Fig. 4d). As the concentration further increase, regular lamellar phase can be observed.

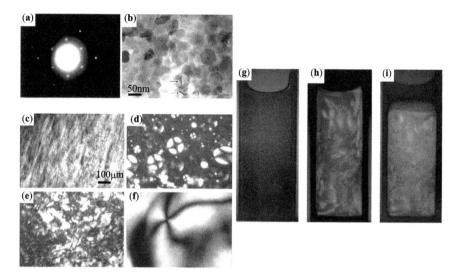

Fig. 3 Nematic phase formed in LDHs aqueous dispersion. **a** Electron diffraction pattern and **b** TEM images of Mg/Al LDH particles. Rare (see arrow) edges of the particles can be seen. **c** 18% (w/w) dispersion. **d** Samples observed after 7 days and **e** 17 days. **f** Schlieren texture with a point defect surrounded by four dark brushes of 22% (w/w) dispersion. **g** Isotropic phase of the 16% (w/w) dispersion. **h** Birefringent nematic phase of the 27% (w/w) dispersion. **i** Isotropic-nematic phase separation of the 27% (w/w) dispersion, as observed 7 days after dilution. Reprinted with permission from (Chemistry of materials **2003**, 15, (17), 3240–3241). Copyright (2003) American Chemical Society

As GO are one atom thick, its aspect ratio is usually very large (700–2600). High aspect ratio leads to much lower isotropic-nematic transition concentration. GO aqueous dispersion can be rather stable for high absolute value of zeta potential. However, adding salts or changing pH will destroy its the liquid crystal structure. Increasing salt concentration will screen the electrostatic repulsion, and absolute value of zeta potential will be decreased. Changing pH to acid will protonated the –COOH groups on the GO [32], leading to reduced hydrophilicity and destabilize the liquid crystal.

Several researches were also performed on liquid crystals formed by chemically functionalized graphene. For reduced GO (RGO), decreased hydrophilicity makes it impossible to form liquid crystal in water. Poulin et al. [33] used bile salts as surfactants to stabilize RGO aqueous dispersion. An isotropic-nematic transition is observed as concentration of RGO increased. Corresponding SAXS 2D pattern also verified the formation of nematic ordering in RGO LC. Physical adsorbed or chemical grafted GO or RGO can also form liquid crystals. For example, Polyvinyl alcohol (PVA) wrapped RGO can be well dispersed in water and form LCs similar to GO [34]. Hyperbranched polyglycerol (HPG) enveloped GO and RGO show good solubility in NMP, and thus the nematic and lamellar phases are observed in the dispersion [35].

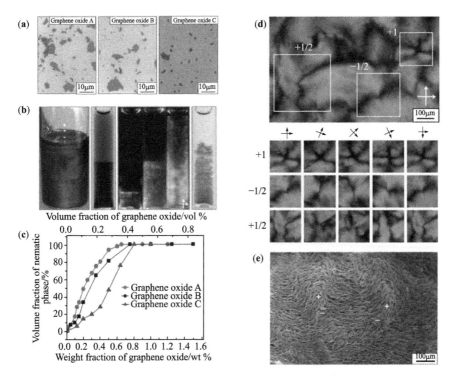

Fig. 4 **a** SEM images of graphene oxide platelets exfoliated from various graphite sources. **b** Left to right: 0.5 wt% graphene oxide dispersion; phase-separated 0.2 wt% dispersion three weeks after preparation; phase-separated dispersions (0.05, 0.2, 0.5 wt%) located between crossed polarizers; coagulated 0.01 wt% dispersion upon adding 50 mM NaCl. **c** Nematic phase volume fraction versus graphene oxide concentration. **d** Typical nematic schlieren texture of a 0.3 wt% dispersion with ±1/2 disclinations and +1 disclination. **e** SEM image of GO liquid crystal in freeze-dried sample (0.5 wt%). Blue and red symbols indicate +1/2 and −1/2 disclinations, respectively. Reprinted with permission from (Angewandte Chemie International Edition **2011**, 50, (13), 3043–3047). Copyright (2011) John Wiley and Sons

1.3 Prospect of Inorganic Liquid Crystals

Considerable progress has been achieved in inorganic liquid crystals, however, more sophisticated understandings on this area are waiting to be explored in theory, experiments, engineering and applications. It is hard to prepare low dimensional inorganic particles (1D rod-like or 2D plate-like), which needs high temperature solid phase reaction. Dispersing/dissolving these inorganic particles in the solvents and preparing stable dispersion/solution is also difficult. For reasons above, only few kinds of inorganic liquid crystals were investigated. Compared with organic liquid crystals, inorganic liquid crystals are similar to rigid polymers and they are all unable to form thermotropic liquid crystals for high melting point, and thus only lyotropic liquid crystals can be achieved. In addition, the polymer dispersions are thermodynamic stable systems, while inorganic liquid crystals are thermodynamic instable for giant phase boundary. According to Gabriel and Davidson [2], surface charged colloidal particles are easier to form inorganic lyotropic liquid crystals. The reasons are listed as below: the surface charge will increase the stability of systems by reducing Van del Waals interaction between particles; The effective volume fraction of particles can be increased because of the double layer, and thus the phase transition concentration will be reduced; Plenty of counter-ions are released in the solution, increasing the system entropy and stability of liquid crystal.

2   The influence of gravity on inorganic lyotropic liquid crystals

Inorganic lyotropic liquid crystals are easy to be influenced by external fields, including gravity, electric field, magnetic field or confining geometries. As diameter of colloid particles are much larger than atoms (10^3 to 10^4 times), and the interactions between particles are relatively weak. The gravity can obviously influence the phase behavior and dynamic of colloidal dispersion [36]. That is to say, gravity has significant influence on the formation of liquid crystals. At low concentration of colloidal dispersion, concentration gradient distributions are proportional to $\exp(-z/l_g)$, where z is vertical height, and l_g is gravitational length. For single particle, gravitational length can be used to the effect of gravity on phase behavior. As the Eq. (1) shows, gravitational length is inverse proportional to density difference and particle volume.

$$l_g = \frac{K_B T}{g \Delta \rho V_{\text{particle}}} \tag{1}$$

where g is gravitational acceleration, $K_B T$ is thermodynamic energy. For example, if one particle has diameter of 1 μm, the density difference to solvent ($\Delta \rho$) is 0.10 g mL^{-1}, the gravitational length of diameter is about 10 μm. The migration generated from gravity is 10 times larger than that from Brownian movement. For a polydisperse system, the chemical potential ($K_B T$) of every components and every interface are all the same, and thus the components with different volume (V_{particle}) have different gravitational length. For dispersions with higher concentration, different type of liquid phases coexists. The particle sediment and fractionation always coexist in colloidal dispersions, and thus rich phase behavior can be observed in settled dispersions.

2.1 Settlement and Fractionation in Inorganic Liquid Crystals Induced by Gravity

Vis et al. [37] observed nematic phase and lamellar phase at low volume fraction in silica coated gibbsite dispersions. The gravitational length of particles calculated by authors is far smaller than 1 mm, which is still large compared to the size of colloid particles. The particles will sediment rapidly under the gravity. The sedimentation of particles will significantly increase the concentration at the bottom of samples, thus facilitating the phase transition. The liquid crystal phase can be formed because of sedimentation, even in systems with low particle concentration. However, it is hard to obtain accurate volume fraction of liquid crystal phase transition.

Mourad et al. [38] succeeded in preparing columnar phase by 2 μm diameter plate-like gibbsite particles. The SAXS results (Fig. 5) showed that upon going from the top toward the bottom of the sample the q-values of the intercolumnar reflections slightly increase, which indicate compaction of the structure and pointed toward the direct role of gravity. In addition, at low q-values up to four concentric Bragg reflections, the reflections are elliptical rather than round, and display elongations similar for all reflections. The "d" value calculated by Bragg peak is smaller than that of ideal columnar phase. Furthermore, observation of texture under polarization microscopy shows the difference of fluctuations in the vertical direction compared to the horizontal direction, which implies the strong effect of gravitational compaction in the structure. The authors predicated that column undulations existed in the system and gravity-brought compaction has drastic influence on the system.

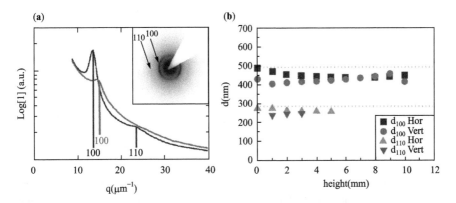

Fig. 5 a Profiles of the X-ray scattering intensity in the horizontal (black curve) and vertical (red curve) direction of the concentrated sample. The inset shows the 2D high resolution SAXS pattern. **b** d_{100} and d_{110} recorded in the horizontal (Hor) and vertical (Vert) direction at different heights above the bottom of the sample. The dashed lines indicate the d_{100} ideal and d_{110} ideal assuming an intercolumnar distance based on the average diameter of the platelets (570 nm) and an ideal hexagonal lattice. Reprinted with permission from (Langmuir **2010**, 26, (17), 14182–14187). Copyright (2010) American Chemical Society

Wijnhoven et al. [39] and Sun et al. [40] observed opposite phenomena on the competition between settlement, gelatinization and liquid phase transition of colloidal system (Fig. 6). The former found that the settlement took place prior to the liquid crystal phase transition, and thus sedimentations can be observed in the bottom of the cuvette. While the latter observed the settlement and liquid crystal phase behavior took place at the same time, and no sedimentations can be observed at the bottom of the cuvette. In order to explain the difference, both of them introduced Peclet number to describe the sedimentation/diffusion balance in gravitational field. Peclet number is the ratio of the time a particle takes to diffuse a distance equal to its diameter D to the time it takes to sediment this distance: $P_e = t_{\text{diff}}/t_{\text{sed}} = (D^2/D')/(D/v_{\text{sed}})$, where D' represents for the diffusion coefficient and v_{sed} the sedimentation velocity. Using the Einstein relation for infinite dilution $D' = k_B T/f$ and $v_{\text{sed}} = m^*g/f$, with f the friction factor and m^* the particle buoyant mass, $P_e = m^*gD/k_B T$. As the gravitational length can be described as $\xi = k_B T/m^*g$, $P_e = D/\xi$, which is a ratio of length scales. Increasing Peclet number lead to increased impact of gravity on particles. In fact, according to the results provided by both of the authors, the conformational transition can take place prior to the impact of gravity, that is to say, the settlement and liquid crystal phase transition should take place at the same time. The presence of sedimentation can be explained as: increasing particle concentration lead to higher probability of particle aggregation from Brownian movement. The aggregates increase the real diameter of particles and thus increasing the P_e. Sun et al. and Wijnhoven et al. observed similar phenomenon in high concentration systems.

van der Beek et al. [41, 42] investigated the influence of ionic strength and gravity on liquid crystal phase behavior of charged gibbsite systems. At relatively high ionic strength, the I–N phase transition was observed; at relatively low ionic strength, the I–C transition occurred. The experimental results are corresponding to the phase diagram obtained by Monte Carlo simulations [43]. After standing for 6 months under the gravitational field, a sample containing an isotropic and a nematic phase in coexistence (6:4) turns to three phases instead of I/N coexistence. The three phases are isotropic, nematic and columnar, from top to bottom, respectively. The particle concentration varied as a function of height, indicating a balance between gravity and osmotic pressure. The author used a simple model to describe this balance qualitatively.

An experimental study on gravity induced liquid crystal formation in gibbsite and silica mixed system was proposed by Kleshchanok et al. [44]. After one year of the sample preparation, all of pure gibbsite samples show coexisted I/N/C phase, which can be explained by gravity and sedimentation of platelets, leading to a density gradient in a capillary of sufficient height. However, for mixed gibbsite/sphere suspensions, the only observed liquid crystalline phase is a columnar phase, in coexistence with an isotropic phase (I/C), in contrast to nematic phase (N) of pure gibbsite systems. The authors ascribed this phenomenon to the strong depletion attraction in the mixture preventing the formation of gravity induced nematic phase. Adding silica sphere significantly broadened the I/C coexistence region.

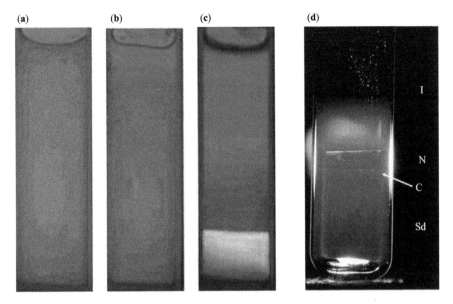

Fig. 6 Phase behavior of 20% (w/w) Mg_2Al LDHs suspension as observed between crossed polarizers. **a** Right after preparation, **b** 39 days and **c** 8 months after preparation, **d** samples (Gibbsite platelets) at 1.5 year after preparation, and observed between crossed polarizers. An isotropic, nematic and columnar phase are visible together with an amorphous sediment. Reprinted with permission from (Langmuir **2007**, 23, (10), 5331–5337). Copyright (2007) American Chemical Society. Reprinted with permission from (Langmuir **2005**, 21, (23) 10422–10427). Copyright (2005) American Chemical Society

Zhang and van Duijneveldt [45] firstly used natural clay sepiolite particles to prepare nonaqueous (toluene) suspension. Clays were firstly treated with DODAB and then steric stabilizer SAP 230 (a poly isobutylene based stabilizer) was grafted onto the surface. In this suspension, isotropic-nematic phase separation can be observed at volume fraction above 0.06. A thin layer of a second birefringent nematic phase slowly forms on top of the initial nematic phase after samples standing for about 1 month, which can be considered as triphasic ($I/N_1/N_2$) equilibrium. The gravitational field has great impact on the triphasic equilibrium of system, for the gravitational length ξ of particles is 0.64 mm and far smaller than the height of cuvette (20 mm). However, it is a pity that authors didn't give further explanation.

Dimasi et al. [46] presented synchrotron X-ray diffraction from gravity dispersed suspensions of Na fluorohectorite over a large NaCl concentration range. Two distinct gel regions were characterized by differences in orientational anisotropy and domain size. The polydispersity in particle size of Na fluorohectorite suspensions is also influenced by gravitational forces, which can sort the particles by size, stabilizing several strata of gels, sols and/or sediments within a single sample tube.

Sun et al. [47] investigated the phase behavior of Mg_2Al LDHs dispersions under microgravity, with the help of SJ-10 satellite. A series of equipments were designed by Li et al. [48] to observe the lyotropic liquid crystals in situ on the satellite (Fig. 7a). The authors firstly performed a pilot study of the liquid crystal phase transition in polydisperse Mg_2Al LDHs under normal gravity. The dispersions are isotropic under the polarizer observation at low concentration (<23 wt%) (Fig. 7c). With increasing concentration (23–30 wt%) of Mg_2Al LDHs, the dispersions show a shear-induced birefringence, indicating a highly-ordered directional alignment of particles. Coexistence of four phases, including an opaque isotropic top phase, a birefringent middle phase, a faint birefringence new phase and a sediment layer of larger platelets, are observed after standing for 5 days (Fig. 7d). When concentration increased up to 32 wt%, three phases coexistence (I/N/S) are observed (Fig. 7e). The authors chose five samples with concentration from 22 to 30 wt% to perform microgravity experiments. The liquid crystal phase transition of Mg_2Al LDHs under microgravity is observed and indicates that the Onsager's theory remained valid under microgravity (Fig. 7f, g). The phase transition concentration increased from 25 wt% under normal gravity to 27 wt% under microgravity, which proves the gravity-brought fractionation facilitate the liquid crystal transition. In addition, relax time of shear-induced birefringence is significantly increased under microgravity, which can be explained by low Peclet number of the Mg_2Al LDHs particles.

2.2 Weakened Impact of Polydispersity in Gravitational Field

In contrast to highly monodisperse rod-like virus [49], the natural or synthetic inorganic colloidal particles are commonly polydisperse [50]. The polydispersity inhibit the formation of ordered structures [51]. However, in the gravitational field, the inhibition of polydispersity on liquid crystal phase transition will be weakened for the fractionation. Vroege et al. [52] observed smectic phase in dispersions of highly polydisperse rod-like goethite nanoparticles. However, Monte carlo simulation showed a terminal polydispersity ($s > 0.18$), above which the smectic phase will be no longer stable [53]. The authors predicted that in gravitational field, particles fractionation induced by gravity will significantly reduce the polydispersity, and thus facilitating the formation of smectic phase. Two systems with $\sigma_D = 0.55$ and $\sigma_D = 0.17$ (σ_D: polydispersity) was investigted. For $\sigma_D = 0.55$, smectic phase and columnar phase in systems are observed, while for $\sigma_D = 0.17$ no columnar phase could be observed in systems. The average length drastically increased to the bottom of the capillary indicating fractionation during sedimentation. Polydispersity is the lowest in the part where only a smectic phase occurred. In contrast, The part where smectic phase coexisted with columnar phase has higher polydispersity, consistent with the notion that the columnar phase can accommodate particles that do not fit into the smectic layers.

For nematic phase, long particles with high excluded volume will be preferred orientation. This tendency will be enhanced because of the sedimentation and lead to extra fractionation. The internal polydispersity of inorganic particles will inhibit the formation of spatial ordered structure (smectic phase or columnar phase). Rod-like particles with low dispersibility will form smectic phase. For particles with

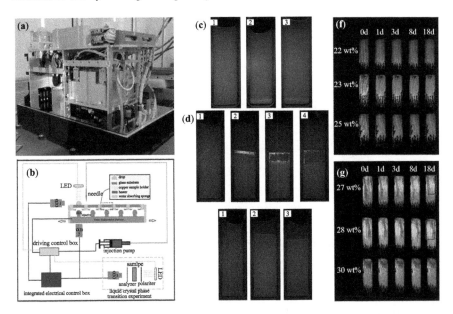

Fig. 7 a The internal structure of the colloidal material box. **b** Experimental principle diagram of the colloidal material box. Reprinted figure with permission from [Microgravity Science and Technology **2016**, 28, (2), 179–188]. Copyright (2016) by the Springer. **c** Polarized light photographs of the 16 wt% Mg$_2$Al LDHs suspension: (1) just prepared, (2) 5 days, (3) 15 days. **d** Polarized light photographs of the 28 wt% Mg$_2$Al LDHs suspension: (1) just prepared, (2) 5 days, (3) 12 days, (4) 15 days. Reprinted figure with permission from [Microgravity Science and Technology **2016**, 28, (2), 95–100]. Copyright (2016) by the Springer. **e** Polarized light photographs of the 32 wt% Mg$_2$Al LDHs suspension: (1) just prepared, (2) 12 days, (3) 15 days. **f** and **g** Polarized optical micrographs of the phase behavior of different concentration samples under microgravity at different time. "0 d" represents "immediately after magnetic stirring". The bead-shaped spheres in the bottom of cuvette are nickel powder fixed by a magnet

higher dispersibility, they tend to form nematic phase rather than unstable smectic phase at lower concentration, or form columnar phase at higher concentration [54]. For example, In systems of plate-like gibbsite particles, the formation of columnar phase is very slow [41]. However, the formation process can be accelerated under a centrifugation force of 900 *g* without arresting the system in a disordered glassy phase, and numerous small crystallites were also observed [55] (Fig. 8).

Petukhov et al. [56] observed of one type of columnar liquid crystal phase formed by thin hard colloidal disks in a dense suspension. The combination of long-range bond-orientational order and short-range translational order between the columns is explained for the size polydispersity of the particles. As Fig. 9 shows, three-phase equilibrium of the isotropic, nematic, and columnar phases is observed between crossed polarizers (Fig. 9a). The columnar phase can be identified by strong Bragg reflections of visible light (Fig. 9c, d). The hexaticlike columnar phase formation is ascribed for the colloidal disks' accommodation of polydispersity at high compression. Close to the N/C transition, the system forms a powder consisting of true

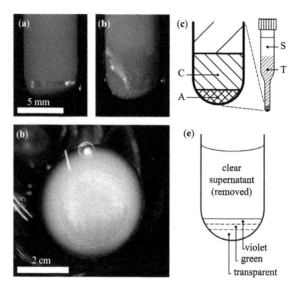

Fig. 8 **a** and **b** Iridescent columnar phase grown in a gravitational field. Samples have been standing for 2 and 4 years, respectively, at $1g$. The scale bar pertains to both images. **c** Sketch of the crystallization tube and its layers: S is the clear supernatant, T is the turbid suspension, C is the well-ordered columnar crystalline layer, and A is the amorphous columnar crystal. **d** At $900g$, using a centrifuge, it takes only 1 day to create a columnar phase. **e** Sketched layers of the sample. Reprinted with permission from (Langmuir **2007**, 23, (23), 11343–11346). Copyright (2007) American Chemical Society

long-range-ordered columnar crystallites, where the relatively large free space between the columns allows for the accommodation of rather highly polydisperse particles. However, for lower ones, due to gravitational compression, little space is available. The geometrical frustration induced by the particle polydispersity suppresses the ordering upon increasing density and favor hexaticlike structuring.

One method to reduce polydispersity of colloidal plates by the synergistic effect of self assembly and gravity was brought by Cheng et al. [57]. The samples were prepared by diluting a concentrated ZrP suspension with deionized water and I/N separation occurred after 5 days. The lower N phase and upper I phases were extracted separately to reduce the polydispersity of the particles. The nematic phases were collected repeatedly for use in the subsequent fractionation. At the concentration of 1.07 wt%, the polydispersity can be reduced from 0.27 to 0.17. This polydispersity reduction method based on the I/N phase transition and gravity can be utilized to select nanoplates with a certain size and to improve size monodispersity.

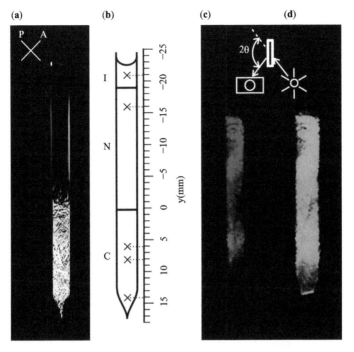

Fig. 9 a Depicts the sample between crossed polarizers (orientation is indicated in the left top) and panel. **b** Identifies the isotropic, nematic, and columnar phase. Crossed mark the positions in the sample where SAXS patterns were taken. **c** and **d** Close-ups of the columnar phase, capturing the Bragg reflections at two different Bragg angles of $2\theta = 140°$ (**c**) and $2\theta = 120°$ (**d**). Reprinted with permission from (Physical Review Letters **2005**, 95, 077801). Copyright (2005) American Physical Society

3 Conclusion

For the large size and weak particle–particle interactions, gravity can significantly influence the phase behavior of inorganic colloidal particles and further influence the liquid crystal phase transition. The phase behavior of liquid crystals without external forces can only be observed under microgravity, which was rarely reported. Gravity-induced fractionation can be observed in liquid crystals systems, which can obviously reduce the impact of polydispersity on the liquid crystal transition. The polydispersity of original dispersion cannot be maintained after standing for long time for gravity. It is necessary to avoid the impact of gravity during the investigation in polydispersity and liquid crystals transition. To provide a microgravity situation, perform the experiments in satellite or space station will be effective, and thus results can be compared to the corresponding ones on earth to understand the impact of gravity. In brief, inorganic liquid crystal transition under microgravity is a promising research aspect in future.

References

1. Steffen W, Köhler B, Altmann M, Scherf U, Stitzer K, zur Loye HC, Bunz UH (2001) Chem-A Eur J 7:117
2. Gabriel JC, Davidson P (2000) Adv Mater 12:9
3. Freundlich H (1916) Zeitschrift für Elektrochemie und angewandte physikalische Chemie 22:27
4. Zocher H (1921) Z Phys Chem 98:293
5. Zocher H, Török C (1967) Acta Crystallogr A 22:751
6. Langmuir I (1938) J Chem Phys 6:873
7. Pelletier O, Bourgaux C, Diat O, Davidson P, Livage J (1999) Eur Phys J B-Condens Matter Complex Syst 12:541
8. Pelletier O, Davidson P, Bourgaux C, Coulon C, Regnault S, Livage J (2000) Langmuir 16:5295
9. Davidson P, Bourgaux C, Schoutteten L, Sergot P, Williams C, Livage J (1995) J Phys II 5:1577
10. Commeinhes X, Davidson P, Bourgaux C, Livage J (1997) Adv Mater 9:900
11. Lamarque-Forget S, Pelletier O, Dozov I, Davidson P, Martinot-Lagarde P, Livage J (2000) Adv Mater12:1267
12. Desvaux H, Gabriel JCP, Berthault P, Camerel F (2001) Angew Chem Int Ed 40:373
13. Camerel F, Gabriel JC, Batail P (2003) Adv Func Mater 13:377
14. Buining P, Philipse A, Lekkerkerker H (1994) Langmuir 10:2106
15. Buining P, Veldhuizen Y, Pathmamanoharan C, Lekkerkerker H (1992) Colloids Surf 64:47
16. Lemaire B, Davidson P, Ferré J, Jamet J, Panine P, Dozov I, Jolivet J (2002) Phys Rev Lett 88:125507
17. Zocher H, Birstein V (1929) Z Phys Chem 141:413
18. Maeda H, Maeda Y (1996) Langmuir 12:1446
19. Emerson W (1956) Nature 178:1248
20. Mourchid A, Delville A, Lambard J, Lecolier E, Levitz P (1995) Langmuir 11:1942
21. Gabriel JP, Sanchez C, Davidson P (1996) J Phys Chem 100:11139
22. Levitz P, Lecolier E, Mourchid A, Delville A, Lyonnard S (2000) EPL (Eur Lett) 49:672
23. van der Kooij FM, Kassapidou K, Lekkerkerker HN (2000) Nature 406:868
24. Van der Beek D, Lekkerkerker H (2003) EPL (Eur Lett) 61:702
25. Rives V (2001) Layered double hydroxides: present and future. Nova Publishers
26. Constantino VR, Pinnavaia TJ (1995) Inorg Chem 34:883
27. Stupp SI, Braun PV (1997) Science 277:1242
28. Liu S, Zhang J, Wang N, Liu W, Zhang C, Sun D (2003) Chem Mater 15:3240
29. Xu Z, Gao C (2011) Nature communications 2:571
30. Kim JE, Han TH, Lee SH, Kim JY, Ahn CW, Yun JM, Kim SO (2011) Angew Chem Int Ed 50:3043
31. Dan B, Behabtu N, Martinez A, Evans JS, Kosynkin DV, Tour JM, Pasquali M, Smalyukh II (2011) Soft Matter 7:11154
32. Kim J, Cote LJ, Kim F, Yuan W, Shull KR, Huang J (2010) J Am Chem Soc 132:8180
33. Zamora-Ledezma C, Puech N, Zakri C, Grelet E, Moulton SE, Wallace GG, Gambhir S, Blanc C, Anglaret E, Poulin P (2012) J Phys Chem Lett 3:2425
34. Kou L, Gao C (2013) Nanoscale 5:4370
35. Hu X, Xu Z, Gao C (2012) Sci Rep 2:767
36. Royall C, van Roij R, Van Blaaderen A (2005) J Phys: Condens Matter 17:2315
37. Vis M, Wensink H, Lekkerkerker H, Kleshchanok D (2015) Mol Phys 113:1053
38. Mourad MC, Petukhov AV, Vroege GJ, Lekkerkerker HN (2010) Langmuir 26:14182
39. Wijnhoven JE, van't Zand DD, van der Beek D, Lekkerkerker HN (2005) Langmuir 21:10422
40. Zhang J, Luan L, Zhu W, Liu S, Sun D (2007) Langmuir 23:5331
41. van der Beek D, Schilling T, Lekkerkerker HN (2004) J Chem Phys 121:5423
42. van der Beek D, Lekkerkerker HN (2004) Langmuir 20:8582
43. Veerman J, Frenkel D (1992) Phys Rev A 45:5632

44. Kleshchanok D, Meijer J-M, Petukhov AV, Portale G, Lekkerkerker HN (2012) Soft Matter 8:191
45. Zhang Z, van Duijneveldt JS (2006) J Chem Phys 124:154910
46. Dimasi E, Fossum JO, Gog T, Venkataraman C (2001) Phys Rev E 64:061704
47. Chen Y, Zhang L, Sun D, Sun Z, Xu S (2016) Microgravity Sci Technol 28:95
48. Li W, Lan D, Sun Z, Geng B, Wang X, Tian W, Zhai G, Wang Y (2016) Microgravity Sci Technol 28:179
49. Lettinga MP, Kang K, Holmqvist P, Imhof A, Derks D, Dhont JK (2006) Phys Rev E 73:011412
50. Palberg T (2014) J Phys: Condens Matter 26:333101
51. Vroege GJ, Thies-Weesie DM, Petukhov AV, Lemaire BJ, Davidson P (2006) Adv Mater 18:2565
52. Van den Pol E, Thies-Weesie D, Petukhov A, Vroege G, Kvashnina K (2008) J Chem Phys 129:164715
53. Bates MA, Frenkel D (1998) J Chem Phys 109:6193
54. Sun D, Sue H-J, Cheng Z, Martínez-Ratón Y, Velasco E (2009) Phys Rev E 80:041704
55. van der Beek D, Radstake PB, Petukhov AV, Lekkerkerker HN (2007) Langmuir 23:11343
56. Petukhov AV, van der Beek D, Dullens RPA, Dolbnya IP, Vroege GJ, Lekkerkerker HNW (2005) Phys Rev Lett 95:077801
57. Chen F, Chen M, Chang Y, Lin P, Chen Y, Cheng Z (2017) Soft Matter 13:3789

SCCO: Thermodiffusion for the Oil and Gas Industry

Guillaume Galliero, Henri Bataller, Jean-Patrick Bazile, Joseph Diaz,
Fabrizio Croccolo, Hai Hoang, Romain Vermorel, Pierre-Arnaud Artola,
Bernard Rousseau, Velisa Vesovic, M. Mounir Bou-Ali,
José M. Ortiz de Zárate, Shenghua Xu, Ke Zhang, François Montel,
Antonio Verga and Olivier Minster

Abstract The accurate knowledge of pre-exploitation fluid compositional profile
is one of the necessary pre-requisites for a successful field plan development of a
petroleum reservoir by the oil and gas industry. Thermodiffusion, leading to a partial
diffusive separation of species in a mixture subject to thermal gradient, is thought
to play an important role in oil and gas reservoir due to the geothermal gradient.

G. Galliero (✉) · H. Bataller · J.-P. Bazile · J. Diaz · F. Croccolo · H. Hoang · R. Vermorel ·
F. Montel
Laboratoire des Fluides Complexes et leurs Réservoirs - IPRA, UMR-5150 CNRS-TOTAL,
E2S - Université de Pau et des Pays de l'Adour, Pau, France
e-mail: guillaume.galliero@univ-pau.fr

F. Croccolo
Centre National d'Etudes Spatiales (CNES), 2, Place Maurice Quentin, 75001 Paris, France

P.-A. Artola · B. Rousseau
Laboratoire de Chimie-Physique, UMR 8000, CNRS, Université Paris-Sud, Orsay, France

V. Vesovic
Department of Earth Science and Engineering, Imperial College London, London, UK

M. M. Bou-Ali
Mechanical and Industrial Manufacturing Department, Mondrago, MGEP Mondragon GoiEskola
Politeknikoa, Arrasate, Spain

J. M. O. de Zárate
Departamento de Estructura de la Materia, Facultad de Física, Universidad Complutense, Madrid,
Spain

S. Xu
Key Laboratory of Microgravity (National Microgravity Laboratory), Institute of Mechanics,
Chinese Academy of Sciences, No. 15 Beisihuanxi Road, Haidian District, Beijing 100190, China

K. Zhang
State Key Laboratory of Enhanced Oil Recovery (Research Institute of Petroleum
Exploration & Development), CNPC, Beijing, China

F. Montel
TOTAL Exploration Production, Pau, France

A. Verga · O. Minster
European Space Agency, ESTEC, Noordwijk, The Netherlands

© Science Press and Springer Nature Singapore Pte Ltd. 2019
W. R. Hu and Q. Kang (eds.), *Physical Science Under Microgravity: Experiments
on Board the SJ-10 Recoverable Satellite*, Research for Development,
https://doi.org/10.1007/978-981-13-1340-0_8

Although major improvements in measuring, simulating and modelling thermodiffusion coefficients have been achieved in the last decades, the improvements are mostly limited to binary liquid mixtures at atmospheric pressure. Thus, the need for accurate data, that would prove invaluable as benchmark reference data for validating models and simulations, was one of the main drivers behind the project "Soret Coefficient measurements of Crude Oil" (SCCO) which used a microgravity set-up implemented in the SJ-10 satellite. This unique project, resulting from a partnership between European Space Agency and China's National Space Science Center enhanced by collaboration among academics from France, Spain, United Kingdom, China and industrialists from France and China, aimed to measure the thermodiffusion coefficients of multicomponent oil and gas mixtures under high pressures. Within this framework, some results on thermodiffusion of one ternary oil mixture and one quaternary gas condensate have been obtained in microgravity and have been qualitatively confirmed by molecular simulations. More precisely, these microgravity results have confirmed on multicomponent mixtures that thermodiffusion leads to a relative migration of the lightest hydrocarbon to the hot region. These results support the idea that, in oil and gas reservoirs, thermodiffusion is not negligible being able to counteract the influence of gravity-driven segregation on the vertical distribution of species.

Keywords Thermodiffusion · Segregation · Multicomponent mixtures · Oil and gas · Molecular dynamics

1 Introduction

The ever increasing world-wide energy demand is putting severe pressure on our ability to manage it effectively, while minimizing adverse climate effects. Although, the renewable energy sources are making strong in-roads, gas and oil remain an important part of the current and near-future energy landscape [1]. Natural gas and crude oil, the two most well-known examples of reservoir fluids, are found, in general, at depths greater than 500 m and require bringing to the surface in a controlled manner. Reservoir fluids, that reside within the pores of petroleum bearing rocks are the remnants of the transformation of organic matter (kerogen) over geological timescales. They are chemically complex mixtures consisting of hundreds of different species, primarily hydrocarbons. In order to optimally exploit the available resources, the petroleum industry creates a considerable demand for reliable values of the thermophysical properties of reservoir fluids over extensive ranges of temperature and pressure. The estimates of thermophysical properties of interest, mainly density and viscosity, can be obtained by a number of standard methods [2].

The petroleum reservoirs are typically subjected to external pressure and/or temperature fields so that gradients in chemical potentials are established leading to diffusive fluxes and resulting in a non-uniform compositional profile. Knowledge of such a profile, especially pre-exploitation profile, is one of the necessary pre-requisites for a successful field plan development of a petroleum containing reservoir.

The non-uniform distribution of species in the reservoir is known to be influenced by a number of phenomena [3]; for instance, in a closed, convection-free reservoir gravitational segregation is assumed to be the most important [4]. However, if there is a temperature gradient within the reservoir, and most reservoir fluids are subject to at least a vertical geothermal gradient of about 0.03 K/m [3], the coupling of heat and mass fluxes [5] leads to a phenomenon known as thermodiffusion, which is also referred to as Soret effect. Thermodiffusion has been shown to strongly influence compositional distribution profile of species [3, 6, 7]. Indeed, there is some evidence based on a number of reservoirs, that thermodiffusion can be as important as gravitational segregation [6, 8–10].

In order to correctly model the thermodiffusion effects one needs a knowledge of thermodiffusion coefficients [5]. Although, there has been a major improvement in the accuracy and reliability in measuring thermodiffusion coefficients in the last decades [11, 12], the improvements are mostly limited to binary liquid mixtures at atmospheric pressure. Recently, the database of the available experimental data has been enhanced by new measurements on ternary mixtures [12, 13] and under high pressures [14–17]. As it is not possible to characterize all the relevant mixtures experimentally, a major effort has been undertaken to develop reliable prediction methods. Progress has been achieved both in the field of thermodiffusion simulation [18] and modelling [12]. Nevertheless, progress has been scant as far as the oil and gas at typical reservoir conditions are concerned, even when the reservoir fluids were represented by synthetic multicomponent mixtures consisting of a few species. The primary reason for this state of affairs can be traced to: (i) complexity of characterizing the thermodiffusion in multicomponent mixtures and (ii) the small magnitude of the thermodiffusion effect under normal laboratory conditions. The latter can be addressed by performing measurements under microgravity conditions [9, 19–22].

The need for accurate data, that can serve as benchmark reference data that would prove invaluable in validating models and simulations, was one of the main drivers behind the project "Soret Coefficient measurements of Crude Oil" (SCCO). The project aimed to measure the thermodiffusion coefficients of six multicomponent fluid mixtures, of interest to reservoir applications, under high pressures used a microgravity set-up implemented in the SJ-10 satellite [23]. The SCCO/SJ-10 project is the result of a unique partnership between the European Space Agency and China's National Space Science Center [24] enhanced by collaboration among academics from France (Université de Pau et des Pays de l'Adour, Université de Paris-Sud and CNRS), Spain (Mondragon Unibertsitatea and Universidad Complutense), United Kingdom (Imperial College London), China (Chinese Academy of Sciences) and industrialists from France (TOTAL) and China (RIPED, CNPC).

In this chapter we aim to summarize some of the results of the project within a context of a short thermodiffusion review and make them available to the larger scientific community.

2 Quantifying Thermodiffusion

As already alluded to in the Introduction, a presence of temperature gradient within a convection-free fluid, generates a composition gradient in a mixture, that results in preferential migration of species towards either cold or hot sections. This separation effect occurs also when fluids are confined in a porous medium. It has been shown that the porous medium has a negligible effect on the magnitude of thermodiffusion, as long as one is not dealing with nano-pores [25]. Thus, the measurements of thermodiffusion coefficients in unconfined fluids can be used for most reservoir applications [26, 27].

The analysis of thermodiffusion is usually performed in the framework of non-equilibrium thermodynamics [5]. In this formalism, the macroscopic relation describing the transport of matter in a two-component mixture can be expressed as

$$\mathbf{J}_1 = -L_{1q}\frac{\nabla T}{T^2} - L_{11}\frac{\mu_{11}^w}{w_2 T}\nabla w_1 \tag{1}$$

where $\mathbf{J}_1$ is the mass flux of component 1, in a centre of mass reference frame, T is the temperature and μ_{11}^w is the derivative of the chemical potential of species 1 relative to its mass fraction, w_1. The L_{ij} are the so-called *phenomenological* or Onsager coefficients, describing the proportionality between fluxes and thermodynamic forces. At the stationary state, i.e. at null mass flux, the partial separation of the species induced by thermodiffusion in binary mixtures is usually quantified by the Soret coefficient, that is defined by,

$$S_T = -\frac{\nabla w_1}{w_1(1-w_1)\nabla T} = -\frac{\nabla x_1}{x_1(1-x_1)\nabla T} \tag{2}$$

where x_1 is the mole fraction of component 1. Note that with the definition given in Eq. (2) the value of the Soret coefficient for a binary mixture is independent of whether mass or molar concentrations are used. By combining Eqs. (1) and (2), an expression of the Soret coefficient in terms of the L_{ij}'s can be given,

$$S_T = -\left(\frac{1}{w_1\mu_{11}^w T}\frac{L_{1q}}{L_{11}}\right)_{\mathbf{J}_1=0} \tag{3}$$

In a multicomponent mixture, it is not possible to quantify thermodiffusion without specifying the reference frame. Although there is still some confusion in the literature, it is lately becoming more customary to quantify thermodiffusion in multi-component mixtures by the so-called (molar) thermal diffusion ratio of each species, which is a dimensionless number defined by,

$$k_{T_i} = -T\frac{\nabla x_i}{\nabla T} \tag{4}$$

with temperature and composition gradients at steady state. The results of the SCCO microgravity experiment will be reported in terms of these (molar) thermal diffusion ratios.

Thermodiffusion has been extensively studied, in particular in binary liquid hydrocarbon mixtures, by experimental, modelling and simulation means. We present in the following a brief review of recent approaches, in particular those developed by the authors in connection with the SCCO project. The reader is referred to the literature for more comprehensive reviews [12, 18, 28, 29] on this topic.

2.1 Experimental Investigations

When a temperature gradient is applied to a fluid, because of thermal expansion, a density gradient appears. If the fluid is a mixture, the Soret effect will induce a compositional gradient which will modify this density gradient. Depending on the relative orientations of the density gradient and the gravity field, convection may occur in terrestrial laboratories. There are two families of experimental set-ups to measure the Soret effect: thermo-gravitational columns, which take advantage of convection, and Soret cells, which work only in a quiescent, purely conductive heat regime. In the framework of the SCCO project, measurements combining these two techniques have been recently performed on binary mixtures of n-alkanes under high pressure [16, 17].

2.1.1 Thermo-gravitational Columns

Following the ground-breaking experiments of Ludwig and Soret in U-shaped glass tubes, the invention of the Clusius–Dickel tube and its application to liquid mixtures [30] was an early milestone in the development of experimental devices capable of accurate and reliable thermodiffusion measurements. The evolution of these devices [31–33] has led to the method usually known as the Thermo-Gravitational Column (TGC) technique. A TGC consists of two vertical plates or concentric cylinders that are separated by a narrow gap, which contains the mixture, across which a temperature gradient ∇T is maintained perpendicular to gravity [34–36]. Coupling of thermodiffusion and solutal expansion induces thermo-solutal convection that highly amplifies the top-bottom gravitational separation. In this technique, it is the so-called thermodiffusion coefficient, D_T, which is directly measured. For a binary mixture, D_T is expressed as the product of the mutual diffusion coefficient D and the Soret coefficient (i.e. $D_T = S_T D$). The theory for concentric TGCs was originally developed by Furry, Jones and Onsager in 1939 [37] and its validity limits have been discussed, among others, by Valencia et al. [38].

TGCs have not only been used to measure thermodiffusion coefficients of binary mixtures of small molecules at normal and high pressures [14], but also for mixtures near the critical point [39]. More recently, the TGC technique has been extended to ternary mixtures [40–43]. TGCs have also been filled with granular matter in order to study the Soret effect in porous media, and to model the separation mechanisms in situations more relevant for geochemical fluids [44, 45].

2.1.2 Soret Cells

A Soret cell is a sample volume bounded by two horizontal parallel plates with a high thermal conductivity, typically copper. The plates have a vertical (z-direction) spacing h and are kept at two different temperatures in order to generate a temperature gradient ∇T parallel to gravity. The volume is laterally confined by a material of low thermal conductivity, frequently glass thus allowing for optical observation in a direction perpendicular to the temperature gradient [11, 12] which facilitates some experimental techniques to measure thermodiffusion, like optical beam bending. In this kind of cells, Soret coefficients (as well as other properties) can also be measured by light scattering from non-equilibrium fluctuations, in which case optical windows in the bounding plates are needed for optical access parallel to the gradient. This approach was first demonstrated by Segre et al. [46] who investigated small-angle Rayleigh light scattering due to temperature and composition fluctuations in a toluene/n-hexane mixture subjected to a stationary temperature gradient.

According to fluctuating hydrodynamics [47, 48], the intensity of scattered light is proportional to $(\nabla T)^2/q^4$ and, hence, it increases very strongly for small scattering wave numbers q. The q^{-4} divergence at $q \to 0$ eventually saturates due to gravity and finite size effects [49–51]. Building on the investigation of non-equilibrium fluctuations in Soret cells, Croccolo et al. [52] have recently obtained Soret and diffusion coefficients of binary mixtures from dynamic shadowgraph experiments. This shadowgraph technique in Soret cells has been extended to high pressures by Giraudet et al. [15].

To apply the dynamic shadowgraph technique for ternary mixtures, theoretical extensions were undertaken. Ortiz de Zárate et al. [53] extended the theory of non-equilibrium concentration fluctuations spectra and their dynamics to micro-gravity conditions, but without accounting for confinement effects. The validation was performed by comparing with the experimental results for a mixture of tetralin/isobutylbenzene/n-dodecane under terrestrial conditions but at large wave vectors, where gravity and confinement effects are negligible [54]. More recently, and in the framework of the SCCO project, gravity has been included in the theory [55]. There are several ongoing experimental verifications in this respect. The incorporation of confinement effects into the theory is currently being developed [56].

2.2 Modelling Developments

In the low density limit it is possible, by means of kinetic theory, to link thermodiffusion to the intermolecular forces [57, 58]. However, and despite numerous efforts and recent progress [12], there is still a lack of a universal microscopic picture explaining thermodiffusion in condensed phases. The lack of an underlying theory is quite general for transport properties [28], but for thermodiffusion it is compounded by its extreme sensitivity to molecular interactions, in particular to cross interactions [59].

When dealing with dense atomic and molecular fluids, there are basically three modelling approaches that have been proposed during the last century. The earliest ones are based on adaptations of the kinetic theory to dense phases [58, 60–62]. They were followed by ones based on pure equilibrium considerations of the phenomena [63] and subsequently by the ones based on liquid states theory [64–66]. There exist also some dedicated empirical approaches which lead to reasonable results when dealing with n-alkanes [59, 61, 67].

In the oil and gas industry, the most popular models are probably those based on equilibrium concepts as they can be linked with thermodynamics quantities readily available in most PVT software packages that make use of Equations of States (EoS). This is somewhat surprising as it is known that such models are not very accurate because of their sensitivity to the choice of the EoS, as thermodiffusion quantification requires second derivatives of thermodynamic properties [68]. Furthermore, these equilibrium type models do not account correctly for the contribution due to irreversibility, although this is not the dominant term in quasi-ideal mixtures, like petroleum fluids [28]. In addition, equilibrium type models cannot take into account properly the mass difference ("isotope") effect which is well correlated with the Soret coefficient in simple linear alkanes [67, 69].

At present, the most promising thermodiffusion models to apply to petroleum fluids are probably those of Artola et al. [64] and of Würger [66] that contain contributions from both equilibrium and irreversibility (through terms related to the mobility of the species). However, these two models still need to be tested systematically against the available data on binary mixtures, in order to better understand how the mobility is related to molecular parameters and thermodynamic conditions. In addition, these models need to be extended to multicomponent mixtures [70], so that they can be applied to real petroleum fluids.

2.3 Molecular Simulations

Equations (2) and (3) offer two different routes to compute the Soret coefficient in binary mixtures by means of molecular dynamics simulations. In approaches using Eq. (2), one mimics a real experiment by introducing a thermal gradient or by imposing a heat flux in the simulation box. Such methods are classified as "boundary driven non-equilibrium molecular dynamics" (BD-NEMD) methods [71, 72].

The other route, through Eq. (3), consists in directly computing the Onsager coefficients L_{ij} either by using equilibrium molecular dynamics (EMD) and the Green-Kubo formalism [73], or by the synthetic non-equilibrium molecular dynamics (S-NEMD) algorithms proposed by Evans [74, 75] or Ciccotti [76]. One drawback in the determination of S_T through the computation of L_{ij} is that additional quantities are needed to relate the L_{ij} to the experimentally accessible S_T, namely, the value of the thermodynamic factor μ_{11}^w. For isotopic or nearly ideal mixtures, one can use the ideal value of μ_{11}^w as a first approximation. However, for a non-ideal mixture, μ_{11}^w differs from the ideal value, strongly affecting the final value of the Soret coefficient. Additional thermodynamic quantities also appear in the expression of L_{1q}, since in the case of mixtures the heat flux J_q is not directly accessible by molecular simulation. In particular, one needs the values of the partial enthalpies of the different components to compute L_{1q} [76]. Therefore, BD-NEMD methods which provide directly Soret coefficients (and thermal diffusion ratios) are becoming the main methods used to obtain the Soret coefficient for realistic mixtures.

Molecular simulation of thermodiffusion has long been limited to the simplest models, i.e. binary Lennard-Jones mixtures. As a result of advances in parallel computing and adapted molecular dynamics simulation packages, it has now become possible to produce quantitative predictions using realistic molecular models, of interest to the crude oil industry. Simon et al. [77] and more recently Antoun et al. [78] employed BD-NEMD to compute the Soret coefficients in alkane binary mixtures. Perronace et al. [79] performed an extensive study of Soret coefficients in n-pentane–n-decane mixtures using EMD, S-NEMD and BD-NEMD simulations. Their results were compared to experimental data and gave a satisfactory agreement. Zhang and Müller-Plathe [80] applied BD-NEMD method to investigate thermodiffusion in benzene–cyclohexane mixtures at three different compositions. In a series of articles, Polyakov and colleagues [81, 82] studied the behaviour of alkane/aromatic mixtures in which they show that the Soret coefficient becomes larger with increasing the degree of branching of the hydrocarbon, thus confirming the experimental findings. Recently and in the context of the SCCO project, molecular simulations have been used to study thermodiffusion in non-binary mixtures [70, 83], as well as in coupling thermodiffusion with gravity segregation so as to mimic the behaviour of a one-dimensional reservoir [9, 10].

3   The SCCO-SJ10 Experiment

The experimental set-up flown in SJ-10 consists of six small and sturdy titanium cells developed for ESA by *Core Laboratories: Sanchez Technologies* (Paris, France) and by QinetiQ Space (Antwerpen, Belgium) and designed to operate at high pressures (HP). These HP cells are built by screwing two titanium end parts to an initially open motorized cylindrical rotating ball valve which, when closed, divides the inner volume into two exactly equal halves. The dimensions of the inner cylinder of the HP cells are: a total length of L = 40 mm and a diameter of D = 6 mm, which becomes a bit narrower in the channel that crosses the ball of the valve, resulting in a total volume of about 1.2 mL. Kalrez 6375 O-ring were selected as sealing materials for the titanium end parts and for the valve sealing stack. HP cells have been certified for pressures up to 600 bars. For temperature control, two aluminium blocks are attached (screwed) at the titanium end parts. Figure 1 shows one of the

Fig. 1 One of the HP cells flown in SJ10 before integration in its triad and in the C-Box. Note the two aluminium blocks attached at the titanium end parts

SJ10-SCCO cells before integration. A cross-section schematic representation of an individual cell can be found in Ref. [23]. Electrical heaters are placed at the two end aluminium blocks and the HP cells, thus fashioned, are attached to two supporting triads and integrated into a hermetically sealed aluminium crate, named C-box, see Fig. 2, which also contains the electronics for temperature control and communications. During the orbital flight the two aluminium blocks of every HP cell were maintained at different temperatures, so as to induce thermodiffusion inside the fluid contained therein. Pt100 sensors with accuracy standard uncertainty of 0.05 K placed at the two ends of each HP cell are used to monitor temperatures through the duration of the experiment. Time-stamped temperature readings are stored in a flash disk also contained in the C-Box. At the end of the experiment the central valves in all HP cells were closed shut separating each fluid sample into two fractions (a "hot" and a "cold" part), which, once recovered after re-entry, were forwarded to the State Key Laboratory of Enhanced Oil Recovery (Research Institute of Petroleum Exploration & Development) for composition analysis. The set-up is similar to the one used during the Foton-M3 mission in 2007 and further details can be found elsewhere [9, 21].

As already described [23] the SCCO-SJ10 experiment has been conducted on six different synthetic mixtures that contain hydrocarbons found in reservoir fluids. The samples were composed of linear alkanes: methane (C1), n-pentane (nC5), n-heptane (nC7) and n-decane (nC10). Binary, ternary and quaternary mixtures containing the

180   G. Galliero et al.

Fig. 2 Left: The six HP cells contained in the C-box. Right: The C-box once closed

aforementioned species have been studied under high pressure in a monophasic state. Table 1 summarizes the composition and pressures used in the six different experimental HP cells. Pressures correspond to the in-flight average temperature of 50.8 °C at which experiments were performed, as further elucidated in Sect. 2 of Chap. 3. The fluids in HP cells A–D are in a compressed liquid state, while HP cells E and F contain compressed gas, which on lowering the pressure produces a small amount of liquid. The latter mimics an important class of fluids of interest to the oil and gas industry, namely gas condensate, which is monophasic in reservoir conditions but diphasic (gas + "condensate", i.e. liquid) at atmospheric conditions. Preparation and injection of the fluids mixture inside the HP cells have been performed in the State Key Laboratory of Enhanced Oil Recovery of the Research Institute of Petroleum Exploration & Development in Beijing as described in Sect. 1 of Chap. 3.

On April 6th at 01:38 local time, a Long March (Chang Zheng) 2D rocket lifted-off from China's Jiuquan Satellite Launch Centre carrying the SJ-10 (Shi Jian 10) research spacecraft. The satellite's scientific payload entailed of a variety of experiments including SCCO. SJ-10 landed on April 18th at 15:04, approximately 12.5 days after its launch, in the Inner Mongolia region north of Beijing. The SCCO C-Box was transported, within a day, to the State Key Laboratory of Enhanced Oil Recovery in Beijing. Additional technical details of the mission are provided in the sections below.

Table 1 Composition and pressure of the six SCCO HP cells embarked in SJ10 (adapted from Ref. [23])

HP cell	Pressure (MPa)	Composition (mole fraction)			
		C1	nC5	nC7	nC10
A	31.1	–	0.5	–	0.5
B	40.2	–	0.5	–	0.5
C	31.0	–	0.3333	0.3333	0.3334
D	40.1	–	0.3333	0.3333	0.3334
E	35.0	0.9649	0.0117	0.0117	0.0117
F	40.0	0.9649	0.0117	0.0117	0.0117

3.1 Fluid Preparation and HP Cell Filling Process

The experiments were performed at pressures given in Table 1 and at the flight operating average temperature of 50.8 °C [23]. In order to achieve these conditions, the HP cells were filled on the ground at 20 °C under a filling pressure that corresponds to the target pressure for each HP cell at the flight operating temperature. The required filling pressure was estimated by means of a Peng-Robinson equation of state [84], as the design of SCCO did not allow for pressure sensors inside the HP cells. Three STIGMA automatic pumps from Core Laboratories were used for sample preparation and HP cell filling procedures. Each pump controlled the injection pressure, the injected volume and temperature. Two different protocols [23], further described below, have been used depending on the mixture type, see Fig. 3.

3.1.1 Binary and Ternary (Liquid) Mixtures Preparation

The four liquid mixtures (A–D) were prepared by first introducing nC5 into pump 1, see Fig. 3 Then, the pre-evacuated pump 2 was filled with, either nC10 or with a nC7–nC10 premix by aspiration and weight control. The desired amount of nC5 was introduced into pump 2 from pump 1 and the mixtures were homogenized by integrated mechanical stirring and then injected into the SCCO HP cells. The cell volume was swept at least three times by the sample. The HP cell pressure was adjusted and controlled by pump 2.

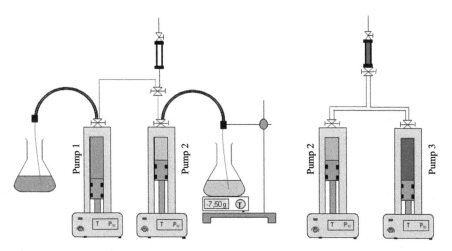

Fig. 3 Sketches of the protocol for the liquid mixture preparation (left figure) and gas-liquid mixtures (right figure) (adapted from Ref. [23])

HP cell ID	T cold side (°C)	T hot side (°C)
A	35.85	65.88
B	35.86	65.88
C	35.84	65.88
D	35.83	65.88
E	35.85	65.88
F	35.87	65.88

Table 2 Average temperatures on the cold and hot plates during operational time of 270 h

3.1.2 Quaternary Mixture (Gas Condensate) Preparation and Filling

The quaternary mixtures of HP cells E and F are single phase at the operating temperature and pressures of Table 1. However, at the filling temperature of 20 °C and corresponding filling pressures they are phase-separated. Hence, it is not possible to use the protocol described in Sect. 3.1.1 and a new protocol was designed. The SCCO HP cells were initially filled with methane (C1) at 20 MPa. The pump 2 was then filled with a C5–nC7–nC10 equimolar ternary mixture using the same procedure as described above, while pure methane was introduced into pump 3 at a pressure of 27.5 MPa. Then, from pump 3 the desired quantity of methane was transferred into pump 2. The quaternary mixture so obtained was homogenized and pressurized to 27.5 MPa and then transferred into the HP cells. The HP cells volume was swept at least five times by the sample. The HP cells pressure was adjusted and controlled by pump 2.

3.2 Conditions During the SJ-10 Orbital Flight

The SCCO experiment was powered on after SJ-10 reached its stabilized orbit, on April 7th at exactly 14:00. The segregation of chemical species occurred in a controlled environment where smooth and seldom orbital manoeuvres resulted in only very small residual accelerations. Inside the C-box, a set of heaters (one per HP cell) and Peltier's (one per triad) elements maintained a relatively stable temperature difference between approximately 36 and 66 °C at each of the six samples. After activation and the initial transitory phase, the target temperatures at the two triads were met in about 40 min. Actual temperatures measured at the HP cell sides are displayed in Fig. 4, illustrating that during the remaining operational time, about 270 h long, the temperatures at the hot and cold sides of each high pressure HP cell were stable to within ±0.05 K. Average values of the temperatures measured at each side of the HP cells, after the short transient, are provided in Table 2.

Tightly attached to its SJ-10 cold plate, the C-Box containing the 6 HP cells kept a rather constant temperature of about 21 °C for the whole experimental run; it decreased to about 19 °C at the end of the orbital flight (see purple curves in

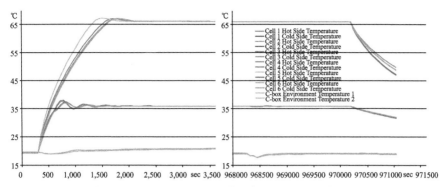

Fig. 4 Evolution of the temperature at the extremities of each HP cell, initial transient (left figure) and final transient (right figure)

Fig. 4). The C-Box is air tight and its inner pressure was continuously monitored. After a short, initial transient the pressure stayed at a relatively constant value of 1.18 ± 0.05 bar.

On April 17th at 21:09, the temperature control at two extremities of the six HP cells was automatically switched-off, following the closure of the intercepting valves about 12 min earlier. The subsequent slow decrease in temperature at both extremities of the HP cells observed in Fig. 4, had no effect on the mixture composition as the fluids remained safely confined within the two separate halves of the HP cells. Typical diffusion coefficients in this kind of hydrocarbon mixtures are at least of the order of 10^{-6} cm^2 s^{-1} [17], which combined with a HP cell length $L = 4$ cm gives a maximum experimental characteristic time, τ_{max}, of about 2.5×10^5 s. The duration of the runs, about $4\tau_{max}$, is so considered long enough for the separation inside the six HP cells to have attained steady state. Twenty-four hours after capsule recovery, the SCCO C-Box was brought back to the State Key Laboratory of Enhanced Oil Recovery in Beijing, where engineers and scientists were ready to split it apart and reach the valuable flown samples of alkane mixtures, in order to start the gamut of tasks for their post-fight sample analysis.

3.3 High Pressure Cell Analysis Process

Two Agilent Gas Chromatographs (GC) 6890 N with Agilent 19091Z-205 HP-1 columns have been used to analyse the six samples; depending on the nature of the mixtures, two different protocols, as described previously [23], have been used.

3.3.1 Binary and Ternary Mixtures

For binary and ternary liquid mixtures (A–D) the HP cells were initially stored, at the CNPC laboratory, at −15 °C in order to limit evaporation of the lightest alkane (nC5). Following this pre-conditioning, the two compartments of each HP cell were emptied using dedicated syringes and the extracted fluids were then stored in Agilent vials with volume redactor. When the extracted volume was insufficient, the liquid samples were diluted in a solvent (carbon disulphide) which did not affect the analysis. Finally, the fluid samples were injected in the gas chromatograph and analysed.

3.3.2 Quaternary Mixtures

For quaternary gas condensate mixtures (E and F) the fluids contained in the two compartments of each HP cell, were initially diluted by using hydrogen, as to avoid phase separation. The resulting diluted mixtures were stored in a dedicated reservoir with a volume of 1 litre and subsequently transferred to the GC for analysis. The reservoirs were heated to the same temperature as the GC line, in order to avoid cold points.

3.4 Results and Discussions

From the gas chromatography analysis of all cold and hot compartments of each HP cell performed after the flight using the protocols above, it appeared that some HP cells have suffered from leakages yielding incoherent results. This conclusion is confirmed independently by accurate weighing of filled HP cells before and after the flight. The leakage may have many root causes, but most of them can be attributed to either unpredictable thermal effects at such high pressures inside the HP cells (more than 300 bar, See Table 1) or to unexpected landing shocks. In particular, HP cells A, B, D and F were found to be partially empty, to the extent that no sensible analysis could be performed and therefore they will not be discussed further. Results on the two exploitable HP cells (C and E), which contained a ternary and a quaternary mixture respectively, which have been presented previously [23], are displayed here in Table 3, for completeness and ease of discussion. The reported compositions in Table 3 have been estimated using normalized peaks-area computation and are an average between two different GC injections for each compartment. The associated error bars correspond to half the difference between the results of the two injections. The composition results were reproducible to within 2% of mole fraction of a given compound. It should be pointed out that the GC protocol has been tested on a HP cell that stayed on earth and was containing the same quaternary mixture as HP cells E and F.

Table 3 Measured GC compositions (in mole fraction) in the two exploitable HP cells after the flight. Initial compositions (before the flight) are shown in Table 1 (adapted from Ref. [23])

HP cell ID	Cold compartment				Hot compartment			
Species	C1	nC5	nC7	nC10	C1	nC5	nC7	nC10
C	N/A[a]	0.3063 ± 0.0151	0.333 ± 0.0033	0.3607 ± 0.0185	N/A[a]	0.3585 ± 0.0019	0.3337 ± 0.0025	0.3078 ± 0.0043
E	0.9612 ± 0.0007	0.0117 ± 0.0002	0.0138 ± 0.0001	0.0135 ± 0.0008	0.0969 ± 0.0056	0.0116 ± 0.0003	0.0096 ± 0.0004	0.0098 ± 0.0049

[a]N/A: Not applicable (Ternary mixture)

As shown in Table 3, compositions of the two compartments (cold and hot) of each exploitable HP cell are noticeably different, as expected. Furthermore, results clearly show that, in both liquid (HP cell C) and gas condensate (HP cell E) mixtures, the lightest species had a tendency to migrate (relatively to the heaviest) to the hot compartment, while the heaviest migrated towards the cold end. Such behaviour is consistent with what has been known qualitatively for a long time [85] and also with more recent experimental results on binary hydrocarbon mixtures [11, 12]. In addition, it appears that the intermediate species, nC7 in the ternary mixture and nC5 in the quaternary mixture, is uniformly distributed showing only weakly migrating patterns. Such a trend is also consistent with what is known regarding mass effect in non-binary mixtures [69, 70]. The observed behaviour is also in agreement with molecular dynamics results on multicomponent hydrocarbon mixtures [9, 10, 58]. More generally, these results confirm that, in oil and gas reservoirs composed of normal alkanes, thermodiffusion tends to counteract (and sometimes overcome) the influence of gravity-driven segregation on the vertical distribution of species [10].

From the compositional difference between the two compartments, and assuming a linear response, it is possible to quantify thermodiffusion in the studied mixtures by means of Eq. (4). Such a computation requires an estimation of the average temperature difference between the two compartments of each HP cell and following the work of Van Vaerenbergh [21], we estimated it to be equal to $12.45 \pm 0.20\,°C$ [23]. As explained above, it is also implicitly assumed that the duration of the experiment (270 h) is long enough for the stationary state to be reached. Results, presented in terms of the thermal diffusion ratio, are provided in Table 4.

Experimental values of thermal diffusion ratios shown in Table 4, confirm quantitatively what is observed in terms of the compositional variations of the cold and hot compartments, i.e. thermodiffusion leads to a relative migration to the hot region of the lightest hydrocarbon in a given mixture. What is more surprising is the magnitude of the thermal diffusion factors [23]. The values obtained for the quaternary mixtures, whose order of magnitude (about 0.1) is consistent with experimental results on non-polar binary mixtures [11, 12]. However, the thermal diffusion ratios for nC5

Table 4 Experimental measurements and molecular simulations results of thermal diffusion ratio of the various species in the two exploitable HP cells (adapted from Ref. [23])

HP cell ID	Thermal diffusion ratio per species (k_{T_i})							
	Experiments				Molecular simulations			
Species	C1	nC5	nC7	nC10	C1	nC5	nC7	nC10
C	–	−1.36 ± 0.47	−0.02 ± 0.14	1.38 ± 0.62	–	−0.09 ± 0.03	−0.09 ± 0.03	0.18 ± 0.04
E	−0.21 ± 0.17	0 ± 0.01	0.11 ± 0.09	0.1 ± 0.08	−0.16 ± 0.04	0.04 ± 0.01	0.05 ± 0.01	0.07 ± 0.02

and nC10 in the ternary liquid mixtures are one order of magnitude larger than those of the quaternary mixture.

To complement these microgravity experiments Boundary Driven Non Equilibrium Molecular Dynamics computations of the thermal diffusion ratios of HP cells C and E fluids have been performed using Mie Chain Coarse Grained molecular models to represent the fluid molecules [86]. Simulations details can be found in Ref. [23].

Simulations results, provided in Table 4, confirm what has been pointed out previously, i.e. microgravity experimental results on the ternary mixture are qualitatively consistent, but quantitatively questionable, whereas experimental results on the quaternary mixture are quantitatively consistent with molecular simulations results, reinforcing thus their reliability.

Dedicated on ground experiments and additional molecular simulations are in progress in order to further analyse these very interesting experimental results obtained in the microgravity environment of a space flight.

Acknowledgements This work has been supported by ESA's Microgravity Applications Programme through the SCCO project. The whole team warmly acknowledges the contributions of QinetiQ Space (Antwerp, Belgium), Core Laboratories: Sanchez Technology (Paris, France) and SISET (Yantai, China) for developing and building both hardware and software of the microgravity experimental set-up flown on SJ-10. The authors also thank TOTAL S.A. and PETROCHINA for permission to publish present and past data. The Spanish teams (Mondragon Unibertsitatea and Universidad Complutense) are thankful to the ATNEMFLU (ESP2017-83544-C3-1-P and ESP2017-83544-C3-2-P) of MINECO. HB and FC acknowledge financial support from the Centre National d'Etudes Spatiales (CNES). The Chinese teams acknowledge financial support from the National Natural Science Foundation of China (Grant No. U1738108) and foundation of SJ-10 recoverable satellite.

References

1. International Energy Outlook (IEO) (2017). www.eia.gov/outlooks/ieo/pdf/0484(2017).pdf
2. Danesh A (1998) PVT and phase behaviour of petroleum reservoir fluids. Elsevier Science, The Netherlands

3. Montel F, Bickert J, Lagisquet A, Galliero G (2007) Initial state of petroleum reservoirs: a comprehensive approach. J Pet Sci Eng 58:391–402
4. Sage BH, Lacey WN (1939) Gravitational concentration gradients in static columns of hydrocarbons fluids. Petr Trans AIME 132:120
5. de Groot SR, Mazur P (1984) Nonequilibrium thermodynamics. Dover, New York
6. Holt T, Lindeberg E, Ratkje KS (1983) The effect of gravity and temperature gradients on methane distribution in oil reservoirs, SPE Paper 11761
7. Whitson CH, Belery P (1994) Compositional gradients in petroleum reservoirs, SPE Paper 28000
8. Ghorayeb K, Firoozabadi A, Anraku T (2003) Interpretation of the unusual fluid distribution in the Yufutsu gas-condensate field. SPE J 8:114–123
9. Touzet M, Galliero G, Lazzeri V, Saghir MZ, Montel F, Legros JC (2011) Thermodiffusion: from microgravity experiments to the initial state of petroleum reservoirs. Comptes Rendus - Mécanique 339:318–323
10. Galliero G, Bataller H, Croccolo F, Vermorel R, Artola PA, Rousseau B, Vesovic V, Bou-Ali M, de Zarate JMO, Xu S, Zhang K, Montel F (2016) Impact of thermodiffusion on the initial vertical distribution of species in hydrocarbon reservoirs. Microgravity Sci Technol 28:79
11. Wiegand S (2004) Thermal diffusion in liquid mixtures and polymer solutions. J Phys: Cond Matter 16:R357–R379
12. Köhler W, Morozov KI (2016) The Soret effect in liquid mixtures—a review. J Non-Equilib Thermodyn 41:151
13. Bou-Ali MM, Ahadi A, Alonso de Mezquia D, Galand Q, Gebhardt M, Khlybov O, Köhler W, Larrañaga M, Legros JC, Lyubimova T, Mialdun A, Ryzhkov I, Saghir MZ, Shevtsova V, Van Vaerenbergh S (2015) Benchmark values for the Soret, thermodiffusion and molecular diffusion coefficients of the ternary mixture tetralin+isobutylbenzene+n-dodecane with 0.8-0.1-0.1 mass fraction. Eur Phys J E 38:30
14. Urteaga P, Bou-Ali MM, Alonso de Mezquía D, Santamaría J, Santamaría C, Madariaga JA, Bataller H (2012) Measurement of thermodiffusion coefficient of hydrocarbon binary mixtures under pressure with the thermogravitational technique. Rev Sci Instrum 83:074903
15. Giraudet C, Bataller H, Croccolo F (2014) High-pressure mass transport properties measured by dynamic near-field scattering of non-equilibrium fluctuations. Eur Phys J E 37:107
16. Lizarraga I, Giraudet C, Croccolo F, Bou-Ali MM, Bataller H (2016) Mass diffusion and thermal diffusivity of the decane-pentane mixture under high pressure as a ground-based study for SCCO project. Micrograv Sci Technol 28:267
17. Lizarraga I, Croccolo F, Bataller H, Bou-Ali MM (2017) Soret coefficient of the dodecane–hexane binary mixture under high pressure. Eur Phys J E 40:36
18. Artola PA, Rousseau B (2013) Thermal diffusion in simple liquid mixtures: what have we learnt from molecular dynamics simulations? Molec Phys 111:3394–3403
19. Legros JC, Van Vaerenbergh S, Decroly Y, Montel F (1994) Expériences en microgravité étudiant l'effet Soret: SCM, SCCO et MBIS, Entropie 198/199:1
20. Georis P, Montel F, Van Vaerenbergh S, Decoly Y, Legros JC (1998) Proc Eur Pet Conf 1:57–62
21. Van Vaerenbergh S, Srinivasan S, Saghir MZ (2009) Thermodiffusion in multicomponent hydrocarbon mixtures: experimental investigations and computational analysis. J Chem Phys 131:114505
22. Khlybov OA, Ryzhkov II, Lyubimova TP (2015) Contribution to the benchmark for ternary mixtures: measurement of diffusion and Soret coefficients in 1,2,3,4-tetrahydronaphthalene, isobutylbenzene, and dodecane onboard the ISS. Eur Phys J E 38:29
23. Galliero G, Bataller H, Bazile JP, Diaz J, Croccolo F, Hoang H, Vermorel R, Artola PA, Rousseau B, Vesovic V, Bou-Ali M, de Zarate JMO, Xu S, Zhang K, Montel F, Verga A, Minster O (2017) Thermodiffusion in multicomponent n-alkane mixtures. NPJ Microgravity 3:20
24. Hu WR, Zhao JF, Long M, Zhang XW, Liu QS, Hou MY, Kang Q, Wang YR, Xu SH, Kong WJ, Zhang H, Wang SF, Sun YQ, Hang HY, Huang YP, Cai WM, Zhao Y, Dai JW, Zheng HQ, Duan EK, Wang JF (2014) Space program SJ10 of microgravity research. Microgravity Sci Technol 26:156–169

25. Hannaoui R, Galliero G, Hoang H, Boned C (2013) Influence of confinement on thermodiffusion. J Chem Phys 139:114704
26. Shapiro A, Stenby EH (2000) Factorization of transport coefficients in macroporous media. Transp Porous Media 41:305–323
27. Platten JK, Costesèque P (2004) The Soret coefficient in porous media. J Porous Media 7:317–329
28. Assael MJ, Goodwin ARH, Vesovic V, Wakeham WA (2014) Experimental thermodynamics volume IX: advances in transport properties of fluids. Royal Society of Chemistry, London
29. Rahman MA, Saghir MZ (2014) Thermodiffusion or Soret effect: historical review. Int J Heat Mass Transf 73:693
30. Clusius K, Dickel G (1939) Das Trennrohrverfahren bei Flüssigkeiten. Naturwissenschaften 27:148
31. Jones RC, Furry WH (1946) The separation of isotopes by thermal diffusion. Rev Mod Phys 18(2):151–224
32. Majumdar SD (1951) The theory of the separation of isotope by thermal diffusion. Phys Rev 81(5):844–848
33. Dutrieux JF, Platten JK, Chavepeyer G, Bou-Ali MM (2002) On the measurement of positive Soret coefficients. J Phys Chem B 106:6104–6114
34. Bou-Ali MM, Ecenarro O, Madariaga JA, Santamaria C, Valencia JJ (1998) Thermogravitational measurement of the Soret coefficient of liquid mixtures. J Phys: Condens Matter 10:3321
35. Bou-Ali MM, Ecenarro O, Madariaga JA, Santamaria C, Valencia JJ (1999) Soret coefficient of some binary liquid mixtures. J Non-Equilib Thermodyn 24:228
36. Bou-Ali MM, Valencia JJ, Madariaga JA, Santamaria C, Ecenarro O, Dutrieux JF (2003) Determination of the thermodiffusion coefficient in three binary organic liquid mixtures by the thermo-gravitational method (contribution of the Universidad del País Vasco, Bilbao, to the benchmark test), Philos Mag 83:2011
37. Furry WH, Jones RC, Onsager L (1939) On the theory of isotope separation by thermal diffusion. Phys Rev 55:1083
38. Valencia JJ, Bou-Ali MM, Ecenarro O, Madariaga JA, Santamaria C (2002) Validity limits of the FJO thermogravitational column technique. In: Köhler W, Wiegand S (eds) Thermal nonequilibrium phenomena in fluid mixtures, vol 233. Springer, Heidelberg
39. Ecenarro O, Madariaga JA, Navarro JL, Santamaria CM, Carrion JA, Saviron JM (1993) Thermo-gravitational separation and the thermal diffusion factor near critical points in binary liquid mixtures. J Phys: Condens Matter 5:2289
40. Bou-Ali MM, Platten JK (2005) Metrology of the thermodiffusion coefficients in a ternary system. J Non-Equilib Thermodyn 30:385
41. Leahy-Dios A, Bou-Ali MM, Platten JK, Firoozabadi A (2005) Measurements of molecular- and thermal-diffusion coefficient in ternary mixtures. J Chem Phys 122:234501
42. Blanco P, Bou-Ali MM, Platten JK, Alonso de Mezquia D, Madariaga JA, Santamaría C (2010) Thermal diffusion coefficients of binary and ternary hydrocarbon mixtures. J Chem Phys 132:114506
43. Alonso de Mezquia D, Wang Z, Lapeira E, Klein M, Wiegand S, Bou-Ali MM (2014b) Thermodiffusion, molecular diffusion and Soret coefficient of binary and ternary mixtures on n-hexane, n-dodecane and toluene. Eur Phys J E 37:106
44. Costesèque P, Fargue D, Jamet P (2002) Thermodiffusion in porous media and its consequences. In: Köhler W, Wiegand S (eds) Thermal nonequilibrium phenomena in fluid mixtures, vol 389. Springer, Heidelberg
45. Costesèque P, Loubet J-C (2003) Measuring the Soret coefficient of binary hydrocarbon mixtures in packed thermo-gravitational columns (contribution of Toulouse University to the benchmark test). Philos Mag 83:2017
46. Segre PN, Gammon RW, Sengers JV (1993) Light-scattering measurements of non-equilibrium fluctuations in a liquid mixture. Phys Rev E 47:1026
47. Ortiz de Zárate JM, Sengers JV (2006) Hydrodynamic fluctuations in fluids and fluid mixtures. Elsevier, Amsterdam

48. Croccolo F, Ortiz de Zárate JM, Sengers JV (2016) Non-local fluctuation phenomena in liquids. Eur Phys J E 39:125
49. Vailati A, Giglio M (1996) q Divergence of non-equilibrium fluctuations and its gravity-induced frustration in a temperature stressed liquid mixture. Phys Rev Lett 77:1484
50. Vailati A, Cerbino R, Mazzoni S, Takacs CJ, Cannell DS, Giglio M (2011) Fractal fronts of diffusion in microgravity. Nat Commun 2:290
51. Ortiz de Zárate JM, Cordon RP, Sengers JV (2001) Finite-size effects on fluctuations in a fluid out of thermal equilibrium. Phys A 291:113
52. Croccolo F, Bataller H, Scheffold F (2012) A light scattering study of non-equilibrium fluctuations in liquid mixtures to measure the Soret and mass diffusion coefficient. J Chem Phys 137:234202
53. Ortiz de Zárate JM, Giraudet C, Bataller H, Croccolo F (2014) Non-equilibrium fluctuations induced by the Soret effect in a ternary mixture. Eur Phys J E 37:77
54. Bataller H, Giraudet C, Croccolo F, Ortiz de Zárate JM (2016) Analysis of non-equilibrium fluctuations in a ternary liquid mixture. Micrograv Sci Technol 28:611
55. Martinez Pancorbo P, Ortiz de Zárate JM, Bataller H, Croccolo F (2016) Gravity effects on Soret-induced nonequilibrium fluctuations in ternary mixtures. Eur Phys J E 40:22
56. Croccolo F, Bataller H (2016) Microgravity in a thin film: how confinement kills gravity. Eur Phys J E 39:132
57. McCourt FRW, Beenakker JJM, Köhler WE, Kuscer I (1990) Non-equilibrium phenomena in polyatomic gases, vol 1. Clarendon Press, Oxford
58. Chapman S, Cowling TG (1991) The mathematical theory of non-uniform gases, 3rd edn. Cambridge Mathematical Library
59. Galliero G, Duguay B, Caltagirone JP, Montel F (2003) On thermal diffusion in binary and ternary mixtures by non-equilibrium molecular dynamics. Phil Mag 83:2097–2108
60. Enskog D (1922) Kungliga Svenska Vetenskapsakademiens. Handlingar 63:5
61. Galliero G, Bugel M, Duguay B, Montel F (2007) Mass effect on thermodiffusion using molecular dynamics. J Non-Equi Thermodyn 32:251–258
62. Villain-Guillot S, Würger A (2011) Thermal diffusion in a binary liquid due to rectified molecular fluctuations. Phys Rev E 83:030501
63. Firoozabadi A, Ghorayeb K, Shukla K (2000) Theoretical model of thermal diffusion factors in multicomponent mixtures. AIChE J 46:892–900
64. Artola PA, Rousseau B, Galliero G (2008) A new model for thermal diffusion, kinetic approach. J Am Chem Soc 130:10963
65. Morozov KI (2009) Soret effect in molecular mixtures. Phys Rev E 79:031204
66. Würger A (2014) Thermodiffusion in binary liquids: the role of irreversibility. J Cond Matt 26:035105
67. Alonso de Mezquia D, Bou-Ali MM, Madariaga JA, Santamaria C (2014) Mass effect on the Soret coefficient in n-alkane mixtures. J Chem Phys 140:084503
68. Gonzalez-Bagnoli MG, Shapiro AA, Stenby EH (2003) Evaluation of the thermodynamic models for the thermal diffusion factor. Phil Mag 83:2171
69. Galliero G, Montel F (2008) Nonisothermal gravitational segregation by molecular dynamics simulations. Phys Rev E 78:041203
70. Artola PA, Rousseau B (2015) Isotopic effect in ternary mixtures, theoretical predictions and molecular simulations. J Chem Phys 143:174503
71. Hafskjold B, Ikeshoji T, Ratkje SK (1993) On the molecular mechanism of thermal diffusion in liquids. Mol Phys 80:1389–1412
72. Rieth D, Müller-Plathe F (2000) On the nature of thermal diffusion in binary Lennard-Jones liquids. J Chem Phys 112:2436–2443
73. Vogelsang R, Hoheisel C (1988) The Dufour and Soret coefficients of isotopic mixtures from equilibrium molecular dynamics calculations. J Chem Phys 89:1588–1591
74. MacGowan D, Evans DJ (1986) Heat and matter transport in binary liquid mixtures. Phys Rev A 34:2133–2142

75. Sarman S, Evans DJ (1992) Heat flow and mass diffusion in binary Lennard-Jones mixtures. Phys Rev A 45:2370–2379
76. Paolini GV, Ciccotti G (1987) Cross thermotransport in liquid mixtures by nonequilibrium molecular dynamics. Phys Rev A 35:5156–5166
77. Simon J-M, Dysthe DK, Fuchs AH, Rousseau B (1998) Thermal diffusion in alkane binary mixtures: a molecular dynamics approach. Fluid Phase Equilib 150–151:151
78. Antoun S, Saghir MZ, Srinivasan S (2018) An improved molecular dynamics algorithm to study thermodiffusion in binary hydrocarbon mixtures. J Chem Phys 148:104507
79. Perronace A, Leppla C, Leroy F, Rousseau B, Wiegand S (2002) Soret and mass diffusion measurements and molecular dynamics simulations of n-pentane–n-decane mixtures. J Chem Phys 116(9):3718
80. Zhang M, Müller-Plathe F (2005) Reverse nonequilibirum molecular-dynaùics calculation of the Soret coefficient in liquid benzene/cyclohexane mixtures. J Chem Phys 123:124502
81. Polyakov P, Müller-Plathe F, Wiegand S (2008) Reverse nonequilibirum molecular dynamics calculation of the Soret coefficient in liquid heptane/benzene mixtures. J Phys Chem B 112:14999–15004
82. Polyakov P, Rossinsky E, Wiegand S (2009) Study of the soret effect in hydrocarbon chain/aromatic compound mixtures. J Phys Chem B 113:13308–13312
83. Galliero G, Srinivasan S, Saghir MZ (2009) Estimation of thermodiffusion in ternary alkane mixtures using molecular dynamics simulations and an irreversible thermodynamic theory. High Temp – High Press 38:315–328
84. Peng DY, Robinson DB (1976) A new two-constant equation of state. Ind Eng Chem Res 15:59
85. Kramers H, Broeder JJ (1948) Thermal diffusion as a method for the analysis of hydrocarbon oils. Anal Chilica Acta 2:687
86. Hoang H, Delage-Santacreu S, Galliero G (2017) Simultaneous description of equilibirum, interfacial and transport properties of fluids using a Mie chain coarse grained force field. Ind Eng Chem Res 56:9213–9226

Ignition and Combustion Characteristics of Overloaded Wire Insulations Under Weakly Buoyancy or Microgravity Environments

Wenjun Kong, Kai Wang, Wei Xia and Shao Xue

Abstract The electric wire, cable and components are the potential igniters, which might cause fire under certain unexpected circumstances. This chapter focuses on the pre-ignition characteristics by overload, the soot emission from the wire insulation during the pre-ignition and ignition stages, the smoke release and distribution characteristics of wire insulation combustion. We reviewed the research work on wire insulations completed by the authors. We first presented the functional simulation methods. The concept of "function simulation" means that the simulation is satisfied in heat transfer sense. A low pressure narrow channel method (LPNCM) was proposed to study fire initiation of wire insulation at microgravity. Then we introduced the experiments completed in microgravity by using the China recoverable satellites of SJ-8 and SJ-10. The experimental hardware were developed to perform the experiments of wire insulation experiments caused by overload on board the SJ-8 and SJ-10 China recoverable satellites, respectively. In the experiments, the pre-fire characteristics including the temperature and radiation characteristics of the wire insulations were presented. For the SJ-10 experiments, the smoke emissions of overloaded wires insulations were investigated. Two smoke emitting modes, namely the end smoke jet and the bubbling smoke jet were identified with polyethylene insulation. The results show that the morphology of pyrolysis front dominated the direction and the range of the end smoke jet. The effects of the insulation thickness and the excess current on the temperature rise were discussed.

Keywords Fire safety · China recoverable satellites · Wire insulation · Function simulation · Overload

W. Kong (✉) · K. Wang · W. Xia · S. Xue
Institute of Engineering Thermophysics, Chinese Academy of Sciences, University of Chinese Academy of Sciences, Beijing, China
e-mail: wjkong@buaa.edu.cn

W. Kong
Beihang University, Beijing, China

K. Wang
Commercial Aircraft Engine Co., Ltd., Aero Engine Corporation of China, Shanghai, China

1 Introduction

Fire safety is one of the most important problems which must be solved properly for the manned space flight since it is closely related to the safety of astronauts and the success of the flight mission.

The well-known fire triangle shows that combustible materials (non-metallic materials), oxygen and igniter are the three necessary elements for fire. Although each one of the three elements is equally important the experiences of manned space flight show that the igniter made much trouble frequently as shown in Table 1 [1]. At least five fire-threatening incidents (STS-6, STS-28, STS-35, STS-40, STS-50) have been reported in the first 50 missions of the US Space Shuttle since 1981. The fire-risk probability is 10%.

Table 1 Shuttle fire-risk experience

Mission	Date	Incident	Result	Response
STS-6	April 1983	Wires fused near material processing unit; crew detected an odor	No atmospheric contamination measured	No alarm
STS-28	August 1989	Cable strain at connector to teleprinter caused insulation failure and electrical short circuit; crew detected a few embers and smoke	Smoke and particle concentration recorded	Circuit breaker did not open; no alarm
STS-35	December 1990	Overheated resistor in digital display unit; crew detected an odor	No atmospheric contamination measured	No alarm
STS-40	June 1991	Refrigerator-freezer fan motor failed; crew noted an irritating odor	Atmospheric contamination identified post-flight	No alarm
STS-50	June 1992	Electronic capacitor in negative body pressure apparatus failed; crew detected an odor	No atmospheric contamination measured	No alarm

These five fire-threatening incidents are all involved electrical component over-heating or electrical short circuits [2, 3]. Heat and increasing temperature due to overload contribute to produce fire. Thus, the electric wire, cable and components are the potential igniters it might cause fire under certain unexpected circumstances.

In order to eliminate these potential igniters as much as possible a series of extremely strict qualification tests have been setup. But all these tests have been conducted on the ground, the microgravity effects were not fully taken into account. Since the natural convection almost vanished at microgravity, the heat transfer of the electric components is quite different at microgravity from on the ground, due to the significantly suppressed natural convection, leading to more heat accumulation in the vicinity of electrical wires and components and higher wire insulation temperatures. The heat loss of the electric components decreased, it might cause the overheating of the electric components and then results in fire. Consequently, it is much more likely to encounter fire threatening scenarios in microgravity due to overheating of electrical wires or components under overload conditions. Therefore, the overheat-ing and damage of wire insulation is one of the main sources of fire incident during manned spacecraft flights.

The ignition of wire insulation by overload in normal atmosphere and gravity has been extensively investigated, with the modes of ignition being classified as (1) arcing, (2) excessive Ohmic heating, and (3) ignition from external heating, and possible contributing factors to ignition systematically discussed [4]. However, there have been limited studies on the ignition of wire insulation in microgravity. The thermal ignition of wire insulation under overload conditions is a slow process, which could take several minutes, several hours, or even longer. It is impossible to create such a long microgravity duration on the ground. Besides, the ground-based tests were limited to very thin fuels, like tissue paper. Practical combustible materials, such as the wire insulation, are thicker and require much longer test time than what is available [5]. On the other hand, space flight experiments in the real microgravity environment are very expensive and not realistic to be conducted on a regular basis.

Anyway, in previous microgravity fire accidents, the fire monitoring and alarm devices did not work, but crew discovered it after they detected an odor or smoke. Based on these, NASA and ESA agreed to strengthen basic research on ignition symptoms in microgravity environments. NASA has funded microgravity research projects to study the combustion characteristics of wire insulations.

Thomas and Donald [6] conducted the first research on the combustion of wire insulations in microgravity in 1971. They found that the Teflon wire insulation could be ignited in 5 s under microgravity in a drop tower. Later, Greenberg et al. [5, 7] provided results of opposed and concurrent flame spreading over polyethylene-insulated nichrome wires by using the Space Shuttle. It was difficult to investigate all the parameters that influence the combustion characteristics of wire insulation, such as oxygen concentration, initial wire temperature, wire diameter, ambient pressure, and dilution gas, because of the limited number of experiments. Kikuchi et al. [8, 9] and Fujita et al. [10, 11] reported results on flame spreading over preheated wire in a quiescent atmosphere in microgravity and investigated the effect of oxygen concentration, dilution gas, wire preheating, and wire thickness by using the 10 s drop tower. Umem et al. [12] proposed a mathematical model to describe the combustion characteristics of ETFE-coated copper wire in microgravity.

We know electrical wires can be involved in a fire by either being subjected to an external heat source or being subjected to internal heating. All above mentioned works conducted by the US and Japan counterparts, only external ignition sources leading to flame spread along wire surface were investigated. In these works, the wire insulation was firstly preheated by electrical overload to a predetermined temperature and then terminated the wire heating and activated the ignitor to produce the flame spreading over the insulation. It is similar to study on flame spread along fuel cylinders, and the results, in fact, cannot fully explain the phenomenon of fire caused by the wire overload and fatigue in microgravity, especially the ignition premonition of the electrical components cannot be obtained. One has to pay much more attentions to study the wire insulations damage by the current overload. It is the most cases of fire involved in wires in microgravity. The understandings of the ignition premonition of wire insulation involved in a fire by being subjected to internal heating has practical signification on the development of fire-prevention measures in manned spacecraft.

Based to our literature search, the combustion characteristics of wire insulation caused by the overload and fatigue of the electrical components in microgravity or weakly buoyant environment were first investigated by the authors here [13–16]. A "function simulation" method [13–15] was proposed and a long time microgravity experiment was conducted on board of the SJ-8 China recoverable satellites in 2006 to study the pre-ignition temperature variations of wire insulations [16].

Nakamura et al. [17, 18] also studied the flame spread of wire insulations in low pressure and showed that the microgravity effect could be simulated by low pressure environment on the ground. After that, Fujita et al. [19] and Takano et al. [20] investigated the flammability of wire insulations by overload with drop tower. These works showed that ignition of wire insulation by current overload was quite a different process in microgravity, when compared with that in normal gravity.

Recently, Takahashi et al. [21] investigated the extinction limits of spreading flames over wires both in normal gravity (1 g) and microgravity (μg). Takahashi et al. [22] studied on unsteady molten insulation volume change during flame spreading over wire insulation in microgravity. Osorio et al. [23] studied the limiting oxygen index and limiting oxygen concentration of ethylene-tetrafluoro-ethylene (ETFE) insulated wires subject to an external radiant flux in both 1 g and μg. Hu et al. [24] investigated flame spread rate over wire at different inclination angles with a high thermal conductivity metal inner core. Fujita [25] reviewed fire safety standards for flammability evaluation of solid material intended for use in a spacecraft habitat.

In 2006, we proposed a satellite project to investigate the pre-ignition characteristics of wire insulation by overload in microgravity, soot emission during the ignition stage, and the smoke distribution of the wire insulation combustion on board SJ-10 satellite [26]. The objectives are to study the smoke emission characteristics of wire insulations and to provide scientific data for the development of fire detection and fire alarm technology in manned spacecraft [27, 28].

In this chapter, we reviewed the research work on wire insulations completed by the authors. We first presented the functional simulation methods, then introduced the experiments completed in microgravity by using the China recoverable satellites of SJ-8 and SJ-10.

2 Research on Fire Initiation in Microgravity by FS Method

The natural convection almost vanished at microgravity. The heat transfer of the electric components is quite different at microgravity from on the ground. The heat loss of the electric components decreased at microgravity, it might cause the over-heating of the electric components and then result in fire. However, this is a gradually developing process. Generally, it takes several minutes, several hours even more. To create such long duration microgravity (10^{-5} g) on the ground is almost impossible.

A "function simulation" method has been introduced to create such an environment on the ground. The term "functional simulation" (FS) means an environment was created on the ground, under the created condition; the heat transfer process is similar to that in microgravity due to the temperature elevation in the wire conductor and insulation by the effects of overload. That is the process of heat transfer in microgravity is simulated in normal gravity. This will be explained more clearly and theoretically in what follows.

The fundamental differential equations governing the incompressible Newtonian fluid two-dimensional "undefined" natural convective heat transfer are:

$$\text{Continuity}: \frac{\partial u}{\partial x} + \frac{\partial v}{\partial y} = 0 \tag{1}$$

$$\text{Momentum}: u\frac{\partial u}{\partial x} + v\frac{\partial u}{\partial y} = \beta \vartheta g + \gamma \frac{\partial^2 u}{\partial y^2} \tag{2}$$

$$\text{Energy}: u\frac{\partial t}{\partial x} + v\frac{\partial t}{\partial y} = \alpha \frac{\partial^2 t}{\partial y^2} \tag{3}$$

where

β Coefficient of volumetric expansion, defined as $\beta = -\frac{1}{\rho}\left(\frac{\partial \rho}{\partial T}\right)_p$,

g Gravitational acceleration

ϑ $T - T_\infty$

ν Kinematic viscosity

α Thermal diffusivity, defined as $\alpha = \frac{\lambda}{\rho C_p}$, λ is thermal conductivity

Based on these basic equations, similarity coefficients can be obtained through similarity analysis. Then the FS condition can be defined by analysis of the similarity coefficients.

For two stable incompressible natural convection fluids, under the conditions that the geometry and boundary conditions are similar, if the ratio of inertial force to viscosity, and the ratio of buoyant force to viscosity are similar at any corresponding point, the corresponding velocity field will be similar.

From the above basic equations, the Grashof number, expressed as $Gr_L = \beta g \vartheta L^3 / \nu^2$, can be conducted by similarity analysis. The Grashof number is representative of the ratio of buoyant force to viscosity. One of the characteristics of the natural convection is that the motive speeds rely on the ratio of buoyancy to viscosity. Thus, the Reynolds number can be expressed as $Re_L = f(Gr_L)$, which means that the Reynolds number is not an independent similarity modulus any more. In other words, the velocity fields for two kinds of stable incompressible natural convection fluids are able to be similar if only the Grashof number is similar at any corresponding point. The Nusselt number ($Nu = \alpha_c l_0 / \lambda_f$) is a dimensionless coefficient describing the heat transfer process under given conditions. It is a standard number to distinguish the heat convection similarity. The Prandtl number ($Pr = \mu c_p / \lambda$) is a dimensionless number of fluid properties. It describes the inherent relationship of velocity distributions and temperature profiles determined by the fluid properties. The Prandtl number for gas far from the critical state is approximately independent of gas temperature and pressure.

Thus, if the corresponding Prandtl numbers and Grashof numbers can be made to have the same values, the corresponding Nusselt numbers will then be the same. Therefore, the FS condition can be determined to be such that the Grashof numbers of the natural convective heat transfer in the two fluids are the same.

In microgravity, the effects of gravity are greatly reduced. If the microgravity environment defines as ten-millionth of Earth's gravitational environment, the FS condition can be determined by the relationship of Grashof number between microgravity and normal gravity. It is,

$$Gr_{\mu g} = \frac{\beta (10^{-5}g)\vartheta L^3}{\nu^2} = 10^{-5} Gr_L \tag{4}$$

As we know that,

$$Gr_L = \frac{\beta g \vartheta L^3}{\nu^2} = \frac{\beta g \vartheta L^3 \rho^2}{\mu^2} \tag{5}$$

According to the state equation for ideal gas:

$$\rho = \frac{p}{RT} \tag{6}$$

Then based on above the Eqs. (4), (5) and (6), we can obtain

$$p_{simu} = \left(\frac{g_{\mu g}}{g}\right)^{1/2} p \tag{7}$$

If the pressure ($p_{simu} = \sqrt{10^{-5}}p = 3.16 \times 10^{-3}p = 320.34$ Pa) in the functional simulation experiment is achieved, it is easily to get $Gr_{simu} = Gr_{\mu g}$. Thus the FS condition is that the pressure in the simulated environment should be set at 320 Pa.

This dimensionless criterion is a property of fluid itself, which describes the inherent relationship between velocity field and temperature field. In the present case the Prandtl number of air is almost a constant which does not vary with its temperature and pressure.

In summary, for the natural convective heat transfer provided the Grashof number equals everywhere, the Nusselt number will also be equal everywhere. That means the simulation of heat transfer is satisfied. So, one can define the criterion of function simulation under the condition keeping the Grashof number equal everywhere.

Based on the concept of FS, a series of experiments have been conducted. Figure 1 shows the systematic diagram of the experiment system. The photograph of experimental set-up is shown in Fig. 2.

Three groups of civil class wire and six groups of astronautic class wire with different insulating materials, thickness of insulation, effective cross-section and rated current have been used in these experiments. Five different levels of environment pressure (100,000, 10,000, 1,000, 100 and 10 Pa) and five different currents (5, 10, 20, 30 and 40 A) have also been used in the experiments. For each wire with certain insulating material, certain thickness of insulation, certain cross-section at different levels of pressure and different currents a set of experiment results can be obtained. A few experiment results will be shown in the present paper. The temperature is the steady equilibrium temperature on the surface of the bare wire. The cross-section of bare wire is 1 mm². The rated current is 20 A. It can be seen from the results that the steady equilibrium temperature greatly increases with the decrease in environment pressure when the current is larger than 20 A. For example, the steady equilibrium temperature is 131 °C when the environment pressure is 100,000 Pa and the current is

Fig. 1 Schematic diagram of experiment system

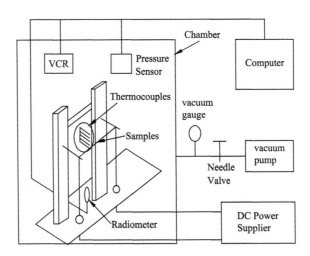

Fig. 2 Photograph of
experiment set-up

30 A. However, the steady equilibrium temperature is 305 °C when the environment
pressure is 100 Pa and current is still 30 A.

The experiment results of red copper wire insulated with polyvinyl chloride are
shown in Fig. 3. The thickness of the insulation material is 0.6 mm. The cross-section
of the bare wire is 0.75 mm^2. The rated current is 16 A. It can be seen from the results,
that the steady equilibrium temperature is 157 °C when the environment pressure is
100,000 Pa and the current is 30 A. But the steady equilibrium temperature becomes
353 °C when the environment pressure is 100 Pa and current is still 30 A. This degree
of overheat is not safe. It might result in the initiation of fire.

The experiment results obtained by the use of FS method proposed in present
chapter indicate that at microgravity the natural convection almost vanished, the heat
loss of the electric components decreased, it might cause the overheat of the electric
components and then result in fire. The experiment results also verified clearly that
the concept FS is a very effective method to simulate the heat transfer process of the
electric components at microgravity.

3   The Low-Pressure Narrow Channel Method

A narrow channel method (NCM) was firstly proposed by Ivanov and Balashov
[29] to study effects of buoyancy on flame spread over non-metal materials by the
change of the value of the Grashof number, Gr, through adjusting the channel height.

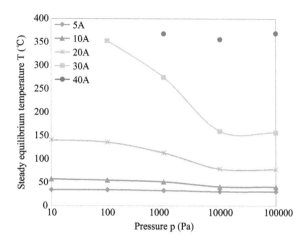

Fig. 3 The steady equilibrium temperature of red copper bare wire with polyvinyl chloride

Zik et al. [30, 31] investigated the flame spread over thermally thin fuels, and observed the instability of fingering flames in the narrow channel. Wichman and Olson [32] and Olson et al. [33] studied the flame spread characteristics of thermally thin fuel by using different channel heights. It is found the observed flame was similar to that in the drop tower with a suitable channel height. Zhang et al. [34–36] further conducted experiments to study microgravity combustion phenomena by the narrow channel. Xiao et al. [37] also studied the effects of channel heights on flame spread rates and the flammability limit with the samples of the thermally thin fuels. The above studies have mainly focused on experimental investigations on the combustion characteristics of the thermally thin fuel using the narrow channel [30–37]. However, most non-metal materials used in the manned spacecraft are the thermally thick. The wire insulation used in this study is also a kind of thermally thick fuel [38]. The literatures of Wang et al. [38] and Xiao et al. [39] suggested that the channel height should be increased to investigate the microgravity combustion properties for the thermally thick fuels by the narrow channel. However if the channel height is increased, the Grashof number is also increased. The weakly buoyancy environment would be lost. Thus we should further study the narrow channel method to simulate the flammability of the thermally thick fuels in microgravity.

The FS method used in the last sections [13–15, 40, 41] is to simulate the natural convective heat transfer at microgravity by using a low pressure method (LPM). It means that the simulation is satisfied in heat transfer sense. The temperature variations of the wire insulation by using the LPM are similar to that at microgravity without combustion occurrence [16]. Once the ignition happens, the chemical reaction rate could be influenced by pressure, and resulted in the different combustion properties under LPM and microgravity. As for the NCM, on one hand, the smaller the channel height, the more similar results of the pre-ignition characteristics of materials to those at microgravity. On the other hand, if the height is too narrow (less than 10 mm), the wall cooling effect and the lack of oxygen transporting would significantly suppress the ignition process. Besides, the conventional narrow channel could not be applied to study the thermally thick materials, like the wire insulation in this paper [38, 39].

To solve those problems, the LPNCM is combined the above two simulation methods to investigate fire initiation of thermally thick materials by firstly increasing the channel height and then reducing the ambient pressure. In this environment the Grashof number equals to that at microgravity. Based on the theory of similarity, the details are described as follows.

The Grashof number for a conventional NCM in 1 atm ambient pressure can be expressed as

$$Gr_\delta = \left(\beta_\delta g_0 \Delta T L_\delta^3 \rho_\delta^2\right)/\mu_\delta^2 \tag{8}$$

where β_δ is the coefficient of volumetric expansion, g_0 is the gravitational acceleration, ΔT is the temperature difference between the wire insulation and the air in the narrow channel, L_δ is the height of the narrow channel, ρ_δ is the air density and μ_δ is the air kinetic viscosity.

The Grashof number for LPNCM can be expressed as

$$Gr_{\delta p} = \left(\beta_{\delta p} g_0 \Delta T L_{\delta p}^3 \rho_{\delta p}^2\right)/\mu_{\delta p}^2 \tag{9}$$

The Grashof number at microgravity can be expressed as

$$Gr_{\mu g} = \left(\beta_{\mu g} g_{\mu g} \Delta T L_{\mu g}^3 \rho_{\mu g}^2\right)/\mu_{\mu g}^2 \tag{10}$$

Therefore the criterion of the LPNCM is

$$Gr_{\delta p} = Gr_\delta = Gr_{\mu g} \tag{11}$$

In order to minimize the influence of buoyancy, the Rayleigh number, Ra, should be less than a critical value Ra_{cr} [42]. Thus the criterion of both LPNCM and NCM for buoyancy to be negligible should be:

$$Gr_{\delta p} = Gr_\delta < Gr_{cr} = Ra_{cr}/Pr \tag{12}$$

where, Pr is the Prandtl number. It is almost a constant which does not vary with its temperature and pressure. So the critical value of the channel height of NCM for buoyancy to be negligible is:

$$L_{\delta,cr} = \sqrt[3]{\mu_\delta^2 Gr_{cr}/\beta_\delta g_0 \Delta T \rho_\delta^2} \tag{13}$$

Thus in NCM, the value of channel height L should be $L_q < L \leq L_{\delta,cr}$. Where, L_q is the quenching distance.

In addition, assuming that the ambient temperature and temperature variation in LPNCM are similar to those both in NCM and µg, β and μ could be treated as the same value. In fact, the changing of environmental temperature in those cases only causes small changes of β and μ, which has been neglected. And this is one assumption of the LPNCM. Besides, substituting the ideal gas equation of state $\rho = p/(RT)$ to Eq. (11), this yields the criterion for the choice of the channel height of LPNCM:

$$L_{\delta p}/L_{\delta} = \sqrt[3]{p_{\delta}^2/p_{\delta p}^2} \tag{14}$$

According to Eq. (14), in LPNCM if the ambient pressure is reduced to $p_{\delta p}$, the channel height should be increased to $L_{\delta p}$. From Eq. (12), the critical value of the channel height of LPNCM for buoyancy to be negligible is:

$$L_{\delta p,cr} = \sqrt[3]{\mu_{\delta p}^2 Gr_{cr}/\beta_{\delta p} g_0 \Delta T \rho_{\delta p}^2} \tag{15}$$

According to the FS method in last section combined with LPM, the critical value of pressure used in LPNCM is obtained as

$$p_{\delta p,cr}/p_{\mu g} = \sqrt{g_{\mu g}/g_0} \tag{16}$$

It is the minimum pressure used in LPNCM. So the pressure range for LPNCM is

$$p_{\delta p,cr} \leq p_{\delta p} \leq p_{\delta} \tag{17}$$

Usually, the value of the ambient pressure is 1 atm for either NCM or μg. The residual gravitational acceleration at microgravity is $10^{-5}g_0$. Thus the critical value of pressure used in LPNCM can be obtained. Substituting this critical pressure into Eq. (15), the critical value of the channel height should be the maximum value used in LPNCM. Meanwhile, substituting the value of p_{δ} into Eq. (15), this yields $L_{\delta p,cr} = L_{\delta,cr}$. Therefore the range of channel height in LPNCM is

$$L_{\delta,cr} \leq L_{\delta p} \leq L_{\delta p,cr} \tag{18}$$

From Eq. (11), for two specific conditions in the LPNCM, the value of $L_{\delta p}$ should be satisfied the following relationship.

$$L_{\delta p1}/L_{\delta p2} = \sqrt[3]{p_{\delta p2}^2/p_{\delta p1}^2} \tag{19}$$

Thus, in the LPNCM, the pressure could be decreased to increase the narrow channel height and L could be chosen according to Eq. (19), which is the principle of the LPNCM. And the specific value of L in the LPNCM should be decided by the detailed experimental conditions, considering the size of thermal thick fuel and the simulated level of gravity. In the following, the effectiveness of LPNCM would be verified by the experimental results.

According to the principle of the LPNCM, a narrow channel setup operated at low-pressure environment (less than 1 atm) has been set up to achieve the weakly buoyancy environment. Figure 4 is the schematics of experimental setup for the tests here. It is consisted with a cabin, a narrow channel system, a vacuum system, a sample setup, an electrical power supply and a measurement system. The experimental cabin is a stainless steel cylinder with a height of 500 mm and an outer diameter of 400 mm. The sample of wire, the narrow channel setup, thermocouples, and the CCD camera were installed inside the cabin. While a vacuum pump, vacuum gages, a constant current supply and data collection system were arranged outside the cabin.

The narrow channel setup is consisted with two pieces of horizontal quartz plate with a thickness of 4 mm, length of 290 mm and width of 140 mm. The channel height H can be adjusted from 10 to 120 mm. The cabin inner ambient pressure p can be adjusted by the vacuum pump. The experiments were conducted in the quiescent environment at fixed pressure p. The temperature histories of the wire insulation and its surrounding environmental temperature variations were measured respectively by K-type thermocouples, all with a diameter of thermocouple wire of 0.2 mm and a sheath diameter of 1 mm. The thermocouple touched the wire insulation surface closely with the hot junction. Five measurement points were arranged around the wire. Three thermocouples were touched the wire insulation with a horizontal distance of 20 mm. Two thermocouples were arranged to measure the surrounding ambient temperature. There are located on the upper and lower sides of a thermocouple touched in the insulation with a vertical distance of 5 mm. This arrangement of measurements can obtain the temperature history of the wire insulation and the representative temperature raising results of its surrounding area. In experiments, polyethylene insulated nickel–chrome wire specimens were used. Each wire is 0.5 mm in inner diameter. The thickness of wire insulation is 0.4 mm. The effective sample length is 70 mm. The rated current for this wire is 1 A. The oxygen concentration inside the cabin was 21% (N_2 diluent) at different pressures.

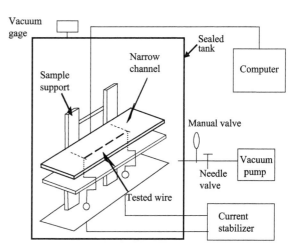

Fig. 4 Schematics of experimental setup by LPNCM

Table 2 Four equivalent conditions of LPNCM

Condition	Narrow channel height, mm	Pressure	Current, A
A	10	1 atm	2, 6
B	13.6	63 kPa	2, 6
C	21.4	32 kPa	2, 6
D	46.4	10 kPa	2, 6

Using the parameters provided in Refs. [34, 35, 39], the critical channel height in NCM can be calculated by Eq. (6) with a result of 10 mm at 1 atm. Thus the experimental conditions in LPNCM can be determined by Eq. (19). The results are listed in Table 2, where typical values of pressure were used based on our previous study [40]. It is seen from Table 2, the pressure and the channel height can satisfy the requirement of Eq. (19). In the experiments, two typical values of current of 2 A and 6 A were used to investigate the effect of current on the fire initiation characteristics of wire insulation. By moderate overload current of 2 A, the temperature variations of the wire insulation were obtained without the wire damaged. While by current overload of 6 A, the wire was breakdown with smoke released from the wire insulation.

3.1 Temperature Variation of Wire Insulation with 2A

Figure 5 shows the variations of the surface temperature of wire insulation at different experimental conditions as shown in Table 2 conducted by the LPNCM with 2 A current. According to the measurement arrangement, there are three thermocouples used to monitor the surface temperature histories of the insulation. It is the temperatures of T_3, T_4 and T_5. The results, shown in Fig. 5, are the arithmetic average of the three temperatures. It is seen from Fig. 5 that the temperature increased rapidly after the power on, about 100 s later the temperature increasing rate slowed down and gradually attained a quasi-steady equilibrium state between the heat produced by the wire and the heat loss from the insulation to the environment. The temperature variation trends are the same as our previous experimental results [16]. It is also seen from Fig. 5 that the temperature rising rates for cases A, B and C were almost the same, while it is different for case D. The temperature rising rate for case D is maximum among them. And the reason for the differences would be discussed later. Furthermore, when compared with the temperature histories of microgravity [5], it can be found that the temperature rising rates for cases B and C are agreement very well with the results conducted in microgravity. Though the initial temperature of wire insulation is 30 °C, 10 °C higher than that in LPNCM, and the wire heating power is 1.27 W [5], a little bit smaller than 1.3 W in LPNCM, the temperature versus time for cases A, B and C, are quite similar with that in microgravity. It proves the validation of LPNCM in simulation of temperature variation characteristics of wire insulation in microgravity.

Fig. 5 The temperature
histories by LPNCM with
moderate overload current of
2 A (the temperature
histories of "μg" are
microgravity results from
Greenberg et al. [5])

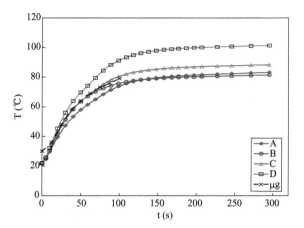

Moreover, we can read the temperature rising time t_p, which is defined as the duration from power on to reach the steady equilibrium temperature, from Fig. 5. Figure 6 shows the variations of t_p under different experimental conditions by LPNCM. As comparison, the results conducted by NCM are also presented there. It is seen from Fig. 6 that the value of t_p decreases quickly with the increase of channel height by NCM, while it keeps similar value in LPNCM for cases A, B and C. In NCM, the effect of buoyancy increases obviously with the increase of the channel height at 1 atm, which leads to the increase of the heat transfer in convection and better cooling conditions for the wire insulation so that t_p of the wire insulation decreases gradually. In LPNCM, the value of t_p for cases A, B and C are almost the same within error limits, because the effect of buoyant convection is still suppressed effectively in the larger channel height with lower pressure environment so that the cooling conditions are similar in those cases. Then, the heat produced by the overloaded current cannot move away through the insulation surface quickly, and thus the value of t_p is similar for those cases. As for comparison, the microgravity result of Greenberg et al. [5] is also presented in Fig. 6. It is seen that the present results of case A, B and C are quite close to that in μg. This again proves that the validation of LPNCM. The parameters for case A, B and C, calculated by Eqs. (10)–(12), satisfies the similarity criterion of LPNCM. While for case D, the value of t_p is increasing to a pretty large value not satisfying the method, which would be discussed later.

We can also obtain the steady equilibrium temperature (T_p) from the temperature histories recorded by different thermocouples with the conditions described in Table 2. Figure 7 shows the steady equilibrium temperature under different experimental conditions. The thermocouples No. 3, 4 and 5 were used to monitor the insulation surface temperature. Point 3 was located at the upper side, while the test point 4 and 5 were located at the down side and with the same vertical height. Thus, the data of T_{4p} shown in Fig. 5 are the average temperature records of T_4 and T_5. In Fig. 7, the solid lines were present results by LPNCM, while the dotted lines were the previous results by NCM from Wang et al. [38]. According to Fig. 7, it is seen that

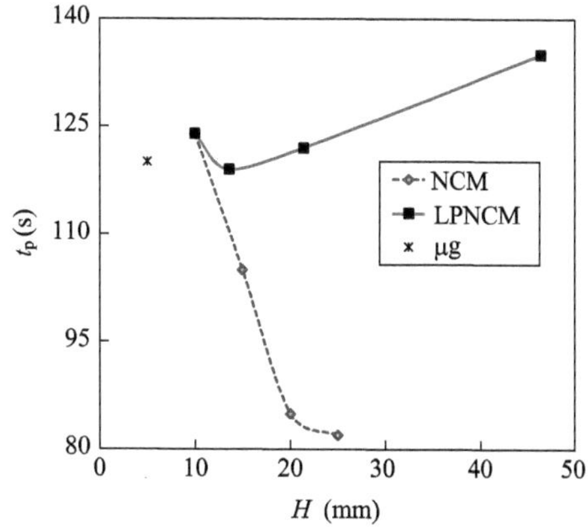

Fig. 6 The contrast of t_p obtained by both LPNCM and NCM with current of 2 A (the t_p of "μg" is the microgravity result from Greenberg et al. [5])

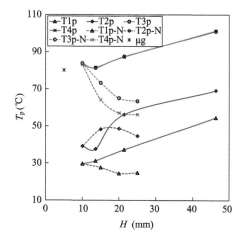

Fig. 7 The steady equilibrium temperature under different experimental conditions of 2 A. The solid lines were present results by LPNCM, meanwhile the dotted lines were the previous results by NCM (the T_p of "μg" is the microgravity result from Greenberg et al. [5])

the results of the insulation surface temperature of T_{3p} and T_{4p} have the same values. This indicates that the temperature of insulation has the same value by LPNCM. For cases A, B and C the temperature differences of T_p are not large. While for case D, T_p increases obviously, and it is far larger than the others. The explanation is as follows. For case D, the pressure was very low (10 kPa), the natural convection was greatly inhibited, and the heat loss of the insulation was decreased, which resulted in the increase of the steady equilibrium temperature. It proves that the effect of lower pressure has overpassed the effect of narrow channel. Thus, the LPNCM is effective to build a weakly buoyant environment and limit the heat transfer of natural convection. However, the environmental pressure should be decreased within a proper range to achieve the similar variation tendency of temperaturein above conditions. For example, to obtain similar temperature histories of T_3 of condition A, the value of pressure is suggested to be not less than 32 kPa.

The reason of the limitation has been provided as follows:

According to the ideal gas equation of state, the mass of air in the narrow channel is $m = pV/(RT)$. Based on Eq. (19), we could obtain the following equation:

$$m_{\delta p1}/m_{\delta p2} = \sqrt[3]{p_{\delta p1}/p_{\delta p2}} \qquad (20)$$

Thus, the total amount of air in the narrow channel decreases with the decrease of pressure. Then if the pressure is too lower (10 kPa in case D), the total amount of air in the narrow channel is decreased too much, which causes the temperature of points near the wire insulation to rise more quickly than that in higher pressure with smaller narrow channel. Then the β and μ of the environment would change greatly so that the LPNCM is inapplicable. That's the reason for the case D not satisfying the method.

It is also seen from Fig. 7 that T_p is decreased with the increase of the channel height in all cases of NCM (after H is greater than 15 mm). This indicates that NCM can be used to simulate microgravity condition only as the channel height is smaller than a specific value (15 mm in this paper). With the increase of the channel height, the buoyancy induced convection is enhanced in NCM. While, for cases A, B and C of LPNCM, T_p is not decreased with the increase of the channel height. Moreover, either T_p of T_{3p} or T_p of T_{4p} in LPNCM is close to that in μg. Thus, LPNCM could effectively suppress the buoyancy induced convection, which could be used to simulate the microgravity environment.

Finally, according to the T_p temperature difference shown in Fig. 7, the uniformity of temperature distribution near the wire insulation could be judged. As for the pair of symmetric points of 3 and 4, the temperature T_p variation curve of T_3 is nearly the same as that of T_4 in the LPNCM, which means that the temperature distribution is uniform in the vertical direction. However, as shown in Fig. 7, the T_p temperature difference in the NCM conditions increases with the increase of the channel height, which means the non-uniformity of temperature distribution in the vertical direction. Thus, when compared with the NCM, the LPNCM improves the uniformity of temperature distribution near the wire insulation.

In summary, the criterion of Eqs. (10)–(12) for LPNCM is experimental proved to be useful in determining the experimental condition. The narrow channel height can be increased from 10 to 21.4 mm by decreasing the pressure from 1 atm to 32 kPa with the LPNCM, which is effective to simulate the temperature variations in microgravity.

3.2 Smoke Emission with 6 A in LPNCM Conditions

As the value of current increases to 6 A, wire insulation will be damaged quickly. A large amount of smoke is emitted without flame. Thus the wire insulation is in the smoke producing stage of fire initiation. The process of smoke producing and moving has been recorded by a CCD camera. The four conditions of LPNCM have been

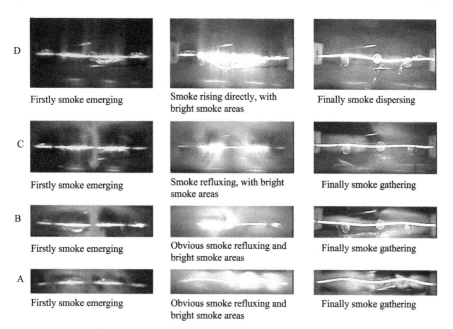

Fig. 8 The smoke emission characteristics by LPNCM with current of 6 A

conducted and repeated three times, with good repeatability. The smoke emission characteristics from wire insulation have been shown in Fig. 8. Three kind of typical stages were chosen and shown in Fig. 8 to describe the smoke emission characteristics of wire insulation as follows. (1) The beginning stage: the smoke begins to produce and its distribution state could be obtained at the early stage. (2) The highlight stage: after a large amount of smoke produced, smoke moves toward outside, with track of circumfluence, due to the confinement by the narrow channel [38]. Meanwhile, there are bright smoke areas along the wire insulation. A highlight smoke distribution area can be observed as strong smoke movement. (3) The evanescing stage: at the end of smoke movement, the accumulated smoke near the wire insulation would evanesce and the distribution state of the left smoke could be observed.

According to Fig. 8, we can analyze the similarity of smoke moving characteristics in four conditions of LPNCM, which has been listed in Table 3. (1) The beginning stage: case A, B and C have the similar smoke moving characteristics—the smoke is emitted from the center area of the wire insulation and then symmetrically moves toward left and right. The shape of the smoke is like a mushroom. However for case D, it has great difference in the smoke moving characteristics. The smoke moves upward directly. There is no clear smoke area under the wire insulation and its distribution of smoke is like an upended curtain in shape. (2) The highlight stage: case A and B have the similar smoke moving characteristics. A group of bright smoke is accumulated near the wire insulation, strong circumfluence movement could be found due to the confinement of the narrow channel, which is similar with that in

Table 3 The contrast of smoke moving characteristics in LPNCM at 6 A

Condition/stage	The beginning stage	The highlight stage	The evanescing stage
A (10 mm, 1 atm)	The smoke is discharged from the center area of the wire insulation and then symmetrically moves toward left and right; the distribution of smoke is like a mushroom in shape	A group of bright smoke is gathered nearby the wire insulation, fierce circumfluence movement could be found due to the confine of narrow channel	The left smoke has the similar moving characteristics, gathering around the wire insulation and moving along it slowly
B (13.6 mm, 63 kPa)			
C (21.4 mm, 32 kPa)		The concentration of smoke begins to decrease, with some circumfluence movement, and a smaller group of bright smoke gathers around the wire insulation	
D (46.4 mm, 10 kPa)	The smoke moves upward directly and its distribution of smoke is like an upended curtain in shape	The concentration of smoke becomes the least and a large amount of smoke move upward, with less smoke circumfluence movement	The smoke evanesces directly and there is no smoke gather around the wire insulation

microgravity [5]. In the condition of case C, the concentration of smoke begins to decrease, with some circumfluence movement, and a smaller group of bright smoke accumulated around the wire insulation. However, for case D, the concentration of smoke becomes small and a large amount of smoke move upward, with less smoke circumfluence movement, which shows small assembling effect of smoke around the wire insulation as comparison with the other conditions. (3) The evanescing stage: At the end of smoke moving, in case A, case B and case C, the residual smoke has the similar moving characteristics, accumulating around the wire insulation and moving along it slowly. Greenberg et al. [5] and Fujita et al. [19] have investigated the movement of smoke from the wire insulation in microgravity after ignition. Results show that smoke resides near the wire insulation for a relatively long period, and then propagates along the wire gradually in microgravity. While smoke would move upward the wire insulation as a result of buoyancy induced flow in normal gravity. Thus, the smoke uniformly moving along the wire in LPNCM has effectively simulated the key movement characteristic of smoke in microgravity. However, in case D, the smoke evanesces directly and there is no smoke accumulated around the wire insulation.

In summary, by the analysis of smoke moving characteristics, case D has great differences with the other conditions, which could not be regarded as an equivalent condition. The others are observed similar smoke moving characteristics at the typical stages. According to our previous results [38], in normal gravity and 1 atm ambient pressure of NCM, if the channel height is greater than 15 mm, it cannot be used to simulate the smoke moving characteristics at microgravity. Besides, results in this paper showed if the pressure reduced to 32 kPa and the narrow channel increased to 21.4 mm, LPNCM can still be used to simulate the phenomena at microgravity, for example it could obtain similar smoke moving characteristics at microgravity.

4 Study on Prefire Phenomena at Microgravity by SJ-8

4.1 Experimental Facility and Process

The experimental setup was designed for microgravity experiments at the China recoverable satellite SJ-8, which offers long microgravity experimental duration and 10^{-4} g microgravity quality and requires experiments that are operator-independent, compact, and constructed to withstand vibrations, large rate of deceleration and meet stringent safety standards. The setup was a stainless-steel cylindrical vessel with an outer diameter of 320 mm, a wall thickness of 2 mm and depth of 450 mm. Figure 9 is a photo of the flight experimental package. It was mainly a test section with a constant current supply to provide overloaded current, a thermocouple temperature measurement system, a radiometer system and a controls system. They were fitted on three levels.

The test sample wires, thermocouples and radiometers were mounted on the first level. The polytetrafluoroethylene-insulated silver-gilt copper wire was used as the test sample. It is an astronautic grade wire cable with black colour. The cable conductor is consisted of 19 pcs single wire. Each wire is 0.12 mm in diameter. The criterion section area of the cable is 0.2 mm². The thickness of cable insulation is 0.1 mm. The rated current for this cable is 2 A. In order to investigate the effects of the overload current on the fire initiation of the wire insulation, two cables were used in the experiments. Each cable was fixed to the sampling holder with a length of approximately 100 mm, and then it was coiled closely around an insulated flame retardant pole. The pole is a silicone rod with diameter of 5 mm. The typical properties of the silicone holder are as follows: thermal conductivity of 0.9 W/m K, surface resistivity of 1.38×1014 Ω, dielectric constant of 2.5 MHz. In the experiments, the temperature histories of the uncoiled wire and the coiled part were measured respectively by K-type thermocouples, all with a sheath diameter of 0.5 mm. For the coiled part temperature measurement, the thermocouple was embedded inside the surface between two coiled wires. For the single wire temperature measurement, the thermocouple touched the wire insulation surface closely with the hot junction. The radiometer was used to measure the radiation emitted from the coiled wire insulations. It is an infrared radiometer with film thermopile. The reason for using the coiled wires for radiation measurement is to increase the detectable area to satisfy the radiometer requirements. Furthermore, we can use one wire to investigate the effects of wire bundles on fire initiation.

The second level of the experimental apparatus was the controls level. It consisted of the central control module, the information module, the relay, the A–D data collection module, and the DC electrical supply, where the DC electricity was supplied by the satellite. All the temperature transducers, radiometers, current transducer and pressure transducer were connected to the AD data acquisition card. All these data were packed up and communicated with the ground control centre.

The constant current supply, providing the overloaded current for the test wire cable, was fitted on the third level.

The pressure in the experimental module at the satellite was lower than the ground ambient pressure. It was about 59 kPa. Thus the low-pressure experiments at normal gravity were conducted to comparing with the microgravity results.

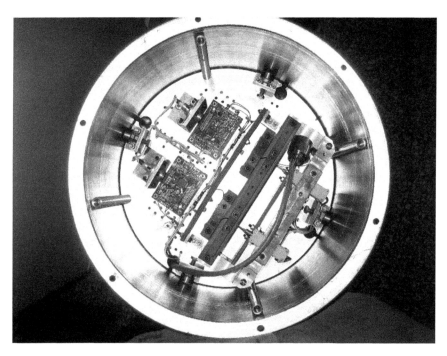

Fig. 9 Photo of the flight experimental package

4.2   Temperature Histories

The temperature histories of the coiled wire insulations at microgravity with over-loaded current of 10 A are shown in Fig. 10. It is seen from Fig. 10 that the temperature increased rapidly after the current inputted, then the temperature increasing rate slowed down and gradually attained a quasi-steady state between the heat produced by the overload and the heat loss to the environment.

The results indicated that the buoyant convection decreased and almost vanished, the heat loss decreased. Thus, the heat produced by the overloaded current cannot move away through the insulation surface immediately, and thus the insulation temperature is gradually increasing. The results also indicted that the heat loss of the electric components decreased at microgravity, it might cause the overheating of the electric components and then results in fire. This is a gradually developing process. It takes several minutes, several hours even more. To create such long duration microgravity on the ground is almost impossible. However space flight experiments at the real microgravity environments are very expensive. It is unsuitable for lots of material tests. Therefore a "function simulation" method is introduced here to create an environment on the ground where the heat transfer of electric wire, cable and components are similar with that at microgravity. So that plenty of experiments can be completed to satisfy requirements of fire safety at microgravity. The suggested experimental method is functionally satisfied in heat transfer sense. The principle and conditions for function simulation are discussed in more detailed in the following.

For flame spread at different pressure environments, the impact of radiative losses can be estimated by the radiative time scale t_{rad}. It can be expressed as $t_{rad} = T_f \rho c_p \Big/ \left(4\sigma a_p \left(T_f^4 - T_\infty^4 \right) \right) \sim \rho / a_p$, where optical thin model is used for

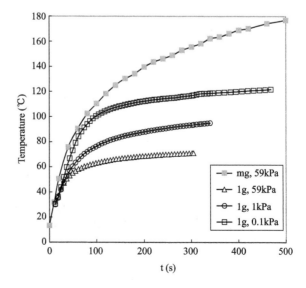

Fig. 10 Temperature histories of the coiled wire insulations at microgravity and normal gravity with current of 10.2 A

the radiative loss from a volume of gas at the flame temperature T_f. Since the Planck mean absorption coefficient of the gas has $a_p \sim P^1$, assuming an ideal gas, we have $t_{rad} \sim P^0$. Thus, the radiative time scale is similar at all pressures. Therefore, in the following analysis to determine the function simulation conditions, the radiative heat transfer is neglected.

For the incompressible Newtonian fluid two-dimensional "undefined" natural convective heat transfer, the Nusselt number can be defined as the ratio of convection heat transfer to fluid conduction heat transfer under the same conditions. It is a dimensionless criterion measuring the convective heat transfer coefficient under certain defined circumstances. It is also a similarity criterion distinguishing whether the convective heat transfer is similar. The Prandtl number is a dimensionless parameter of a convective system that characterizes the regime of convection. This dimensionless criterion is a property of fluid itself, which describes the inherent relationship between velocity field and temperature field. In the present case the Prandtl number of air is almost a constant which does not vary with its temperature and pressure. The Grashof number is a dimensionless number in fluid dynamics which approximates the ratio of the buoyancy to viscous force acting on a fluid. The characteristic of natural convection is that the fluid velocity depends only on the ratio of buoyancy to viscous force. Meanwhile, for natural convection the Reynolds number is no longer an independent similarity criterion. Only the Grashof number equals everywhere the velocity field will be similar. Thus, for the natural convective heat transfer provided the Grashof number equals everywhere, the Nusselt number will also be equal everywhere. That means the simulation of heat transfer is satisfied. So, one can define the criterion of "function simulation" under the condition keeping the Grashof number equal everywhere.

We know the Grashof number is $Gr = (\Delta\rho / \rho)gL^3 / v^2 \sim P\Delta\rho g L^3 / \mu^2$. If the reduced gravity of 10^{-4} g is utilized and density variation induced by temperature is assumed to be independent of pressure, the relationship of reduced pressure for function simulation P_s and the pressure at microgravity P_{mg} can be expressed as: $P_s = 10^{-4} P_{mg}$. Thus by using reduced ambient pressure in normal gravity can realize the identical Grashof number at microgravity.

Figure 10 shows experimental results under reduced pressures of 59, 1 and 0.1 kPa. It is seen from Fig. 10 that the temperature varying tendencies are all the same under function simulations and microgravity environment. The temperature increased rapidly after the current inputted, and then the temperature change tendency was slowed down to attain their maximum value, after that the temperature remained almost unchangeable. We define these maximum values of temperature as the steady equilibrium temperature of the wire insulations under the experimental conditions. It is seen from Fig. 10, the steady equilibrium temperature greatly increased with the decrease of the environment pressures in normal gravity. In the experiments, the steady equilibrium temperature was 70 °C at pressure of 59 kPa, while it was 120 °C at pressure of 0.1 kPa. This is because that the decrease of the environment pressure leads to the decrease of the buoyancy convection, and thus the heat loss of the wire insulation decreased. Therefore, the temperature of the wire insulations increased with the decrease of the pressures.

From the discussions of functional simulation, to achieve the presented micro-gravity experimental condition, the reduced pressure should be around 5.9 Pa. This is not achieved in Fig. 4's reduced pressures experiments completed in normal gravity. Thus the microgravity data do not fit to any of the reduced pressure data.

The steady equilibrium temperature T_s at quasi-steady state can simply estimate from the balance of heat fluxes between conduction, convection and radiation. The results can be expressed as: $T_s \sim \alpha^{-1} \sim Nu^{-1}$. Where α is the heat transfer coefficient. From the classical heat transfer textbook, we know the Nussle number and Grashof number have relation of $Nu = f(Gr^n)$. Thus the steady equilibrium temperature can be expressed as: $T_s \sim Gr^{-n}$.

Table 4 shows the experimental temperature and rough estimated results by the exponent (n) relation of $T_s \sim Gr^{-n}$. Where $n = 0.12$.

It is seen from Table 4, with $n = 0.12$, the estimated results by using of the relation $Nu = f(Gr^n)$ agreed very well with the experimental results. Actually, the exponent (n) in relation $Nu = f(Gr^n)$ is about 0.12 can be found in many literatures on heat transfer from a cylinder [43]. Using this value, predicting the microgravity data is possible by extrapolation to the 10^{-4} g in the g term in Grashof number. In Table 4 the pressure of 5.9 Pa is the functional simulation condition of the microgravity environment of 10^{-4} g. The predicted steady equilibrium is slightly larger than the microgravity experimental data. The reason is that at microgravity experiments the steady equilibrium has not achieved yet. Thus it is reasonable to declare that the predicted reduced pressure value is agreed with the microgravity experimental results. Thus the heat transfer of buoyancy convection explains the observed trend very well.

The temperature histories of a single wire and the coiled wire measured by ther-mocouples in the microgravity environment. The cable wire coiled closely around the insulated flame retardant pole, which is used to simulate the parallel arrangements of cables required by ASME cable standard. It was observed that the insulation tem-perature change tendency is the same for the single wire and coiled wire, but the temperature increasing rate and the final reached maximum temperature during the experiments are quite different. The insulation temperature behaves firstly increasing quickly and then the temperature increasing rate slow-down, where the temperature increasing rate can be inferred from the results. For the single wire, the insulation temperature increased from 13.2 to 43.5 °C in the time duration of 140 s, and thus the

Table 4 Experimental and estimated results of steady equilibrium temperature	Conditions	Experimental results, °C	$T_s \sim Gr^{-n}$, °C
	P = 59 kPa	70	61.22852
	P = 1 kPa	95	99.87482
	P = 0.1 kPa	125	131.6607
	P = 5.9 Pa		184.9072
	Microgravity 10^{-4} g	176	

temperature increasing rate is 0.217 °C/s. While in the same time duration, the insulation temperature increased from 13.2 to 124.9 °C for the coiled wire, and results in a temperature increase rate of 0.798 °C/s. In the whole experimental duration, the final insulation temperature of the coiled wire was 234 °C, while the final insulation temperature for the single wire was 88 °C. It infers that the heat loss decreased quickly in the wire coiled case than that in the single wire case at microgravity. Therefore, from our experimental results, it is concluded that it is easier to cause the over heat of the electric components and then results in fire for the wire bundle situation in the microgravity environment.

4.3 Effects of Overloaded Currents

In the space experiments, two kinds of overloaded currents were used to investigate the effects of overloaded currents on the temperature profiles of the wire insulations. The currents varied from 9.9 to 10.2 A. The current difference is 0.3 A. The imperceptible difference in manufacture may lead to a small difference in resistance of the wire or cable. These differences may cause disaster under overloaded conditions, and thus the purpose of using a small current difference in the experiment was to investigate the trivial difference in overloaded current may cause what happen to the wire insulation. The results conducted in microgravity condition show that the temperature histories of the single wires under overloaded currents of both 9.9 and 10.2 A were almost the same. It means that the difference of 0.3 A of the overloaded currents cannot bring significant difference for the single wire in the microgravity condition. Whereas for the coiled wire the difference of overloaded current, even as small as 0.3 A, resulted in the observed difference of the temperature profiles. It is seen from the results, the insulation temperature variation can be divided into two stages. In the first stage, the temperature increased quickly, and then the temperature variation changed to the second stage, where the temperature increase rate slowed down. In the second stage, the insulation temperature increased gradually. It is observed that the effects of currents were mainly in the first stage. The insulation temperature increasing rate increased with the increase of the overloaded currents, and thus the attained maximum temperature in the first stage increased with the increasing the overloaded current. In the second stage, however the temperature increasing rate was almost the same for the different currents used in the experiments. The final insulation temperature increased with increasing the overloaded currents.

4.4 The Radiation Characteristics

Radiation is one of the most important factors involved in fire initiation of wire insulation caused by overload. The radiation signal can be used for fire monitor. The radiation energy emitted from the insulation was measured by an infrared radiometer.

The infrared lens is a germanium crystal. It can detect the low-temperature radiation energy. The radiometer is specially designed for the experiments. The temperature range is from -10 to 500 °C with 1.5% accuracy. It constructed to withstand vibrations and satisfied other stringent requirements for space experiments. Due to the radiation energy being not intuitive, the output of the radiometer was transferred to the temperature reading. Hereafter we denoted it as radiation temperature. The radiation temperature of the radiometer was calibrated by the blackbody calibration system.

It is seen from the results that the radiation temperature increased with the increasing the overloaded currents. During the experiments, the radiation temperature increased continually. It implies that the steady equilibrium between the heat produced by the overload and the heat loss to the environment is never achieved at microgravity. It is in reason to infer that the overheating of the electric components might finally result in fire if the current never be switched off. The change tendency of the radiation temperature was similar to that of the thermocouple temperature readings. However the thermocouple temperature reading is greater than the radiation temperature in approximately 50 °C. The results demonstrated that the temperature inside the coiled wire was greater than the radiation temperature emitted by the insulation surfaces.

The experimental hardware has been developed to perform the experiments of fire initiation of wire insulation caused by overload on board the China recoverable satellite. The present results show that the temperature of the wire insulation increased continually in the microgravity environment, while in normal-gravity, the temperature might finally attain quasi-steady equilibrium even in low-pressure environment. The steady equilibrium temperature increased with the decrease of the pressure due to the suppression of the buoyancy convection in low-pressure environment. The results imply that the low-pressure environments can be used to simulate the microgravity environments for the study of the fire precursor of wire insulation. The simulation is satisfied in heat transfer sense by selecting different pressure to simulate different gravity level. Nevertheless the results indicate that the microgravity experiments cannot be replaced absolutely by the simulation method in normal-gravity. The effects of wire bundle were investigated at microgravity. The results showed that the insulation temperature of the coiled wire was much greater than that of single wire. Therefore, it is easier to cause the overheating of the electric components and then results in fire for the wire bundle situation in the microgravity environment. The effects of overloaded currents on the insulation temperature and radiation characteristics of the coiled wire were presented. The results showed that temperature increased with the increasing current. For coiled wire the radiation temperature was less than the insulation temperature obtained by thermocouples. The results demonstrated that the temperature inside the coiled wire was greater than the radiation temperature emitted by the insulation surfaces. The experiment results indicate that at microgravity the natural convection almost vanished, the heat loss of the electric components decreased, it might cause overheating of the electric components and then results in fire. The results show that the fire risks of wire overload in the microgravity environment are much more serious than that in normal-gravity.

5   Wire Insulations Experiments in Microgravity by SJ-10

5.1   Setup of Experiment on SJ-10

The long-term overloaded wire insulation combustion (LOWIC) space experiment was performed in one payload of the 24th recoverable satellite of China, the SJ-10 satellite [26, 44, 45], carried by the CZ-2D carrier rocket at 1:38′04″ am, April 6, 2016. The payload was sent into the 250 km earth orbit, offering high-quality (better than 10^{-3} g, @ ≤ 0.1Hz) microgravity for long experimental duration [46].

Figure 11a shows the setup. It was mainly a test section with a constant current supply to provide overloaded current, a thermocouple temperature measurement system, a soot volume fraction measurement system and a controls system. We applied the laser extinction method to analyze the smoke emission in high-concentration zone. A laser generator (wavelength: 635 nm, installed with a spatial filter), two mirrors, two convex lens, a beam splitter lens, and two 8-bit black and white CCD cameras (resolution: 1024×776 pixel, acquisition rate: 25 Hz, observing diameter: 78 mm) were fixed on an optical platform. The CCD camera 2 captured images directly, and the CCD camera 1 recorded tests through the light extinction path. We left a tiny angle between the direct optical path and the light extinction path. We installed a pressure sensor (Honeywell ASDX030P-AAA5, range: 0–1.5 atm, accuracy: 1 kPa), an oxygen sensor (range: 0–100%, accuracy: ±0.01%), and seven K-type nickel chromium-nickel silicon thermocouples (diameter: 0.2 mm, range: 0–900 °C, accuracy: ±0.4%t)to record parameters. The 28 V power was supplied by the satellite and the current was changed by modifying the duty cycle of the wire [47].

Figure 11b shows the sample configuration of the LOWIC experiment. Seven insulated Cr20Ni80 wires (core diameter: 0.5 mm, total resistance: 0.356 Ω) were held and fastened in a wire holder (three in a horizontal column and four in a vertical column). The effective length of cores was 70 mm, and the insulation length was 40 mm. The two columns were 16 mm apart from each other. In each column, paralleling wires were 15 mm apart from adjacent ones. Seven thermocouples were arranged near the midpoints of wires and labeled with the sequence numbers of corresponding wires.

Figure 11c illustrates the operating processes and the power consumptions of the payload with different conditions. In testing, the power source orderly energized the labeled wires with constant currents. At least 24 h interval was set between each two tests to eliminate the influence of previous tests. To compare the temperature evolutions with/without insulation layer, the bare wires #2, #3 and #6 were also energized after insulated-wire tests. Every testing procedure needed 2 h, during which the tested wire was electrified for 660 s and the CCD cameras and LED lights were turned on 15 s before electrifying the wire.

Figure 12 shows the static analysis of the payload box. The experiment apparatus was designed to be compact and operator-independent, withstanding vibration and

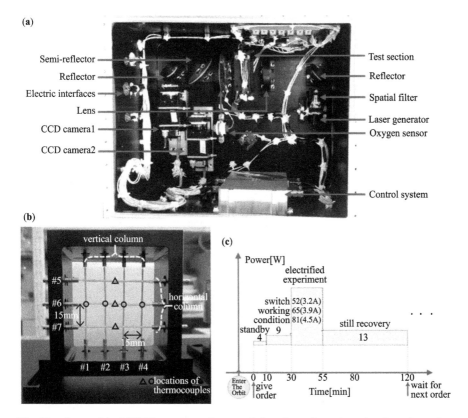

Fig. 11 **a** Setup of the LOWIC experimental system. **b** Sample configuration; triangles—the positions of thermocouples placed in horizontal column; circles—the positions of thermocouples placed in vertical column, **c** the operation process and power consumptions of the payload

high deceleration rate. The payload box was a $500 \times 400 \times 300$ mm rectangle vessel made of aluminum alloy, and weighted 25.5 ± 0.255 kg with all facilities. The average of total power consumption was 22.7 W and the maximum was 81 W.

5.2 Sample Selection

The tested samples are summarized in Table 5. The polyethylene (PE), though differentiated from the low flammable poly tetra fluoroethylene (PTFE, also known as Teflon) extensively used in space, was selected as the insulation material in the present study. Since the first space experiment on the wire combustion [5, 7] many researchers have extensively studied the combustion of wires using PE insulation. Such easy-melting and highly-flammable materials enable us to observe the smoke behavior of the overloaded wire safely when conducting experiment by moderate

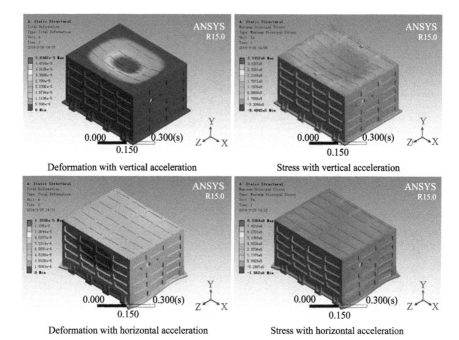

Deformation with vertical acceleration

Stress with vertical acceleration

Deformation with horizontal acceleration

Stress with horizontal acceleration

Fig. 12 Static analysis of the payload box with 90.88 m/s^2 acceleration along the Z-axis (the deformation was shown as 3000 times larger than the result)

Table 5 Parameters of the experiment

Test	Current, A^{-1}	Thickness, mm^{-1}	Material	Wire
a	3.2	0.20	PE	#1
b	3.2	0.40	PE	#2
c	3.2	0.50	PE	#3
d	3.9	0.40	PE	#4
e	4.5	0.40	PE	#5
f	4.5	0.15	PTFE	#6
g	4.5	0.40	PVC	#7
h	3.2	Non-insulated		#2
i	3.2			#3
j	4.5			#6

current (3.2–4.5 A in this study) on the satellite. In the present study, the excess current and thickness were selected as tested parameters to further explore the characteristic of wire insulation in space.

On the ground buildings and electric devices, the less expense and easily-manufacturing of polyvinyl chloride (PVC) makes it be extensively used as insulation for electric wires. In space, the high flame resistance and low chemical activity of PTFE could expand the safe operating condition of electric wires, making it as a typical material used as insulation in spacecraft. So far, the behaviors of overloaded electric wires insulated by these two typical materials have not been compared with PE-insulated wire. In the present study, wires insulated by PVC and PTFE (wire#6 and wire#7) were tested after the PE-insulated samples to observe the smoke emitting on the surface of insulation. Wire#6 was adapted as the most commonly used sample in China's spacecraft, with 0.15 mm thickness of insulation. Wire#7 was coated by PVC insulation with a thickness of 0.4 mm, the same as the insulation thickness in test a, d, and e.

5.3 Processing of Images

The laser extinction is a commonly used fire-detection methods in space, and owning to the high-luminosity of the images captured directly, the CCD camera 2 only could make valid observation at the beginning period of smoke production. Although pyrolysis and combustion products of PE were not a single component, the laser extinction method still could be used to qualitatively record the smoke production and its following diffusing track. To simplify the computation, in this study, the component of smoke was assumed as soot, regarded as spheroids obeying the Beer-Lambert Law [48, 49]. After the data of CCD camera 1 filtered by Gauss denoising, the smoke concentration was calculated by the three-point Abel inversion. The image processing was as follows:

5.3.1 Preparation for Calculation

- Select photos before electrifying as background picture
- Get the average gray value of each background pixel
- Calculate the average gray values of the experimental pixels and those of the four subsequent pictures
- Divide the experimental values by the background values, getting the ratio picture
- Gauss denoise the ratio picture
- Normalize the ratio value of boundary, getting sharp borders of wire, flame and smoke
- Abel inverse the accumulated values in the camera direction, obtaining the smoke distribution on the cross section.

5.3.2 Calculation the Smoke Distribution

According to the Bouguers-Lambert-Beer law, the light intensity captured by camera is:

$$I = I_0 exp\left(-\int_0^L K_e dx\right) \tag{21}$$

where I_0 is the original light intensity, and K_e represents the extinction coefficient. L is the total distance of light movement. The smoke particles are regarded as soot particles, and on this scale, the albedos and radiation emission of particles could be neglected. Therefore, K_e is equal to the absorb coefficient K_a, which could be given as:

$$K_a = \frac{\pi^2}{\lambda} E(m) N \int_0^x P(D) D^3 dD \tag{22}$$

In Eq. (22), λ is the wavelength of laser light, and m represents the complex refractive index. The refractive index m is fitted by $m = 1.755 + 0.576i$, according to $E(m)$ is given by:

$$E(m) = Im\left|\frac{m^2 - 1}{m^2 + 2}\right| \tag{23}$$

In hence, the smoke volume fraction f_v can be calculated by:

$$f_v = \frac{\pi}{6} N \int_0^x P(D) D^3 dD \tag{24}$$

where $P(D) dD$ is the occurrence possibility of particles diameters between D and $D + Dd$. After the original data filtered by Gauss denoise method, the extinction index K_e can be calculated by Abel inversion. Consequently, the smoke volume fraction is given as:

$$f_v = \frac{\lambda K_e}{6\pi E(m)}. \tag{25}$$

5.4 Smoke Emission

5.4.1 Wire with PE Insulation

Definition of Smoke Jetting Angle

Figure 13 shows the laser-illuminated images of the overloaded wires in normal gravity. It can be observed that at the end of vertical insulations, a jet angle could be observed between the insulation surface and the tangent of the high-concentration boundary. In low-pressure case, the high-concentration smoke zone expanded, and the jet angle increased from 0° to around 70° when the pressure decreased [50].

Smoke Emission at Wire Ends

Figure 14 shows the smoke emission of #1–3 wires at variable moments, with different insulation thicknesses with the same excess current. Figure 15 shows the smoke emissions of wire#4 and #5, with 3.9 A and 4.5 A current, respectively. Due to the absent buoyancy, the smoke distributing zone at two ends of the insulation were mirror-symmetrical. Such interesting space phenomenon is rarely observed on the ground where we could only gauge jet angles at the lower end of insulations. As the Fig. 14 illustrates, the high-concentration smoke zone experienced a notable shape change. Initially, the high-concentration smoke lied in a jet paralleled to the core. As time lapsing, the high-concentration smoke zone formed a half-ellipsoid surrounding the insulation end, and eventually turned into a stable shape. Meanwhile, the outer surface of insulation reached the pyrolysis temperature, and initiated smoke production. The variance of the smoke jet angle with wire#3 is shown in Fig. 16.

This process could be schematically illustrated by Fig. 17. At an early stage of overloading (0–30 s, in present study), only oxidative pyrolysis happens due to its relatively low activation energy. The heat was only supplied by the core and

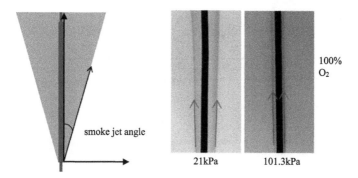

Fig. 13 Definition of the jet angle and its laser-illuminate images in normal gravity [9]

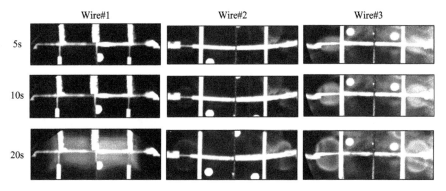

Fig. 14 Smoke emissions with different insulation thicknesses in space, wire#1–3

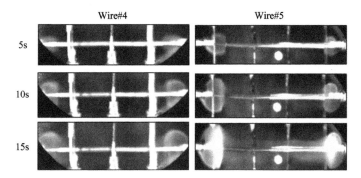

Fig. 15 Smoke emissions with different currents in space, wire#4 and #5

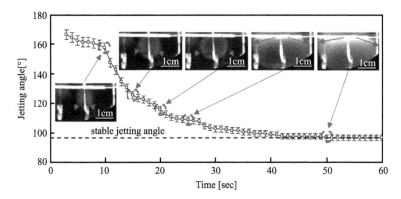

Fig. 16 Variation of the smoke jet angle with wire#3

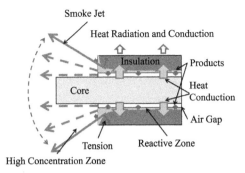

Fig. 17 Schematic illustration for the evolution of smoke jetting angle

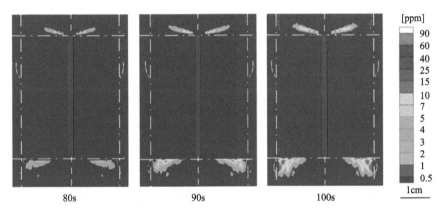

Fig. 18 Smoke distribution of the smoke jet at the ends of wire#3, computed by the laser extinction method; dotted lines—the cores; red rectangle—the insulation being tested

the reactive zone lied in the inner surface of insulation, where the PE could reach the oxygen and the core both. The pyrolysis front, located on the inner surface of swelled insulation, produced smoke escaping along the length of the wire. Then, the fuel consumption and the evolution of temperature field expanded the reactive zone. The space between the insulation end and the core was also expanded by the tension caused by the jet flow. The jet angle enlarges and eventually reached a fixed value after the reactive zone expended to the outer surface. Meanwhile, the outer insulation started to produce smoke. The insulation emitted the smoke with an unchanged morphology of the pyrolysis front. With a fixed jet angle, the smoke aggregated along the main jet, thus the CCD camera 1 could capture the smoke emission at the wire end, as Fig. 18 shows. The notable main jet constantly produces smoke till the intensive pyrolysis on the inner surface ruptures the insulation layer.

The velocities of the high-concentration smoke boundary caused by the oxidative pyrolysis from the outer surfaces of insulation were calculated. As shown in Fig. 19, the average moving rate of the high-concentration smoke boundary are in the scale of 0.3–0.5 mm/s. Smoke emission with the velocity in this scale only could be observed

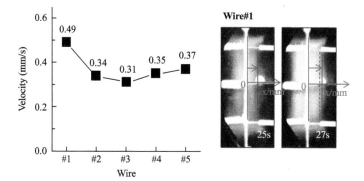

Fig. 19 Average velocity of the advancing high-concentration smoke boundary caused by reaction on the outer insulation surface in the middle of wire, 0–16 s since the beginning of the outer reaction in the case of wire#1–4, 0–14 s in the case of wire#5

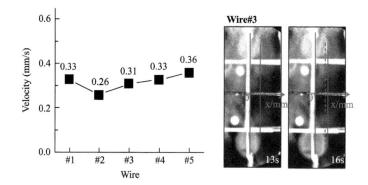

Fig. 20 Average velocities of the advancing high-concentration smoke boundary at the end of wire, 0–26 s since the beginning of end smoke jet in the case of Wire#1–4, and 0–16 s in the case of wire#5

in microgravity due to the absence of buoyancy. The velocity simply decreased with the insulation thickness and increased with the current. This velocity may be affected by many factors, such as the temperature field, the oxygen transportation etc. which requires further study in the future. Figure 20 shows the average velocity of the high-concentration boundary at the wire end moving along the direction perpendicular to the length of wire, where a same scale with the moving rate shown in Fig. 19 was found.

As shown in Fig. 21, the observable evolution of the smoke emitting stages significantly varied with the insulation thickness and the excess current. Large insulation thickness and current helped to produce visible smoke jet earlier. The evolution time of the end smoke jet became shorter with thicker insulation and larger excess current, as the stage 2 in Fig. 21 shows, indicating the end smoke jet is easier to achieve its stable sate. This might be because the reactive zone is more easily to be expanded to the outer surface of insulation if the insulation is thin and the current supplies larger

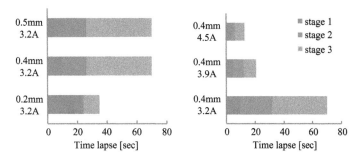

Fig. 21 Evolution of the smoke emitting stages at the wire end; stage 1—from the beginning of overload to the time the end smoke jet been captured by CCD camera 2; stage 2—from the beginning of the visible end smoke jet to the time it turned stable; stage 3—from the beginning of the stable smoke jet to the time smoke been captured by CCD camera 1

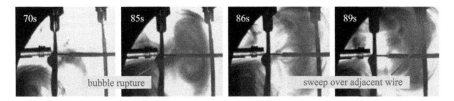

Fig. 22 Laser illuminated images of a typical bubbling jet process with wire#1

heat. Likewise, the space between the core and the insulation will be easier to be enlarged by the gas flow. In the end of stage 3, the smoke could be detected by CCD camera 1 decades second later than the end smoke emission turned stable because the laser extinction method could only observe high-concentration smoke.

Bubbling Smoke Jet

A bubbling smoke jet mode was also observed later. Caused by the heat provided by the core, the non-oxygen pyrolysis produced a bubble of smoke between the insulation layer and the core, as Fig. 22 shows. Unlike the wire end, the inner surface in middle section was isolated from the oxygen. The non-oxygen pyrolysis occurred when the temperature was high enough. The gathered smoke might raise the pressure, then, as the stress increased through smoke accumulation, there could be an insulation layer rupture. Subsequently, as the red arrow indicated, the pyrolysis product swept over the adjacent cables, showing the non-oxygen pyrolysis may cause dangerous interaction between wires. In this study, all bubbling jets happened later, after the end jets turned stable.

5.4.2 Wire with PTFE and PVC Insulations

Figure 23 shows the smoke emitting behavior of wire#6. The PTFE-insulated wire failed to achieve any intensive reaction with air in the moderate-current experiment. In the early stage of overload, the wrapping insulation produced thin smoke and detached orderly from the wire end to the middle, as Fig. 23a shows. As shown in Fig. 23b, the detachment of insulation turned slow later, creating a chance for the insulation to be sufficiently heated, where the slow rupture of the insulation occurred. After heated for 660 s, the insulation has not completely detached with the core.

Figure 24 shows the smoke emission of wire#7. The PVC-insulated wire produced heavy smoke with 4.5 A current. As shown in Fig. 24a, the concentration of smoke was much higher than that of wire#6. Afterwards, the high-concentration zone of smoke lost its symmetry within 10 s, as illustrated in Fig. 24b. An intensive smoke jet emerged and created strong disturbance of the ambient environment. After 660 s overload, the PVC insulation remained a bulged remnants around the metallic core.

(a)   (b)

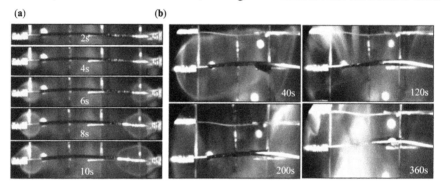

Fig. 23 **a** Symmetrical smoke emission and ordered detachment of PTFE insulation in early stage. **b** Slow rupture of PTFE insulation in late stage

(a)   (b)

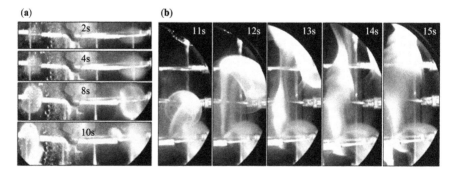

Fig. 24 **a** Symmetrical smoke emission of PTFE insulation in early stage. **b** Asymmetrical smoke jet of PTFE insulation at the wire end

5.5 Temperature Variation

5.5.1 Simplified Model

A simplified model is applied to describe the temperature rise of wire. When over-loaded, the constant current will heat the wire up. The temperature of wire rises and will eventually reach a nearly steady-state value. Only heat transfer occurs before any reaction starts.

Other assumptions are made as follows:

(1) The temperature field of the wire is uniform and thermally-thin theory is applied.
(2) The wire is long and the heat was conducted only along the radial direction.
(3) Through the whole process, all thermal properties and the electric resistance of wire are constant.
(4) All parameters of the ambient medium are constant.

The energy equation is:

$$mc\frac{dT}{dt} = I^2R - KA(T - T_0) \tag{26}$$

where m, A, and R are the mass, area of outer surface and electric resistance of wire per unit length, with the specific heat capacity c. The integrated heat transfer coefficient on the outer surface is K. On the ground, the heat transfer coefficient K depends on many factors, such as the natural convection, forced convection, heat conduction, and thermal radiation. T and T_0 represent the temperatures of wire and ambient environment respectively. The time since the overload is t. Therefore we get excess temperature as:

$$\theta = T - T_0 = \frac{I^2R}{KA}\left(1 - \exp\left(-\frac{KAt}{mc}\right)\right) + (T_1 - T_0)exp\left(-\frac{KAt}{mc}\right) \tag{27}$$

where T_1 is the initial temperature of wire, and set $\theta_\infty = \frac{I^2R}{KA}$, $\theta_0 = T_1 - T_0$, $t_m = \frac{mc}{KA}$, thus we get:

$$\theta = \theta_\infty\left(1 - \exp\left(-\frac{t}{t_m}\right)\right) + \theta_0 exp\left(-\frac{t}{t_m}\right) \tag{28}$$

For the temperature rising rate f:

$$f = \frac{d\theta}{dt} = \frac{\theta_\infty - \theta_0}{t_m}\exp\left(-\frac{t}{t_m}\right) = \frac{I^2R - \theta_0 KA}{mc}exp\left(-\frac{KAt}{mc}\right) \tag{29}$$

Therefore if the wire undergoes different excess currents, we get:

$$\frac{\partial f}{\partial I} = \frac{2RI}{mc} exp\left(-\frac{KAt}{mc}\right) \tag{30}$$

And the temperature rising rate changes with the integrated heat transfer coefficient as:

$$\frac{\partial f}{\partial K} = \frac{-(\theta_0 Amc + (\theta_\infty - \theta_0)KA^2 t)}{(mc)^2} \exp\left(-\frac{KAt}{mc}\right) \tag{31}$$

We can easily find the temperature rising rate increases with the excess current I while decrease with the integrated heat transfer coefficient K on the outer surface of wire. Though in actual pre-ignition scenarios, the heat transfer is much more complicated than the model demonstrated above, it helps us predict the tendency of the temperature when the condition changes.

Figure 25 shows the temperature records when wire#2 and wire#3 is electrified and provides comparison with the bare-wire tests. Instead of monotonically rising as the bar-wire temperature did, the temperatures of insulated wires showed transient fluctuations. The integrated heat transfer coefficient K in Eq. (9) changed here, and the current I remained constant through the overload. Other thermocouples also recorded similar fluctuations, showing the temperature fluctuations were not caused by the thermal expansion of heated core.

5.5.2 Confirmation of Bubbling Jet

A one-dimension analysis of heat transfer could help analyze the temperature fluctuations further. As Figure 25 shows, in the tests of wire#2 and #3, the fluctuations

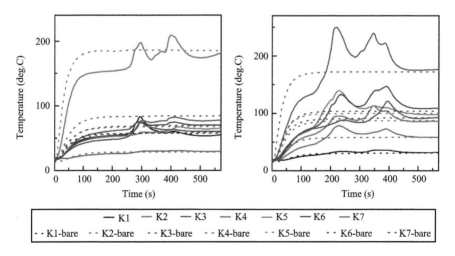

Fig. 25 The temperature records in insulated/bare wire tests, with 3.2 A current

occurred later than the temperatures in bare-wire tests reached a steady state. In steady-state heat transfer, the temperature of metallic core maintained constant (when the temperature rising rate was less than 0.1 °C/s, set in the present study), the heat was conducted through the insulation layer with a constant heat flux per unit time. Before any insulation rupture or bubble smoke jet occurred, the insulation still acted as a heat conductor. Here we get:

$$q = \frac{T_i - T_0}{R_{tm}} = const \tag{32}$$

where q is the heat supplied by the core per unit length, and R_{tm} is the thermal resistance per unite length of the media between the core and the thermocouples. T_i is the temperatures on the inner surface of insulation, and T_0 is the value recorded by thermocouple. For the thermal resistance, we get:

$$R_{tm} = \frac{1}{2\pi \lambda_{pe}} \ln\left(\frac{d_i + \delta}{d_i}\right) + \frac{1}{2\pi \lambda_{air}} \ln\left(\frac{d_o}{d_i + \delta}\right) \tag{33}$$

where λ_{pe} and λ_{air} are the thermal conductivities of PE and air respectively. d_i is the diameter of the core and δ is the insulation thickness. d_o indicates the distance from the thermocouple to the center axis of core. Usually, λ_{pe} is almost ten times larger than λ_{air}. The temperature records of insulated tests should be higher than those in bare-wire tests. Before temperature fluctuating, the present study recorded lower temperatures with insulations than those with bare cores. This implies the heat was not totally conducted to the outer surface of insulation and some amount of it may be consumed by temperature rising, phase transition and pyrolysis of the insulation. When the fluctuation occurred, thermocouples even recorded higher temperatures than in bare-wire tests, reflecting the high temperatures of insulation debris and pyrolysis products inside the insulation layer.

5.5.3 Typical Temperature Evolutions of PE-Insulated Wires

Compared with bare-wire tests, the temperature history of PE-insulated wire could be summarized into three stages:

(1) In the early stage, the insulation layer acted as a heat insulator, slowing down the temperature rise recorded by the thermocouple. According to Eq. (9), since the mc of PE is higher than that of air, the existence of insulation decreased the temperature rising rate f and therefore made the temperature rose more slowly than that of bare wire.

(2) In the middle stage, temperature fluctuations occurred due to the bursting high-temperature smoke and the rupture of insulation. The pyrolysis products and insulation debris caused fluctuating temperatures which were even higher than those in the bare-wire tests.

(3) In the late stage, when the degradation of insulation completed, the temperature records reached the balance values, which were nearly the same as those in bare-wire tests.

5.5.4 Effect of Thickness on PE Insulation

Figure 26 shows the temperature histories of wire#1–3 in normal gravity and microgravity. In microgravity, the temperatures rose faster and achieved a higher steady-state value. The temperature histories show more different when the insulation thickness changed. In 0–80 s, the temperatures increase less with thicker insulations. This may be because the larger mc of PE decrease the f in Eq. (29). Besides, the temperature fluctuated more times with thicker insulation, showing times of bubbling smoke jets were needed to completely rupture the insulation.

The first peaks of fluctuations with 0.2 mm, 0.4 mm and 0.5 mm came in 111 s, 294 s, and 219 s respectively, indicating there may be a tradeoff between the smoke production and the stress tension on the inner surface of insulation. The maximum temperature increases with insulation thickness. It may be because higher tension the insulation could undergo made it possible to form high-temperature and high-pressure smoke inside the bubble.

5.5.5 Effect of Current on PE Insulation

Figure 27 illustrates the temperature histories of 0.4 mm PE-insulated wires with different current. In normal gravity, though with different current, the temperature histories fail to show notable differences since the thickness of the natural convection boundary very thin and its cooling effect became stronger when the density

Fig. 26 The temperature histories of the wire#1–3, electrified by 3.2 A current

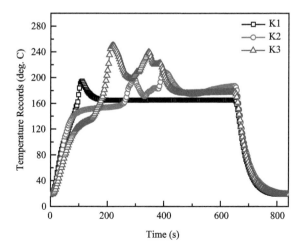

Fig. 27 The temperature histories of the wires#2, 4, and 5, with 0.4 mm insulation

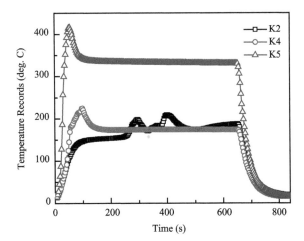

difference of ambient air became larger. In microgravity, the temperature rises with different current could be easily differentiated with each other. In the early stage, the temperature rising rate significantly increased with the current. It showed the similar tendency as the Eq. (30). The first peaks of temperature fluctuations also came earlier with larger current.

The extents of overheat, defined as the differences between the local extreme temperature and the balanced temperature, could measure the additional temperature rise caused by burst pyrolysis products and hot insulation debris. The overheat extent increased from 28 °C to 85 °C when the current increased from 3.2 to 4.5 A, implying hotter smoke inside the bubble with larger current. The frequency of fluctuation decreased with the current and the peak of it became sharper. This may showed that with large current, stronger smoke production and less bubbling jets needed to rupture the insulation layer. The 4.5 A excess current managed to induce a stable and visible flame, whereas only flash were observed in other tests. We should note that after the intensive rupture of wire, the cores were usually inevitable bent, which was caused by the heat and the strong insulation rupture. This bending will change the distance between the wire and its thermocouple, bringing detecting errors.

5.6 Environment Parameters

Figure 28 shows the variation of pressure inside the payload box when testing samples. When electrifying the tested wire, the heat supplied by current heated the air up and produced gaseous products of pyrolysis, leading to higher records of the pressure sensor. After all tests, the ambient pressure decreased by less than 0.3PSI. Figure 29 illustrates the relative concentration of oxygen during the tests. The wire#5 achieved to sustain a visible flame in the early stage of overload, both the pressure and oxygen

Fig. 28 The pressure inside
the payload box during tests

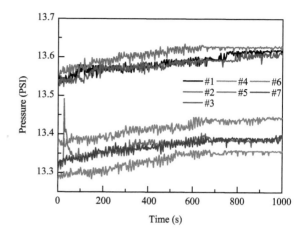

records abruptly increased and subsequently decreased at that time. When the experiment ended, the oxygen concentration decreased by less than 0.0035 mol fraction
in total, lying in the permitted range of error

5.7 Conclusions of SJ-10 Experiments

In the space experiment, the smoke emission of the overloaded wire was investigated
in long-term microgravity for the first time. The smoke emission firstly started in wire
ends, decades second before the temperature record reached a balanced value. The
evolution of the high-concentration smoke zone showed the geometry of pyrolysis
front dominated the direction of end smoke jet. The laser extinction method detected

Fig. 29 The relative oxygen
records inside the payload
box during tests, setting the
oxygen record before any
test as 1.0

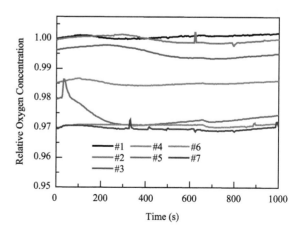

the end smoke jet until the jet boundary reached a stable angle and the smoke accumulated along a main jet. In the middle wire section, bubbling smoke jets occurred and caused fluctuations of temperature records when rupturing the insulation. The bubbling smoke jet may be attributed to the non-oxygen pyrolysis on the inner surface. It produced high-temperature smoke that eventually burst out from the melt insulation. The smoke emission of overloaded wire with PVC and PTFE insulation were observed. The wrapping PTFE insulation was orderly detached from the metallic core when overload, without intensive smoke emission. The high-concentration zone of smoke at the end of PVC-insulated wire lost its symmetry within 10 s and created a heavy smoke jet.

The temperature records also showed unique phenomena in microgravity and also helped to confirm the bubbling smoke jet with PE insulation. In the early stage, the temperature rose slowly with thicker insulation. When thicker insulation being ruptured, the stronger and more frequent temperature fluctuations were observed. With a lager current, the fluctuations came earlier and with sharper fluctuating peaks. The fluctuation times decreased under the similar condition, indicating the insulation was ruptured faster and more completely.

References

1. Robert F (1994) Risks and issues in fire safety on the space station, NASA TM-106430
2. Friedman R (1998) Fire safety in extraterrestrial environments. NASA TM-207417
3. Limero T, Wilson S, Perlot S, James J (1992) The role of environmental health system air quality monitors in space station contingency operations. SAE Trans 101:1521
4. Babrauskas V (2003) Information on specific materials and devices. In: Babrauskas V (ed) Ignition handbook: principles and applications to fire safety engineering, fire investigation, risk management and forensic science. Fire Science Publishers, Issaquah, pp 774–795
5. Greenberg PS, Sacksteder KR, Kashiwagi T (1994) Wire insulation flammability experiment: USML-1 1 Year post mission summary. NASA CP 3272 2, p 631
6. Thomas HC, Donald AP 1971 Burning of Teflon-insulated wires in supercritical oxygen at normal and zero gravities. NASA TM-2174
7. Greenberg PS, Sacksteder KR, Kashiwagi T (1995) Wire insulation flammability. NASA CP 10174, p 25
8. Kikuchi M, Fujita O, Ito K, Sato A, Sakuraya T (1998) Experimental study on flame spread over wire insulation in microgravity. Proc Combust Inst 27:2507
9. Kikuchi M, Fujita O, Ito K, Sato A, Sakuraya T (2000) Flame spread over polymeric wire insulation in microgravity. Space Forum 6:245
10. Fujita O, Kikuchi M, Ito K, Nish K (2000) Effective mechanisms to determine flame spread rate over thylene-tetrafluoroethylene wire insulation: discussion on dilution gas effect based on temperature measurements. Proc Combust Inst 28:2905
11. Fujita O, Nish K, Ito K (2002) Effect of low external flow on flame spread over polyethylene-insulated wire in microgravity. Proc Combust Inst 29:2545
12. Umem A, Uchi M, Hira T (2002) Physical model analysis of flame spreading along an electrical wire in microgravity. Proc Combust Inst 29:2535–2543
13. Kong WJ, Lao SQ, Zhang PY (2006) Study on wire insulation flammability at microgravity by functional simulation method. J Combust Sci Technol 12(1):1–4
14. Chen LF, Xin Z, Kong WJ (2006) Functional simulations of the fire precursor of the wire insulation in quiescent microgravity environment. Chin J Space Sci 26(3):235–240

15. Kong WJ, Wang BR, Law SQ (2007) Study on fire precursor of wire insulation in low-pressure environments. J Eng Thermophys 28(6):1047–1049
16. Kong WJ, Wang BR, Zhang WK (2008) Study on prefire phenomena of wire insulation in microgravity. Microgravity Sci Technol 20:107–113
17. Nakamura Y, Yoshimura N, Matsumura T et al (2008) Flame spread over polymer-insulated wire in sub-atmospheric pressure: similarity to microgravity phenomena. Prog Scale Model, 17–27
18. Nakamura Y, Yoshimura N, Ito H et al (2009) Flame spread over electric wire in sub-atmospheric pressure. Proc Combust Inst 32:2559–2566
19. Fujita O, Kyono T, Kido Y (2011) Ignition of wire insulation with short-term excess electric current in microgravity. Proc Combust Inst 33(2):2617–2623
20. Takano Y, Fujita O, Shigeta N (2013) Ignition limits of short-term overloaded electric wires in microgravity. Proc Combust Inst 34(2):2665–2673
21. Takahashi S, Ito H, Nakamura Y, Fujita O (2013) Extinction limits of spreading flames over wires in microgravity. Combust Flame 160:1900–1902
22. Takahashi S, Takeuchi H, Ito H, Nakamura Y, Fujita O (2013) Study on unsteady molten insulation volume change during flame spreading over wire insulation in microgravity. Proc Combust Inst 34(2):2657–2664
23. Osorio AF, Mizutani K, Fernandez-Pello C, Fujita O (2015) Microgravity flammability limits of ETFE insulated wires exposed to external radiation. Proc Combust Inst 35:2683–2689
24. Hu LH, Zhang YS, Yoshioka K et al (2015) Flame spread over electric wire with high thermal conductivity metal core at different inclinations. Proc Combust Inst 35(3):2607–2614
25. Fujita O (2015) Solid combustion research in microgravity as a basis of fire safety in space. Proc Combust Inst 35(3):2487–2502
26. Hu WR, Zhao JF, Long M, Zhang XW, Liu QS, Hou MY, Kang Q, Wang YR, Xu SH, Kong WJ, Zhang H, Wang SF, Sun YQ, Hang HY, Huang YP, Cai WM, Zhao Y, Dai JW, Zheng HQ, Duan EK, Wang JF (2014) Space program SJ-10 of microgravity research. Microgravity Sci Technol 26:159–169
27. Xue S, Kong W (2019) Smoke emission and temperature characteristic of the long-term overloaded wire in space. J Fire Sci 37(2): 99–116
28. Xue S, Kong W (2018) Overload characteristic of typical electric wire insulation material in microgravity. In: Keynote lecture, 12th Asian microgravity symposium, Zhuhai, China, 12–16 November 2018
29. Ivanov AV, Balashov YV (1999) Experimental verification of material flammability in space. NASA/CR-1999-209405
30. Zik O, Olami Z, Moses E (1998) Fingering instability in combustion. Phys Rev Lett 81(18):3868–3871
31. Zik O, Olami Z, Moses E (1998) Fingering instability in solid fuel combustion: the characteristic scales of the developed state. The Combustion Institute, pp 2815–2830
32. Wichman IS, Olson SL (2003) Flamelet formation in hele-shaw flow. NASA/CP-2003-212376/REV1, pp 29–31
33. Olson SL, Miller FJ, Jahangirian S, Wichman IS (2009) Flame spread over thin fuels in actual and simulated microgravity conditions. Combust Flame 2009:1214–1226
34. Zhang X (2007) Simulation of flame spread over thin solid fuel under microgravity using narrow channel on ground. Chin J Theor Appl Mech 39(4):466–472
35. Zhang X (2008) Opposed-flow flame spread over a thin solid material in narrow channels in normal and microgravity. J Eng Thermophys 29(2):347–350
36. Zhang X, Yu Y (2008) Comparability of flame spread over thin solid fuel surface under different gravities. J Combust Sci Technol 14(4):289–294
37. Xiao Y, Hu J, Wang SF (2010) A narrow channel experimental study on flammability characteristics of thermally thin fuels under simulated microgravity conditions. J Astronaut 31(7):1877–1882
38. Wang K, Wang BR, Ai YH, Kong WJ (2012) Study on the pre-ignition characteristics of wire insulation in the narrow channel setup. Sci China Technol Sci 55:2132–2139

39. Xiao Y, Ren T, Wang SF (2010) Flame spread over thermally thick fuels in narrow channel apparatus. J Eng Thermophys 31(8):1423–1426
40. Wang K, Ai YH, Wang BR, Kong WJ (2012) Study on pre-fire characteristics of wire insulation by overload in weakly buoyant environment. J Eng Thermophys 33(4):689–693
41. Wang K, Wang BR, Kong WJ, Liu FS (2014) Study on the pre-ignition temperature variations of wire insulation under overload conditions in microgravity by the functional simulation method. J Fire Sci 32(3):257–280
42. Bergman TL, Lavine AS, Incropera FP (2011) Fundamentals of heat and mass transfer, 7th edn. Wiley, Jefferson City, USA
43. Gebhart B, Jaluria Y, Mahajan RL, Sammakia B (1988) Buoyancy-induced flows and transport. Hemisphere, Washington, DC
44. Hu WR, Tang BC (2017) Kang Q progress of microgravity experimental satellite SJ-10. Aeron Aero Open Access J 1(3):00016
45. Zhao HG, Qiu JW, Tang BC et al (2016) The SJ-10 recoverable microgravity satellite of China. J Space Explor 5(1):101
46. Wang Y, Zhao H, Zhang Y et al (2016) Establishing and evaluation of the microgravity level in the SJ-10 recoverable satellite. Aeros China 17(4):3–13
47. Kong WJ, Wang BR, Xia W (2016) Experimental facility for wire insulation combustion in SJ-10. Physics 45(4):219–224
48. Greenberg PS, Ku JC (1997) Soot volume fraction imaging. Appl Opt 36(22):5514–5522
49. Iuliis SD, Barbini M, Benecchi S et al (1998) Determination of the soot volume fraction in an ethylene diffusion flame by multiwavelength analysis of soot radiation. Combust Flame 115(s1–2):253–261
50. Xia W, Wang K, Wang BR et al (2016) Study on combustion characteristics of overloaded wire insulation at the early stage of fire in weak buoyancy. J Eng Thermophys 37(4):876–882

Flame Spread in Low-speed Forced Flows: Ground- and Space-Based Experiments

Shuangfeng Wang and Feng Zhu

Abstract Flame spread across solid materials in low-velocity flow regime is of fundamental interest and practical importance, whereas experiments for this type of combustion have been a challenge. The present article introduces recent progress of this research field in the light of the flame spread experiments that have been performed at the Key Laboratory of Microgravity, CAS. The experimental methods employed to offer a slow convective flow include the narrow channel apparatus in normal gravity, ground-based microgravity free drops, and orbital space flights. The highlighted topics involve the confinement effects on flame spread, concurrent flame behaviors near the quenching limit, flame spread and extinction over thick solids in opposed and concurrent flows, and dynamics of spreading flame over thick solids in step-changed flow.

Keywords Flame spread · Low-velocity flow · Microgravity · SJ-10 Satellite of China · Narrow channel apparatus

1 Introduction

Flame spread over the surface of a solid combustible material constitutes a unique configuration of propagating diffusion flame combustion, and is generally considered as a significant model of real life fire processes. Compared with conventional homogeneous diffusion flames in gases, the flame spreading over solids is different in that the fuel is originally in condensed phase and the pyrolysis process of the solid is needed to generate fuel vapor. Thus the flame spread phenomena involve inherently complex interaction of heat and mass transport processes, and the

S. Wang (✉) · F. Zhu
Key Laboratory of Microgravity (National Microgravity Laboratory),
Institute of Mechanics, Chinese Academy of Sciences, Beijing 100190, China
e-mail: sfwang@imech.ac.cn

School of Engineering Science, University of Chinese Academy of Sciences,
Beijing 100049, China

© Science Press and Springer Nature Singapore Pte Ltd. 2019
W. R. Hu and Q. Kang (eds.), *Physical Science Under Microgravity: Experiments
on Board the SJ-10 Recoverable Satellite*, Research for Development,
https://doi.org/10.1007/978-981-13-1340-0_10

237

chemical reaction in both gaseous and condensed phases. In addition, as the flame spreads across the solid surface, heat transfer takes place at least in the parallel and perpendicular directions simultaneously, i.e., a multi-dimensional system is required to describe the spreading flame problem. Notwithstanding much progress over the years, a thorough understanding of solid fuel flame spread is not in hand, with the complexity of the problem partly explaining the research status.

In order to generalize experimental results and to develop simplified theories, the flame spread over solids is traditionally classified into two sub-categories: opposed flow spread when a flame spreads against the oxidizer flow, and concurrent or flow-assisted spread when a flame spreads in the same direction as the oxidizer flow. On the other hand, there exist two limiting cases with regard to the thickness of the solid fuel: thermally thin when the fuel is heated uniformly throughout its entire thickness by the spreading flame, and thermally thick when solid heat conduction fails to penetrate the entire fuel thickness such that significant temperature gradients develop in the normal direction to the surface. The division of flame spread between distinct sub-categories is meaningful in the sense that the dominant mechanisms are different in the model problems, and most studies have fallen into the four regimes.

It is well known that the flame spread behavior in opposed or concurrent flows depends strongly on the magnitude of the flow velocity. The influence of high-velocity and moderate-velocity has been investigated by numerous experimental and theoretical works. For gas velocities smaller than that induced by buoyant flow, however, flame spread experiments have been a challenge. This situation stems from the fact that it is difficult to offer a very small velocity convective flow in normal gravity and a microgravity environment is needed to eliminate the complications of buoyant flow. Particularly, for a thick solid, the ignition and subsequent transition to flame spread has a relatively long time scale, which precludes the use of ground-based facilities such as drop towers for a microgravity test. From the viewpoint of fundamental understanding of flame spread, microgravity environment is of interest to researchers because a purely forced oxidizer flow can be generated and therefore flame behaviors in low-velocity flow regime could be examined. Studies on this type of combustion, however, have been motivated primarily by the practical concerns of fire safety for inhabited spacecraft in the past decades, since spacecrafts are designed to maintain a low velocity atmospheric circulation (of the order of 10 cm/s).

The present article surveys recent experimental results of flame spread and extinction phenomena over solid fuels, which have been obtained by the authors' research team. Since several excellent reviews have addressed the solid flame spread process, this review is prepared with a narrow scope to focus on spreading flames in the low-velocity flow regime, whereas the highlighted topics are of scientific interest and practical importance. Readers are referred to Fernandez-Pello [1], Wichman [2], Sirignano and Schiller [3], and the book by Quintiere [4] for a comprehensive review of the science of flame spread, while readers who desire a review of flame spread in microgravity are referred to T'ien et al. [5] and Fujita [6].

2 Effects of a Confined Space on Microgravity Flame Spread

Most of previous studies on microgravity flame spread have been performed in open areas. A realistic fire could start and develop in constrained spaces [7, 8]. Meanwhile, the available test section sizes for microgravity experiments are usually rather limited, implying that the influence of a confined space should be taken into account to interpret the experimental results. Only few studies have treated geometry effects on flame spread in microgravity [9–11]. Although they demonstrated clearly the modifications of flame characters in a confined space, the problem deserves more effort. In particular, the quantitative effect of a finite size space on opposed-flow flame spread remains to be examined systematically considering the varied space dimensions.

In our study, effects of confined spaces on flame spread over thin solid fuels in a low-speed opposing air flow have been investigated by combined use of microgravity experiments and numerical computations [12]. To examine the effect of a confined space, the flow tunnel had a variable height in a range of 1.5–5 cm (in experiments) or 6 cm (in computations), while the velocity of the imposed flow was fixed at 5 cm/s. Thin cellulosic paper was used as the solid fuel. The microgravity experiments were conducted in the 3.6 s drop tower at the Key Laboratory of Microgravity, CAS, and flame data were gathered for spread rate, flame length, and flame appearance for various tunnel heights. The observations were compared with the numerical results that based on a two-dimensional flame spread model. Simulations also examined the flow field in the tunnels and the heat release rate from the flame. The results were used to explain the observed flame behaviors in confined spaces.

The drop-tower experiments showed that the transition of the opposed-flow flame to a steady-state microgravity flame was completed within approximately 1 s from the drop start. Presented in Fig. 1 are side-view images of the microgravity flames spreading in flow tunnels of different heights. In the most narrow tunnel (tunnel height, H, is 1.5 cm), the visible flame is very small. The brightest part of the flame is at or near the leading edge and the flame is almost entirely faint blue, indicating an absence of significant soot production. As the tunnel height increases slightly to 2 cm, the length and luminance of the flame increase dramatically, with a bright yellow trail visible behind the blue leading edge. Near the trailing edge, the flame becomes dimmer and thinner, and a glow of soot is noted. The overall flame curves away from the fuel plane and then keeps nearly parallel with the fuel plane. When H increases to 2.5 and 3 cm, the overall flame length increases further. However, a more significant change is observed to occur in the trailing portion of the flame, which curves back toward the fuel plane near the trailing edge. In an even higher tunnel ($H = 3.5$ and 4 cm), the flame tail becomes nearly flat again, curving very slightly back towards the fuel plane near the trailing edge. There is evidence that the tunnel height reaches a critical value at approximately 3 cm, beyond which the changes of the flame appearance are gradual and monotonic with increasing tunnel height. In fact, combustion tests have also been carried out in a 1 cm high tunnel. The flame

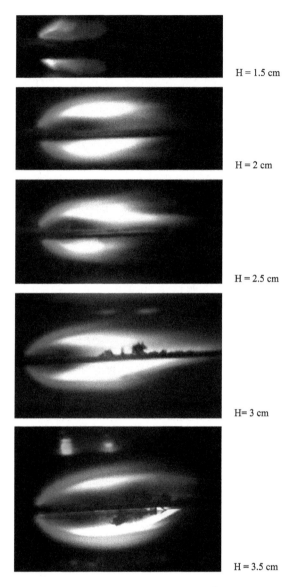

H = 1.5 cm

H = 2 cm

H = 2.5 cm

H= 3 cm

H = 3.5 cm

Fig. 1 Images of steady microgravity flames spreading in flow tunnels of different heights. The air flow enters from left at 5 cm/s and 1 atm. (reprinted from reference [12]. Copyright 2015, with permission from Springer Nature)

extinguishes shortly after ignition, and a self-sustained spreading flame could not be achieved in the slow air flow of 5 cm/s under microgravity conditions. In addition to heat loss, such a flame-spread limit may be caused by the limited oxygen supply into the tunnel.

The tunnel size has a significant influence on the magnitude of flame spread rate. It is shown that the spread rate curve is non-monotonic with respect to tunnel height. When H is increased from 1.5 cm, the spread rate increases until a maximum is reached at approximately $H = 3$ cm. The flame spread is 75% faster in the 3 cm tunnel than in the 1.5 cm tunnel (1.325 cm/s as against 0.758 cm/s). A reverse flame spread trend is observed when $H > 3$ cm, i.e. the spread rate decreases with increasing tunnel height. In this regime, an interesting observation is that an invariable spread rate may be approached in a flow tunnel with large height since the spread rate seems to decrease asymptotically with H. Flame spread rates in a 20 cm diameter tunnel have been measured by Olson [13] for a cellulosic paper whose area density is similar to that of the present fuel. In an air flow of 5 cm/s, the spread rate was determined to be 1.08 cm/s, comparable to the present test value of 1.14 cm/s which is obtained at $H = 5$ cm. It implies that the confinement effects on flame spread become less important when the test tunnel is relatively large. Compared with experiments, the computations produce higher spread rates especially in the low range of H. The non-monotonic trend of flame spread rate versus tunnel height, however, is well predicted. The entire flame length curve is also non-monotonic with tunnel height, while a maximum value occurs at approximately $H = 3$ and 4 cm from the experiments and computations, respectively.

Since the flame spread rate depends on heat release rate from the flame [4], the computed result for the total heat release rate is examined as a function of the tunnel height. It is shown that the tunnel size has a significant influence on the chemical energy release. The total heat release rate is highest in the 3 cm high tunnel, and it decreases in narrower or higher tunnels. Thus, a non-monotonic trend of heat release rate vs. tunnel height is obtained. This trend can satisfactorily explain the variation of the flame spread rate with tunnel height, although it should be regarded as a result of the flow modification in confined spaces.

Figure 2 shows the computed flow velocity contours around the spreading flame for five different tunnels. Since there is thermal expansion near the flame and the expansion is constrained in the direction perpendicular to incoming flow (y direction), the streamwise flow (in the x direction) is accelerated towards the downstream region. This flow acceleration effect is observed in all the tunnels studied. But it is most pronounced in intermediate-height tunnels ($H = 3$–4 cm) because the degree of the acceleration depends on both the flame size and the tunnel dimension. Note that the larger downstream velocity produces a longer flame. Thus, the trend of flow acceleration behind the flame is consistent with the non-monotonic variation of flame length with tunnel height. With the velocity component in the y direction, there is an additional observation about the confinement influence. At the leading edge of the flame, the flow is directed outwards due to the blowing of pyrolysis gases from the fuel. Above the flame, however, the flow deflects back from the tunnel wall. This inward flow pushes the flame towards the fuel surface, and transports more oxygen into the flame (compared with the no inward flow case which is assumed to happen in a large tunnel). Because of the enhanced combustion reaction in the flame, the heat flux from the flame to the fuel is increased, so the flame spread becomes faster. Once again, such a flow trend is most pronounced in intermediate-height tunnels, explaining the trend of flame spread rate with tunnel height.

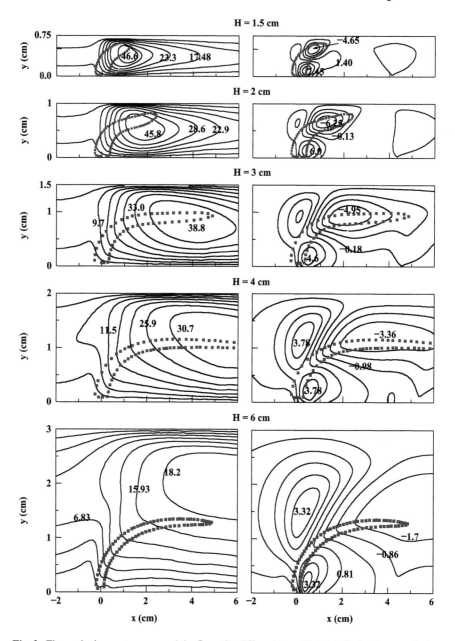

Fig. 2 Flow velocity contours around the flame for different tunnel heights. Left column: velocity component in the streamwise direction (x direction); Right column: velocity component in the direction perpendicular to the incoming flow (y direction). (reprinted from reference [12]. Copyright 2015, with permission from Springer Nature)

3   Near-Limit Instabilities of Concurrent Flame Spread Over Thin Solids

As an alternative mode of solid combustion, smoldering wave shares much common characteristics with spreading flame so far as their structural properties and the involved physical mechanisms are concerned [14]. Followed the experimental studies of smoldering combustion by Olson et al. [15], Zik and Moses [16, 17], and Zik et al. [18], Olson et al. [19] conducted experiments on opposed-flow flame spread over thin solid fuel in actual and simulated (by employing a narrow channel apparatus) microgravity conditions. Similar to the smolder case, the initially continuous flame front was observed to break into separate flamelets when the imposed oxidizer flow velocity was reduced near to the quenching limit, leaving a fingering pattern on the fuel surface. Zhang et al. [20] suggested that diffusive-thermal instability is the mechanism that is responsible for the fingering patterns observed in the flame spread mode. So far the cellular and fingering instabilities identified in experiments, either in smolder or flame spread mode, have been confined to the opposed flow regime. In a numerical study of diffusive-thermal instability of forward smolder waves, Lu and Dong [21] predicted that fingering instability can develop near the quenching limit where heat loss should play a dominant role.

The experimental work of Wang et al. [22], which was partly aimed at providing experimental data in the context of concurrent flame spread, has been inspired by the numerical results of Lu and Dong [21]. The objective of the work was to obtain an overall picture of the dynamical behavior of concurrent flame spread near the quenching limit. In order to suppress buoyant convection, a Narrow Channel Apparatus (NCA) was developed for experimental study of concurrent flame in very slow oxidizer flows. For all of the tests, thin filter paper was used as the solid fuel sample, which had an effective width of 150 mm and a length of 254 mm.

The experimental results show that, when the imposed oxidizer flow velocity is above an oxygen concentration dependent threshold, the flame spread is usually characterized by a continuous flame front and a uniform spread rate, and is therefore deemed to be stable. If the oxidizer flow velocity is reduced below the threshold, the flame front breaks into separate flamelets, thus marking the onset of flame instability. Two distinct types of instabilities are identified, namely, fingering or cellular instability, which is characterized by cellular flame fronts with fingered burned-out pattern trailing behind, and traveling wave instability, which is characterized by transverse creeping motion of the flamelets along the unburned fuel edge. Both types of instabilities are usually accompanied by recurrent flamelet growing and splitting during the flame spread process. A critical oxygen concentration, which is approximately 19% for the fuel tested, is identified as a separation between these two instability regimes. Specifically, for oxygen concentrations below the critical value, the instability is of fingering or cellular type, whereas for supercritical oxygen concentrations traveling wave instability prevails.

Shown in Fig. 3 is a typical evolution process of flame spread that demonstrates fingering or cellular instability. In an atmosphere with oxygen concentration of 15%

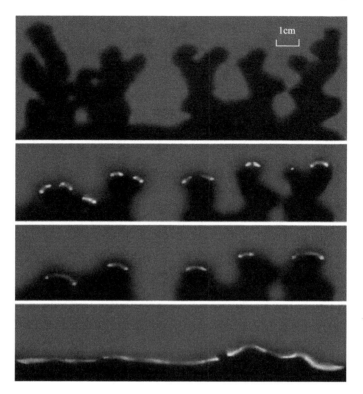

Fig. 3 Flame evolution process demonstrating fingering or cellular instability. (reprinted from reference [22]. Copyright 2016, with permission from Taylor & Francis)

and gas flow velocity 6 cm/s, shortly after ignition, the initially continuous flame front breaks into several discrete flamelets, which spread downwind together and leave behind a fingering burned-out pattern. It is evident that each flamelet assumes an arc shape, with the convex surface facing downstream. Moreover, the arc-shaped flamelet grows in width as it moves forward, and once its width exceeds a critical value, the flamelet splits in the middle into two children flamelets, resulting in a bifurcation of the trailing finger tip. However, all of the second-generation children flamelets go out eventually. A test with an even lower velocity, 5 cm/s, has been carried out subsequently, but it turned out that the initiated flame front failed to spread forward and extinguished soon after ignition. This suggests that the quenching limit for the current oxygen concentration should lie between 5 and 6 cm/s.

When the gas flow velocity is increased to 7 cm/s at the same oxygen concentration (15%), the initially continuous flame front is seen to break into three isolated flamelets, of which only one is survived. Subsequently, this single flamelet undergoes a series of very regular, period-doubling-like bifurcation processes, which are characterized by recurrent flamelet growing and splitting, along with random merging and extinction of the children flamelets. Eventually, the survived children flamelets succeed in burning through the fuel sample, leaving behind a tree-like burned-out pattern. Such a period-doubling-like bifurcation mode arising in concurrent flame spread is qualitatively reminiscent of the front splitting mode identified in the experiments of reverse smolder [15–18] and opposed flow flame spread [19, 23], although the gas flow directions are opposite in the two situations. This suggests that the currently identified fingering or cellular regime of concurrent flame spread may also be a manifestation of the diffusive-thermal mechanism.

Figure 4 presents a sequence of images for concurrent flames spreading in air (oxygen concentration 21%) with flow velocity 3 cm/s, showing the evolution of typical traveling wave instability. As can be seen, after an initial developing stage, the continuous flame front is survived by a single flamelet. In the subsequent development, this single flamelet experiences recurrent growing and splitting, spawning a train of children flamelets that distribute along the unburned fuel edge. However, different from the fingering or cellular regime, each of the flamelets undergoes an essentially transverse creeping motion, traversing the fuel edge often in a back and forth manner. As a consequence, the unburned fuel maintains a relatively smooth edge at all times and there is no leftover trailing behind the flamelets.

By conducting systematic experiments that cover comprehensive ranges of oxygen concentration and flow velocity, we have plotted the flammability map and stability diagram of concurrent flame spread. As shown in Fig. 5, the flammability boundary exhibits a U-shaped curve, with the interior corresponding to the flammable region. This observation is consistent with the computational results of Kumar et al. [24]. The left branch of the flammability boundary corresponds to the quenching limit, near which heat loss plays a significant role in flame spread; the right branch is believed to correspond to the blow-off limit, which, due to the focus of the study (flame behaviors near quenching limit), has not been delineated except for very low oxygen concentrations. The merging point of the two branches defines an absolute oxygen limit, below which concurrent flame spread cannot be sustained for any imposed flow velocities.

The flammable region in Fig. 5 can be divided into three sub-regions, which correspond to distinct flame spread regimes. Specifically, the marginal stability boundary separates a region that is characterized by essentially continuous flame fronts spreading at uniform speeds, from regions where the flames suffer instabilities. The marginal stability boundary lies very close to the quenching limit for higher oxygen concentrations and progressively gets farther as the oxygen concentration goes down towards the absolute oxygen limit. Herein we could come to a conclusion that the flame spread is more susceptible to flame instabilities in low oxygen environments. The second dividing line corresponds to a critical oxygen concentration, which is approximately 19% for the tested fuel. It divides the entire unstable region into two parts: below the critical oxygen concentration, the flame instability is of fingering or cellular type, whereas above the critical oxygen concentration, the flame instability is of traveling wave type.

Wang et al. [22] asserted that the two kinds of instabilities identified in concurrent flame spread are diffusive-thermal in nature and may be classified into the category of near quenching limit instability of non-adiabatic diffusion flames. In addition, an

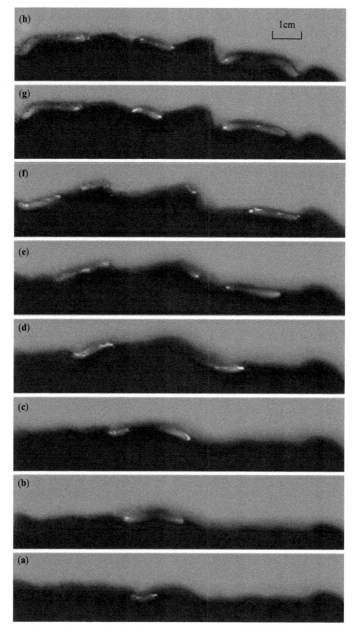

Fig. 4 Flame evolution process demonstrating traveling wave instability. (reprinted from reference [22]. Copyright 2016, with permission from Taylor & Francis)

Fig. 5 Flammability map and stability diagram of concurrent flame spread. (reprinted from reference [22]. Copyright 2016, with permission from Taylor & Francis)

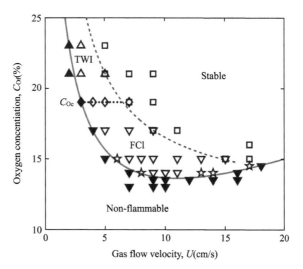

attempt had been made to gain insight into the physical mechanisms of the flame instabilities by exploiting the similarities between concurrent flame spread and forward smolder combustion. It was found that the conditions for the onset of fingering instability in flame spread are in accord with the stability analysis results of smolder waves [21].

4 Opposed versus Concurrent Flames Spreading Over a Thick Solid

The forward heat transfer is mainly by gas-phase conduction in opposed-flow flame spread. In contrast, convection prevails for concurrent spread. Since convective heat transfer is usually more effective than conduction, concurrent spread is generally expected to be more rapid and hazardous than opposed spread. As the velocity of the convective oxidizer flow decreases, however, the relative effectiveness of the heat transfer modes may be changed. For thermally-thin fuels, both microgravity experiments and numerical simulations [13, 24–26] have revealed that the opposed flame can spread faster than the concurrent flame if the flow velocity is low enough. Due to constraints on the available test time in drop towers, most research on microgravity flame spread has dealt with thermally-thin fuels [5], and the number of data points has been limited for flames spreading over thick fuels in the low-velocity flow regime. Such that our understanding of the near-limit flame spread over thick fuels is far from mature. The problem requires further research on both opposed and concurrent flame spreads.

By employing a Narrow Channel Apparatus to suppress buoyant flow, Zhu et al. [27] conducted systematical experiments to observe the flame spread and extinction processes over a thick PMMA sample (with a thickness of 10 mm) in opposed and concurrent flows. The tests focused on low-velocity flow regime ($U \leq 15$ cm/s) and hence complemented experimental data previously reported for high and moderate velocity regimes. The flame spread rates were measured as a function of the velocity and oxygen concentration of the forced gas flow. The experimental results were analyzed in the framework of existing theoretical models of flame spread, and the data for opposed flames were compared with those for concurrent flames. Flammability maps were constructed for both opposed and concurrent flames. A comparison of flammability limits between the two flame spread modes revealed their relative flammability in low-velocity flows.

Flame spread rates are shown in Fig. 6 as a function of flow velocity for both opposed and concurrent flames. In the low-velocity flow regime considered, the flame spread rate increases monotonically as the gas velocity increases in both flame spread modes. In opposed flows, the flame spread rate also depends on oxygen concentration, and it increases with increasing oxygen concentration. In concurrent flows, however, the flame spread rate appears practically independent of the oxygen concentration. A more interesting observation is that, at a given flow velocity, the opposed flame spreads much faster than the concurrent flame, and this trend remains unchanged throughout the flow velocity range in the experiments. In particular, at very low gas velocities (near quenching limits) or at high oxygen concentrations, the spread rate in opposed spread may be one order of magnitude larger than that in concurrent spread. Also plotted in Fig. 6 are the microgravity data on opposing flame spread rate over PMMA plates 20 mm thick, which were measured in sounding rocket experiments by Olson et al. [28] at 50% O_2 and $U = 1, 5$, and 10 cm/s. They are substantially below the NCA results for the same oxygen level, while a similar trend is exhibited as the flow velocity increases. This discordance may be contributed to the fuel sample size adopted in the experiments. In the microgravity tests the sample is rather narrow (6.35 mm). Since lateral heat loss from the flame exerts an influence over a distance of the order of cm in low-velocity flow [11], the flame is cooled

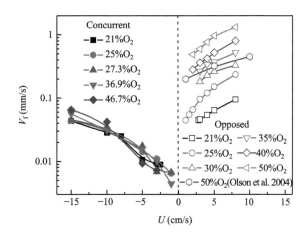

Fig. 6 Flame spread rate as a function of flow velocity for opposed and concurrent spreading flames at different oxygen concentrations. The velocity of opposed flow is defined as positive, and concurrent flow as negative. (reprinted from reference [27]. Copyright 2016, with permission from Springer Nature)

everywhere across its width, resulting in a reduced spread rate. On the contrary, in our experiments the sample has a much larger width (50 mm), and for the center portion of the flame where the spread rate is measured the side heat loss effect can be avoided. Note that the effect of the sample width on flame spread rate measurements has been extensively observed for thin solid fuels [11, 29, 30].

For opposed flame spread in moderate- and high-velocity flow regimes, the flow effects have been shown to be contained a non-dimensional controlling parameter, the Damköhler number [31, 32]. In Fig. 7, the flame spread rate data reported in Fig. 6 for opposed flame are presented versus the Damköhler number of Bhattacharjee et al. [32], Da_{EST}, together with the experimental results of Fernandez-Pello et al. [31] for high flow velocities. Herein the spread rates are normalized with a theoretical spread rate, $V_{f,EST}$, predicted by an extended simplified theory (EST) of opposed flame spread [32]. The formula for $V_{f,EST}$, based on EST, overcame the main drawbacks of the de Ris formula [33] and performed significantly better than the latter. However, it is noted that the assumptions of infinitely fast chemistry and neglect of radiation are retained in EST, and such that the derived spread rate formula is for the thermal regime (i.e., flame spread in moderate-velocity flow). In Fig. 7, it is clearly seen that the non-dimensional spread rates, obtained under extensive environmental conditions, collapse onto a universal curve over the entire range of the Damköhler number. The shape of the collapsed curve suggests that three distinct segments may be identified with respect to Da_{EST}. For intermediate Damköhler numbers, $10^5 < Da_{EST} < 10^6$, $V_f \approx V_{f,EST}$ ($V_f/V_{f,EST} \approx 1$), and the flame spread is in the thermal regime; for low Da_{EST}($<10^5$), V_f is depressed below $V_{f,EST}$ ($V_f/V_{f,EST} < 1$) by the finite-rate chemical kinetics, and the flame spread is in the kinetic regime; the third regime, characterized by large Da_{EST} ($>10^6$), is indicated by the present experimental data, where the flame spread is in the low-velocity quenching regime and the spread rate is lowered ($V_f/V_{f,EST} < 1$) primarily by radiative heat loss.

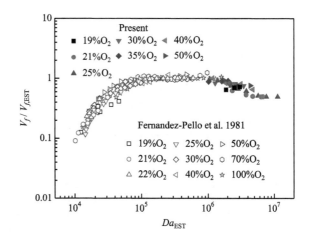

Fig. 7 Correlation of non-dimensional flame spread rate, $V_f/V_{f,EST}$, with Damköhler number, Da_{EST}, for opposed flame spread. (reprinted from reference [27]. Copyright 2016, with permission from Springer Nature)

Fig. 8 Experimental spread rate for concurrent spreading flame in a comparison with theoretical prediction. (reprinted from reference [27]. Copyright 2016, with permission from Springer Nature)

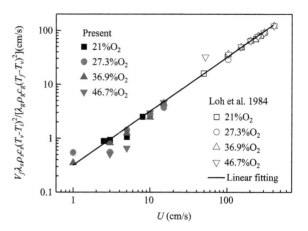

For concurrent flame, the theoretical models [1, 34, 35] predicted that the spread rate is linearly proportional to the flow velocity: $V_f \lambda_s \rho_s c_s (T_v - T_i)^2 / [U \lambda_g \rho_g c_g (T_f - T_v)^2]$ = const. Where T_f is the adiabatic flame temperature, T_v the pyrolysis temperature of the solid, T_i the initial temperature of the solid, λ the conductivity, ρ the density, and c the specific heat, and the subscripts s and g denote the gas and sold phase respectively. In Fig. 8, for concurrent flame, a plot of $V_f \lambda_s \rho_s c_s (T_v - T_i)^2 / [\lambda_g \rho_g c_g (T_f - T_v)^2]$ is presented versus U for the spread rate data reported in Fig. 6. The experimental data of Loh and Fernandez-Pello [34] for high flow velocities are also given in Fig. 8, and the linear regression line corresponds to their theoretical model. It is seen that the measured flame spread rates agree with the prediction in general, indicating that the heat transfer model captures the controlling mechanisms of flame spread. On the other hand, the experimental results at low flow velocities lie below the model prediction. This deviation may be caused by the neglect of radiative loss in the theoretical model, whereas radiation becomes important in low-intensity flame spread [36].

Using the NCA test results and other published data on opposed flame spread [28, 31], the flammability boundaries for flames over PMMA are plotted for both opposed and concurrent flow configurations (Fig. 9). For the opposed case, the flammability boundary can be obtained over a wide range of flow velocity, showing a typical U-shaped. The left branch of the flammability boundary corresponds to quenching extinction limit in low-velocity flows, and the right branch corresponds to blow-off extinction limit in high-velocity flows. The absolute oxygen limit, indicated by the bottom of the U-shaped curve, is determined to be approximately 18.5% O_2. For the concurrent flame, only a left flammability boundary is shown in Fig. 9 due to the absence of experimental data for high-velocity flows. The absolute oxygen limit for concurrent flame spread is estimated to be around 14% O_2, substantially below that for opposed flame. As can be seen in Fig. 9, the concurrent spread has a wider flammable range than the opposed spread. For concurrent spread, a narrow flammable region is added beyond the flammability boundary of opposed spread. Particularly, when oxygen concentration is reduced below the absolute oxygen limit for opposed spread, a flame can be sustained only in concurrent spread.

Fig. 9 Flammability
boundaries for PMMA in
opposed and concurrent
flame spread modes
(reprinted from reference
[27]. Copyright 2016, with
permission from Springer
Nature)

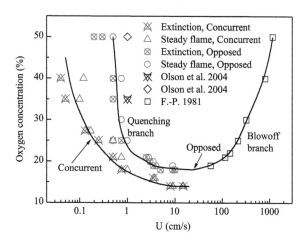

5   Spread and Dynamic Transition Behaviors of Flames Over a Thick Solid in Microgravity

In this section, some preliminary results of a space-based microgravity experiment, which was conducted aboard the SJ-10 satellite of China in April 2016 to investigate flame spread behaviors over a thermally-thick PMMA in low-velocity flows [37, 38], will be described. We made an observation of flame behaviors in step-changed opposing flows. In other words, the focus was on the spread of steady flames, as well as the flame response to a change of imposed flow.

A payload specifically designed (Fig. 10) have been developed for the space experiment "Ignition and Burning of Solid Materials in Microgravity". The microgravity flame spread experiments are conducted in a 39-L combustion chamber, in which eight flow tunnels are installed. The flow tunnels have an identical cross section of 95 mm × 95 mm and a length of 120 mm, and four of them are used for experiments of flame spread over thick PMMA plate. Each of the four tunnels accommodates an aluminum sample holder. A PMMA sample, with a size of 61.6 mm long × 50 mm wide × 10 mm thick, is mounted flush with the sample holder. Five thermocouples (type-R, 0.075 mm in diameter) are located along the centerline of the sample to measure the gas and solid phase temperatures during the test. A resistively heated wire is used as the igniter, which is embedded in the sample 15.8 mm away from the downstream edge. Such a design, igniting at the middle of the sample, is intended to observe a potential combination of opposed and concurrent flame spreads, although previous studies have focused primarily on either opposed or concurrent flame. Under the tested conditions, however, opposed-flow flame spread occurred merely. The forced gas flow in the tunnel is induced by a fan fixed at the downstream end of the tunnel. The bulk flow velocity is calibrated using a Laser Doppler Velocimeter (LDV) on ground, and can be adjusted in a range of 0–12 cm/s. At the upstream end of the channel, an aluminum honeycomb is installed

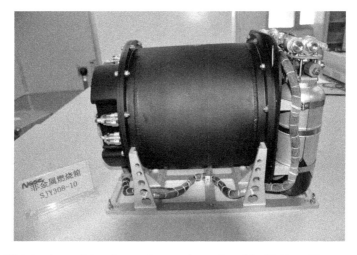

Fig. 10 Flight hardware of the microgravity experiment aboard the SJ-10 satellite

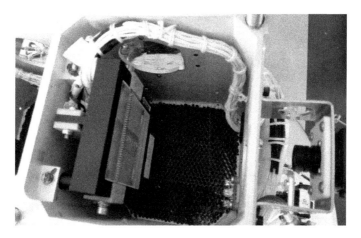

Fig. 11 Cross-sectional view of one flow tunnel

to straighten the flow. On two orthogonal walls of the flow tunnel, observation windows are presented. Shown in Fig. 11 is the cross-sectional view of one flow tunnel, thereby the test section configuration is demonstrated.

In order to establish the specified oxidizer atmosphere in the combustion chamber, a gas control system is integrated with the payload. The gases are filled into the chamber from two bottles, which are charged with 21 and 50% O_2–N_2 mixtures, respectively. At the beginning of each test, residual gas in the chamber is vented to the vacuum of outer space, and then gases from the bottles are charged into the chamber one after the other according to a specified pressure ratio. By blending the two mixtures, the desired O_2–N_2 mixture with preset oxygen concentration is

Table 1 Prescribed time durations of gas flow velocities in each experiment

Oxygen concentration \ Gas flow speed \ Time	9 cm/s	6 cm/s	3 cm/s	0
40%	60 s	50 s	150 s	> 15 min
35%	55 s	40 s	90 s	> 15 min
30%	80 s	80 s	120 s	> 15 min
25%	90 s	90 s	150 s	> 15 min

achieved in the chamber. The chamber pressure is monitored by a pressure transducer. The oxygen concentration is measured by an oxygen transducer in real time. Oxygen concentrations specified for the four flame spread tests are 25%, 30%, 35%, and 40%, respectively, whereas the pressure is at 101 kPa. The microgravity experiments are performed one by one during the orbital flight mission. The experimental procedure is automatically controlled. Firstly, an initial flow with a velocity of 9 cm/s is established in the tunnel. The igniter is then energized for 20 s. In the subsequent duration, the gas flow is adjusted three times with a step change, decreasing to 6 cm/s, 3 cm/s, and finally to 0. Each flow velocity is kept for a prescribed period as listed in Table 1. The flame spread process is simultaneously recorded by two color CCD cameras at a framing rate of 25 fps from top view and side view.

Figure 12 shows the opposed-flow flame spread processes when the flow undergoes stepped changes from 9 to 3 cm/s at 40%, 35%, and 30% O_2, respectively. Depending on the ambient oxygen concentration, the flame initially established over the fuel takes two different forms. At 40 and 35% O_2, it appears as a uniform flame, which extends across the sample width with a continuous leading front; at 30% O_2, a small, three-dimensional flame is achieved, which is named flamelet. As the flow velocity is decreased from 9 to 6 cm/s at 40 and 35% O_2, the flame front remains continuous after a rapid transition process. When the flow is further reduced to 3 cm/s in either oxygen atmospheres, the uniform flame changes its apparent form into flamelets after a relatively long transition process, noting that two separated flamelets forms at 40% O_2 and a single flamelet at 35% O_2. The flamelets go to extinction when the flow is stopped finally and thus a quiescent environment is achieved. At 30% O_2 the initial flamelet survives as the flow decreases from 9 to 6 cm/s, and the transition process is rather short. But it extinguishes when the flow decreases to 3 cm/s. The observed flame spread modes are summarized in Table 2. In the 25% O_2 test, the flame is observed to fail to spread following the ignition transition.

It is seen in Fig. 12 that the flame can respond to the sudden step change of the gas flow velocity to re-establish a new steady state. The transition process between two similar flame modes is relatively short. The transition from a uniform flame to flamelets, however, lasts much longer, and the transition process is characterized by flame oscillations. Figure 13 gives a flame image sequence showing one cycle of the flame oscillation when the flow is reduced from 6 to 3 cm/s at 40% O_2. First a uniform flame shrinks into individual flames separated by non-burning fuel, and

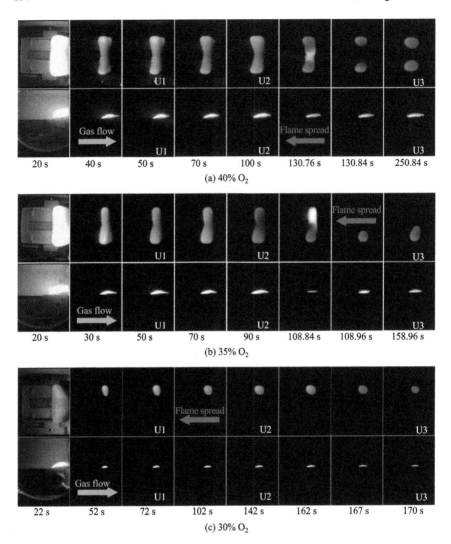

Fig. 12 Flame spread process after the igniter is turned off at 40%, 35%, and 30% O_2 respectively. U1 = 9 cm/s, U2 = 6 cm/s, and U3 = 3 cm/s. Time is from the moment the igniter is energized

Table 2 Flame spread modes at different oxygen concentrations and flow velocities

Oxygen concentration \ Gas flow speed / Flame mode	9 cm/s	6 cm/s	3 cm/s	0
40%	Uniform flame	Uniform flame	Two flamelets	Extinction
35%	Uniform flame	Uniform flame	One flamelet	Extinction
30%	One flamelet	One flamelet	Extinction	\
25%	Fail to spread	\	\	\

Fig. 13 Sequence of front-view images of flame oscillations during flamelets formation at an oxygen concentration of 40% after the flow speed is decreased from 6 to 3 cm/s. The images are separated by an equal time interval 0.08 s

then the flamelets re-connect to form a united flame. This process persisted for about 40 s, and finally spreading flamelets are formed. At 35% O_2, flame oscillations are also observed as the flow velocity is decreased to 3 cm/s, whereas only one flamelet is formed after the transition process. Flame oscillations are also observed when the flamelets go to extinction following sudden decrease of flow velocity. Figure 14 illustrates the dynamic process of flame extinction at 40%, 35%, and 30% O_2. Note that the transition process lasts about 1 s, much shorter than the uniform to flamelets transition.

In Fig. 15, the variations of flame (or flamelet) leading edge position and flame length are shown as a function of time for different oxygen concentrations. As can be seen, except the transition process following the step change of flow velocity, the relative flame position data show a linear variation with time, indicating a constant flame spread rate; but the flame length continually increases with time even at specified oxygen concentration and flow velocity, indicating that the flame does not reach a completely steady state. The leading edge standoff distances are also determined. A constant value can be obtained under a specified oxygen and flow condition after the transition process. The spreading flames after the transition could be regarded as steady because opposed flame spread is controlled primarily by processes in the vicinity of the flame leading edge.

The RGB two-color pyrometry method [39–42] is utilized to resolve the temperature field of the spreading flames. Based on the flame leading edge temperature determined, we carry out an energy balance analysis at the fuel surface:

$$\lambda_g \frac{dT_g}{dy} = \lambda_s \frac{dT_s}{dy} + \dot{m}'' L_v + \varepsilon \sigma T_s^4 \tag{1}$$

where, λ_g, λ_s, ε, T_g, T_s, L_v, σ, and $\dot{m}''$ are gas thermal conductivity, solid phase thermal conductivity, the radiative emittance of the fuel surface, the temperature of the gas and solid, latent heat of vaporization, Stefan-Boltzmann constant and mass burning rate, respectively. The left hand term of Eq. (1) is the conductive heat flux from the flame to the surface. As the energy input at the surface, it does not include the radiative feedback from the flame because this feedback may be ignored for near-limit flames [28]. The three terms on the right hand of Eq. (1) constitute the output of the energy, and they represent the in-depth conduction into the solid, the heat to vaporize the fuel, and the surface radiative heat loss, respectively. The values of each

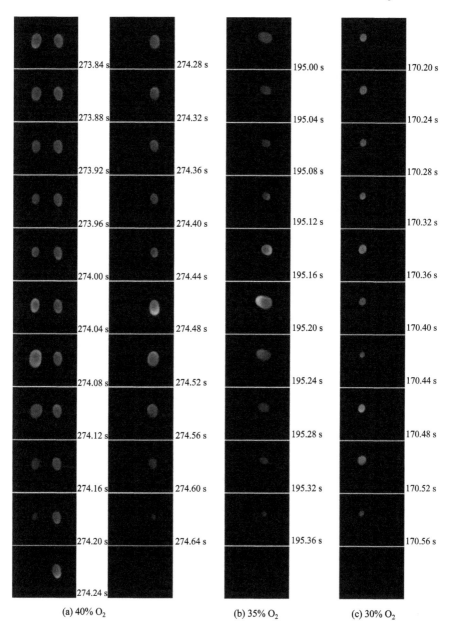

(a) 40% O_2 (b) 35% O_2 (c) 30% O_2

Fig. 14 Sequence of front-view images of flame variation process before extinction. At 40% and 35% O_2, flame spreads in the quiescent environment before extinction, while at 30% O_2, flame extinguishes with a flow speed of 3 cm/s

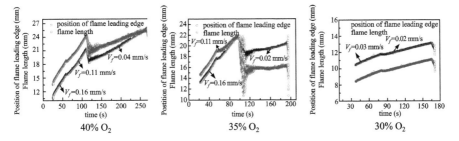

Fig. 15 Relative positions of flame leading edge and flame lengths as a function of time at different oxygen concentrations

heat flux term in Eq. (1) are calculated for flames at different oxygen concentrations and flow velocities. The results are shown in Fig. 16 as a function of flame spread rate. The conductive heat from the flame is clearly seen to decrease with decreasing flame spread rate, resulting from the reduced flame temperature and increased flame standoff distance. The heat loss through solid phase conduction exhibits a similar trend, whereas the surface radiative heat loss remains essentially unchanged. As a result, the heat absorbed by the solid fuel for vaporization, which can be regarded as the net heat flux utilized for flame spread, is observed to decrease as the flame spread slows down. We proceed to examine the heat utilization ratio and the heat loss ratio:

$$F_{utilization} = [\dot{m}''L_v] \Big/ \left[\lambda_g \frac{dT_g}{dy} \right] \tag{2}$$

$$F_{loss} = \left[\lambda_s \frac{dT_s}{dy} + \varepsilon\sigma T_s^4 \right] \Big/ \left[\lambda_g \frac{dT_g}{dy} \right] \tag{3}$$

Fig. 16 Heat flux terms in Eq. (1) as a function of flame spread rate

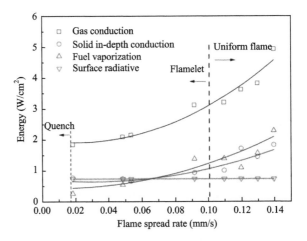

Fig. 17 Heat loss and reutilization ratios as a function of flame spread rate

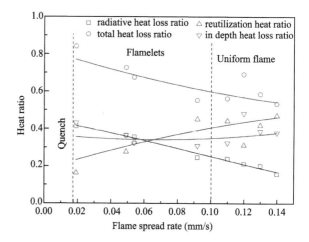

The ratios are plotted in Fig. 17 as a function of flame spread rate. Overall, the total heat loss ratio increases with the reduced flame spread rate, while the reutilization ratio has an opposite variation trend. At the boundary between uniform flame and flamelets, the loss ratio is about 55%. For flamelets with an extremely low spread rate at extinction, the loss ratio exceeds 80%, which is in accordance with the finding of Olson et al. [28].

6 Summary

The problem of flame spread over solid materials in low-velocity flow regime is of fundamental interest and practical importance for spacecraft fire safety. To test this type of flame spread, a microgravity environment is desired to eliminate the complications of buoyant flow. For thermally-thick solids, long-duration microgravity experiments in space are particularly needed, while the opportunity to conduct space-based experiments has been scarce. In view of these constraints, the Narrow Channel Apparatus (NCA) has been developed to suppress buoyant flow in normal gravity, and thus provide a feasible method for flame spread experiments in slow gas flows.

The present article surveys the major research works that have been recently performed at the Key Laboratory of Microgravity, CAS, to study flame spread in low-velocity flows. The topics involve the confinement effects on flame spread, concurrent flame behaviors near the quenching limit, flame spread and extinction over thick solids in opposed and concurrent flows, and dynamics of spreading flame over thick solids in step-changed flow. The experimental results described herein are obtained from normal gravity NCA tests, microgravity free drops, and space experiments aboard the SJ-10 satellite of China.

Effects of a confined space on flame spread over thin solid fuels in a low-speed opposing air flow is examined by combined use of drop-tower experiments and numerical computations. Results show that the height of the flow tunnel has significant influence on flame behaviors in the low flow velocity regime. The flame spread rate curve is non-monotonic with respect to tunnel height, with the fastest flame occurring in the 3 cm high tunnel. The flame length and the total heat release rate from the flame also change with tunnel height, and a faster flame has a larger length and a higher heat release rate. A confined space is observed to modify the flow around the spreading flame. The flow accelerates in the streamwise direction. Above the flame, the flow deflects back from the tunnel wall. This inward flow pushes the flame towards the fuel surface, and increases oxygen transport to the flame. The flow modification the changes heat release rate of the flame and explains the variations of spread rate and flame length with tunnel height. The results suggest the confinement effects should be taken into account to assess accurately the fire hazard aboard spacecraft.

Flame instabilities near the low flow velocity quenching limit of concurrent flame spread over a thin fuel are experimentally studied by employing a narrow channel apparatus. Depending on the magnitude of the ambient oxygen content and/or the imposed flow, we identify two distinct kinds of instabilities, namely fingering or cellular instability, and traveling wave instability. The former is characterized by cellular flame fronts, and the later by transverse motion of flamelets along the unburned fuel edge. The flammability map and stability diagram of concurrent flame spread are constructed using oxygen concentration and flow velocity as coordinates. The flammable region is divided into three sub-regions that correspond to different flame spread regimes. The marginal stability boundary separates a region where continuous flame spreads from regions where the flames suffer instabilities. Meanwhile, a critical oxygen concentration (approximately 19% O_2 for the tested fuel) divides the entire unstable region into fingering instability and traveling wave instability. It is believed that that the two kinds of instabilities in concurrent flame spread are diffusive-thermal in nature. The physical mechanisms of the flame instabilities may be revealed by exploiting the similarities between concurrent flame spread and forward smolder combustion.

Systematical experiments in a narrow channel apparatus are performed for flame spread and extinction phenomena over a thick PMMA, with emphasis placed on a comparison of the two flame spread modes in low-velocity flows. At a given flow velocity, it is found that the opposed flame spreads much faster than the concurrent one. In general, the spread rate data for both opposed and concurrent flames can be correlated by corresponding prediction models. However, the experimental results in low-velocity gas flows are observed to deviate from theoretical predictions due to the neglect of radiative heat loss in the theoretical models. Flammability limits are presented for both flame spread configurations, showing that concurrent spread has a wider flammable region than opposed spread. An additional flammable area for concurrent spread exists beyond the flammability boundary of opposed spread. Particularly, the oxygen concentration limit for concurrent flame spread is approximately 14% O_2, much lower than that for opposed flame spread (18.5% O_2). It should be noted that more microgravity experiments are needed to validate the NCA test results.

For a thick solid (PMMA), steady spread, extinction, and dynamic transition behaviors of spreading flame in step changed low-speed opposing flows are investigated by conducting microgravity experiments aboard the SJ-10 satellite. Depending on the oxygen concentration and flow velocity, the spreading flames are observed to take two different forms: uniform flame, and flamelets. A uniform flame sustains at high oxygen concentration. At lower oxygen concentration and flow velocity, however, only flamelets can survive. It is noted that the uniform flame and flamelets can respond to the sudden step change of the gas flow to re-establish a new steady state. The transition process between two similar flame modes, i.e., from uniform flame to uniform flame, or from flamelets to flamelets, is relatively short. By contrast, the transition process from a uniform flame to flamelets lasts much longer, which is characterized by flame oscillations. An energy balance analysis at the fuel surface shows that the heat absorbed by the solid fuel for vaporization decrease as the flame spread rate slows down. The total heat loss ratio increases with decreasing flame spread rate, and excessive heat loss results in the flame extinction ultimately.

Acknowledgements The present works are supported by the National Natural Science Foundation of China (grant No. U1738117) and the Strategic Priority Research Program on Space Science of CAS (grant Nos. XDA04020410, XDA04020202-10). The authors would like to thank all cooperators for their contribution to these studies. Special gratitude is due to Prof. Wen-Rui Hu for his persistent support of the first author's research work.

References

1. Fernandez-Pello AC (1984) Flame spread modeling. Combust Sci Technol 39:119–134
2. Wichman IS (1992) Theory of opposed-flow flame spread. Prog Energy Combust Sci 18:553–593
3. Sirignano WA, Schiller DN (1997) Mechanisms of flame spread across condensed-phase fuels. In: Dryer FL, Sawyer RF (eds) Physical aspects of combustion, a Tribute to Irvin Glassman, Gordon and Breach, pp 353–407
4. Quintiere JG (2006) Fundamentals of fire phenomena. Wiley, Chichester, West Sussex, England
5. T'ien JS, Shih HY, Jiang CB, Ross HD, Miller FJ, Fernandez-Pello AC, Torero JL, Walther D (2001) Mechanisms of flame spread and smolder wave propagation. In: Ross HD (ed) Microgravity combustion: fire in free fall. Academic Press, San Diego, pp 299–418
6. Fujita O (2015) Solid combustion research in microgravity as a basis of fire safety in space. Proc Combust Inst 35:2487–2502
7. Friedman R (1994) Risks and issues in fire safety on the space station. NASA TM-106430
8. Friedman R (2000) Testing and selection of fire-resistant materials for spacecraft use. NASA TM-209773
9. Shih H-Y, T'ien JS (1997) Modeling wall influence on solid-fuel flame spread in a flow tunnel. AIAA paper 97-0236
10. Nakamura Y, Kashiwagi T, McGrattan KB, Baum HR (2002) Enclosure effects on flame spread over solid fuels in microgravity. Combust Flame 130:307–321
11. Zhang X, Yu Y (2011) Experimental studies on the three-dimensional effects of opposed-flow flame spread over thin solid materials. Combust Flame 158:1193–1200

12. Wang SF, Hu J, Xiao Y, Ren T, Zhu F (2015) Opposed-flow flame spread over solid fuels in microgravity: the effect of confined spaces. Microgravity Sci Technol 27:329–336

13. Olson SL (1991) Mechanisms of microgravity flame spread over a thin solid fuel: oxygen and opposed flow effects. Combust Sci Technol 76:233–249

14. Ohki Y, Tsugé S (1974) On flame spreading over a polymer surface. Combust Sci Technol 9:1–12

15. Olson SL, Baum HR, Kashiwagi T (1998) Finger-like smoldering over thin cellulosic sheets in microgravity. Proc Combust Inst 27:2525–2533

16. Zik O, Moses E (1998) Fingering instability in solid fuel combustion: the characteristic scales of the developed state. Proc Combust Inst 27:2815–2820

17. Zik O, Moses E (1999) Fingering instability in combustion: an extended view. Phys Rev E 60:518–531

18. Zik O, Olami Z, Moses E (1998) Fingering instability in combustion. Phys Rev Lett 81:3868–3871

19. Olson SL, Miller FJ, Jahangirian S, Wichman IS (2009) Flame spread over thin fuels in actual and simulated microgravity conditions. Combust Flame 156:1214–1226

20. Zhang Y, Ronney PD, Roegner EV, Greenberg JB (1992) Lewis number effects on flame spreading over thin solid fuels. Combust Flame 90:71–83

21. Lu Z, Dong Y (2011) Fingering instability in forward smolder combustion. Combust Theor Model 15:795–815

22. Wang SF, Wang SD, Zhu KC, Xiao Y, Lu ZB (2016) Near limit instabilities of concurrent flame spread over thin solid fuel. Combust Sci Technol 188:451–471

23. Olson SL, Miller FJ, Wichman IS (2006) Characterizing fingering flamelets using the logistic model. Combust Theor Model 10:323–347

24. Kumar A, Shih HY, T'ien JS (2003) A comparison of extinction limits and spreading rates in opposed and concurrent spreading flames over thin solids. Combust Flame 132:667–677

25. Olson SL, Kashiwagi T, Fujita O, Kikuchi M, Ito K (2001) Experimental observations of spot radiative ignition and subsequent three-dimensional flame spread over thin cellulose fuels. Combust Flame 125:852–864

26. Kashiwagi T, McGrattan KB, Olson SL, Fujita O, Kikuchi M, Ito K (1996) Effects of slow wind on localized radiative ignition and transition to flame spread in microgravity. Proc Combust Inst 26:1345–1352

27. Zhu F, Lu ZB, Wang SF (2016) Flame spread and extinction over a thick solid fuel in low-velocity opposed and concurrent flows. Microgravity Sci Technol 28:87–94

28. Olson SL, Hegde U, Bhattacharjee S, Deering JL, Tang L, Altenkirch RA (2004) Sounding rocket microgravity experiments elucidating diffusive and radiative transport effects on flame spread over thermally thick solids. Combust Sci Technol 176:557–584

29. Altenkirch RA, Eichhorn R, Shang PC (1980) Buoyancy effects on flames spreading down thermally thin fuels. Combust Flame 37:71–83

30. Shih HY, T'ien JS (2003) A three-dimensional model of steady flame spread over a thin solid in low-speed concurrent flows. Combust Theory Model 7:677–704

31. Fernandez-Pello AC, Ray SR, Glassman I (1981) Flame spread in an opposed forced flow: the effect of ambient oxygen concentration. Proc Combust Inst 18:579–589

32. Bhattacharjee S, West J, Altenkirch RA (1996) Determination of the spread rate in opposed-flow flame spread over thick solid fuels in the thermal regime. Proc Combust Inst 26:1477–1485

33. de Ris JN (1969) Spread of a laminar diffusion flame. Proc Combust Inst 12:241–252

34. Loh HT, Fernandez-Pello AC (1984) A study of the controlling mechanisms of flow assisted flame spread. Proc Combust Inst 20:1575–1582

35. Fernandez-Pello AC (1979) Flame spread in a forward forced flow. Combust Flame 36:63–78

36. Zhao XY, T'ien JS (2015) A three-dimensional transient model for flame growth and extinction in concurrent flows. Combust Flame 162:1829–1839

37. Hu WR, Zhao JF, Long M, Zhang XW, Liu QS, Hou MY, Kang Q, Wang YR, Xu SH, Kong WJ, Zhang H, Wang SF, Sun YQ, Hang HY, Huang YP, Cai WM, Zhao Y, Dai JW, Zheng HQ, Duan EK, Wang JF (2014) Space program SJ-10 of microgravity research. Microgravity Sci Technol 26:159–169

38. Zhao HG, Qiu JW, Tang BC, Kang Q, Hu WR (2016) The SJ-10 recoverable microgravity satellite of China. J Space Explor 4(3):101/1–9
39. Shimoda M, Sugano A, Kimura T, Watanabe Y, Ishiyama K (1990) Prediction method of unburnt carbon for coal fired utility boiler using image processing technique of combustion flame. IEEE Trans Energy Convers 5:640–645
40. Lou C, Zhou HC, Yu PF, Jiang ZW (2007) Measurements of the flame emissivity and radiative properties of particulate medium in pulverized-coal-fired boiler furnaces by image processing of visible radiation. Proc Combust Inst 31:2771–2778
41. Hossain MM, Lu G, Yan Y (2012) Measurement of flame temperature distribution using optical tomographic and two-color pyrometric techniques. In: IEEE international instrumentation and measurement technology conference proceedings. IEEE, Graz, Austria
42. Müller B, Renz U (2001) Development of a fast fiber-optic two-color pyrometer for the temperature measurement of surfaces with varying emissivities. Rev Sci Instrum 72:3366–3374

Experimental Study on Coal Combustion at Microgravity

Hai Zhang, Bing Liu, Yang Zhang, Qing Liu, Cheng Zuo, Jialun Chen, Yu Qiao and Minghou Xu

Abstract Coal is the primary energy supply worldwide and will remain dominant in energy consumption structure in China for a long period. The fundamental studies on coal combustion are still of great demand due to the increasing pursuit of the higher efficiency and lower emission. Microgravity (μg) environment minimizes the buoyancy effect that the coal particles are subjected to during the combustion under normal gravity, and thus provides an ideal environment to discover new phenomena and obtain more accurate data. The results are useful to explore the basic combustion principles and the strength of the buoyancy effect to coal combustion, and to validate and develop coal combustion models for the ground coal utilization. The paper reviews the experimental studies on the coal combustion at microgravity conducted using drop towers. Furthermore, it introduces the state-of-art coal combustion study using Chinese SJ-10 satellite. With the careful design of the microgravity experimental system, space experiments on both single coal particles and pulverized coals were conducted in the first time. The entire burning process for coal particles, including the high volatile content bituminous coal and low volatile content anthracite coal were observed. The differences in volatile release mode, ignition temperature and delay time, volatile flame evolution, flame shape, and char burnout time for the coal particles burning at microgravity from counterparts at normal gravity were reported. With the new observation and data from drop tower and SJ-10 satellite experiments, more accurate models could be developed for the ground coal utilization.

Keywords Microgravity · SJ-10 satellite · Coal · Combustion · Ignition · Buoyancy effect

H. Zhang (✉) · B. Liu · Y. Zhang · Q. Liu
Key Laboratory for Thermal Science and Power Engineering of Ministry Education, Department of Energy and Power Engineering, Tsinghua University, Beijing 100084, China
e-mail: haizhang@tsinghua.edu.cn

C. Zuo · J. Chen · Y. Qiao · M. Xu
State Key Laboratory of Coal Combustion, Huazhong University of Science and Technology, Wuhan 430074, Hubei, China

© Science Press and Springer Nature Singapore Pte Ltd. 2019
W. R. Hu and Q. Kang (eds.), *Physical Science Under Microgravity: Experiments on Board the SJ-10 Recoverable Satellite*, Research for Development,
https://doi.org/10.1007/978-981-13-1340-0_11

1 Introduction

Coal is a widely used fossil fuel, occupying an important position in the energy consumption structure for many countries around the world. According to IEA's report, between 1972 and 2016, the world's total primary energy supply increased by almost 2.5 times from 5523 Mtoe to 13761 Mtoe, while coal was the second dominant fuel other than oil, accounting for about a quarter of the total amount [1, 2]. During the same period, coal supplied one third of all energy used worldwide and contributed about 40% of the electricity generation.

In China, coal has been being the dominant primary energy supply for decades. Shown in Fig. 1, between 2005 and 2014, the coal production amount kept increasing and then slightly dropped in 2015 and 2016. In 2017, it slightly increased again by 3.52 Bton, accounting for slightly less than half of the world coal consumption. The detected coal reserve is about 1600 Bton, accounting for about 96% of total reserve fossil fuel. Thus, it is forecasted that coal can be still used as the primary energy supply for at least 300 years for China.

Coal is primarily used for the generation of electricity and commercial heat. In the last ten years in China (Table 1), the thermal installed capacity (95% were coal-fired units) kept increasing to 1106 GW, though its fraction decreased from 77.4% in 2007 to 62.2% in 2017. Even in 2017, the electricity generated by thermal power still accounted for 73.4% of the total generated electricity.

Though renew energy supply grows rapidly, coal will keep as a major primary energy supply globally. In China, coal will serve as the dominant primary energy supply for several decades more due to the limited sources or reserve of renewable energy. It is expected in 2030, coal-fired units will increase to 1350 GW, still accounting for 45% of the total installed capacity in China.

Combustion is the main way for coal utilization, producing power and heat. When coal is burnt in the furnace of a boiler, the chemical energy stored in the coal turns into the thermal energy of hot flue gas. Then, the hot flue gas transfers heat to water and

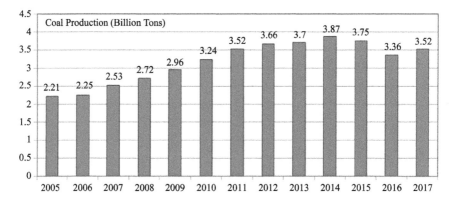

Fig. 1 The annual coal production in China during 2005 and 2017

Table 1 Thermal installed capacity and electricity generation in China in 2007–2017

Year	Total capacity MW	Total electricity TWh	Thermal capacity MW	Thermal capacity ratio, %	Thermal electricity TWh	Thermal electricity ratio, %
2007	713,290	3255.9	554,420	77.4	2698.0	82.9
2008	792,530	3433.4	601,320	75.9	2779.3	80.8
2009	874,070	3650.6	652,050	74.6	2981.42	81.7
2010	962,000	4141.3	706,630	73.5	3325.3	80.3
2011	1,055,760	4717.6	765,460	72.5	3793.5	80.4
2012	1,144,910	4977.4	819,170	71.5	3910.8	78.6
2013	1,247,380	5361.6	862,380	69.1	4143.9	77.3
2014	1,360,190	5545.9	915,690	67.3	4173.1	75.2
2015	1,525,270	5739.9	1,005,540	65.9	4230.7	73.7
2016	1,650,000	5990.0	1,050,000	63.4	4332.2	72.3
2017	1,777,030	6275.8	1,106,040	62.2	4611.5	73.4

steam running inside metal tubes. In a power plant, the high temperature high pressure steam will drive a steam turbine then an electric motor to convert the thermal energy into mechanical energy and electricity respectively. During this energy conversion process, a few kinds of air pollutants are formed, including SO_2, NOx, particulate matter, and CO_2. Hence, coal firing is also regarded as the main contributor of the air pollutants, severely threating the environment and the health of human beings.

From above introduction, we can easily to understand that coal combustion is extremely important for the ordinary living, economic development, and environmental protection. Consequently, the researches on coal combustion have been conducted for over 100 years worldwide. In the past three decades, especially in the new century, the coal combustion technologies have been developed rapidly in China. Since 2011, China has kept ranking the top one in thermal installed capacity for power generation in the world. A number of newly-built thermal power units reach the world's leading level in efficiency and emission control.

However, due to the complexity of coal and coal combustion process, there are still many aspects not well known, from single coal particle combustion to coal firing in the large furnace of a commercial boiler. With the increasing pursuit of the higher efficiency and lower emission, both fundamental studies and technology development on coal combustion are still of great demand. In addition, the coal combustion research is also a must for the safety of the combustion system.

The fundamental studies on coal combustion characteristics not only could discover basic phenomenon from which the combustion and pollutant formation mechanisms can be revealed, but also could provide the key data for industrial use and to develop models to simulate the entire coal combustion process. The associated data include the chemical kinetic parameters, ignition temperature and delay time, burnt out time, and pollutant formation rate, etc. They are certainly a base for the design of burners and boiler, and the development of the combustion technology. Without properly grasping the combustion characteristics of a coal, the boiler could be designed too big to cause extra cost or reduce the flame stability, or too compact to induce slagging or overheating of the heating surfaces.

A number of studies on the coal combustion characteristics have been carried out and several experimental methods have been developed. The commonly used ones include the thermogravimetric analyzer (TGA) [3, 4] and drop tube furnace (DTF) [5, 6] in laboratory. TGA measures the mass loss of testing sample during the combustion process with a constant heating rate, and obtain the curves of temporal mass variation (i.e., TG curve) and the gradient of the temporal mass variation (i.e., DTG curve). Form the TG or DTG curves, the ignition temperature, ignition delay time, and kinetic parameters such as activation energy can be derived. However, the drawbacks of the TGA method are the slow heating rate and the packing of coal particles. The TGA has a heating rate of 10–100 K/min, much lower than the real one in the boiler of 10,000–100,000 K/s. In addition, coal particles are packed in the reactor during the experiment. The particles interact each other and the gas diffusion among the particles is difficult to evaluate and keep consistent. Thus, the data obtained by TGA method are not distinct ones, and they are not able to reflect the industrial process.

Differently, DTF could reach a high heating rate close to the reality with coal particles dispersedly burning in a high temperature furnace. However, as the particles moving from the top of the furnace, they are actually subjected to forced convection even in a laminar flow or natural convection, and the particles are still burning in a non-isotropic environment. For coal cloud experiments, it is very difficult to get a uniform dispersion of the particles. Thus the coal combustion behaviors could be also different from the ideal ones that are used in the simulation. Form another view, in normal gravity (1-g), the coal combustion is subjected to the influence of buoyance-induced natural convection, which could change the heat transfer, and moreover sweep the surrounding volatile, CO produced by char combustion and hot products. Obviously, the observation and data obtained in 1-g experiments are "contaminated" by the gravity.

On the other hand, in modeling, the gravity term and thereby the induced buoyancy effect is mostly not considered as this non-linear term simply destructs the one-dimensional assumption. One dimensional (1-D) model is the base of coal modeling and combustor simulation. The contradiction between the 1-g experimental conditions and those used in theoretical analyses and mathematical models is a main driving force for the study of coal combustion at microgravity (μg).

Microgravity (μg) provides an ideal environment to discover new phenomena, to explore the instinct mechanisms and to provide data accurately validate the combustion theory since the effect of buoyancy can be minimized. As a result, combustion process can be simplified with increasing the combustion time scale [7–9]. Through the comparison between the microgravity experimental results and those corresponding ones obtained from 1-g experiments, the buoyancy effect could be examined and quantitatively described. With theoretical studies, data correction, and modeling

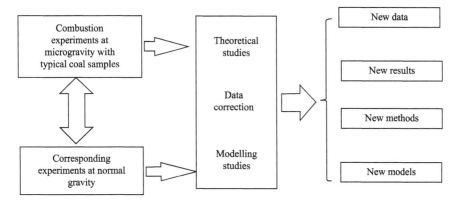

Fig. 2 The purposes of study of coal combustion at microgravity

studies, new and more accurate data, results, analytic method and models can be obtained for the ground coal combustion research (Fig. 2).

There are a few kinds of facilities used for coal combustion at μg. The commonly used one is drop tower. Basically, the experimental rig will do a free drop in the drop tower. Depending on the height of the tower, the μg time varies from 1 to 10 s, usually with couple of seconds. The μg level for a drop tower is usually 10^{-4} to 10^{-3} g_0. This level satisfies for the ignition experiments for a mm-sized coal particle and the burning experiments for the pulverized coal in general. However, the entire combustion process of the mm-sized coal particles, especially for those low volatile coals could be more than 20 s or even longer. Then, facilities with longer μg time are required. So far, such experiments were only done by Chinese researchers on the SJ-10 satellite.

In the next section, the studies on coal combustion at μg with a drop tower and SJ-10 satellite will be introduced respectively.

2 Studying Coal Combustion at Microgravity Using a Drop Tower

2.1 The Experiments on Ignition of Single Coal Particles

Understating the ignition process of coal particles is important for burner and combustor design, and unwanted-fire prevention. For simplification, single-particle approaches were often taken [10–19]. Though coal ignition has been studied widely before and a large amount data, such as the ignition temperature were obtained. However, nearly all existing data still cannot be deemed as the instinct ones, upon which mathematical models can be well validated. One of the important reasons is that the corresponding experiments were conducted on the earth at normal gravity (1-g). At 1-g, the ignition process is subjected to the influence of the buoyancy-induced natural convection, i.e., buoyancy effect, due to the large temperature difference, and thereby the large density difference between the burning volatile matter and ambient gas. The natural convection can sweep the surrounding volatile cloud away [16, 17], noticeably affecting the ignition temperature and time. The buoyancy effect is expected to more significant for larger size particles, as the Grashof Number Gr ($g\rho^2\beta\Delta T d_p^3/\mu^2$) increases [17]. To minimize such an effect, microgravity (μg) experiments are desired.

So far, very few experimental studies on single coal ignition or combustion at μg have been conducted. Pioneering study was conducted by Poland scholar Gieras et al. [13, 14] in 1980s. They assessed flame propagation from a burning coal particle to unburnt one at μg using a ~1.4 s small drop tower. The experiments were conducted with an array of mm-sized single particles with a certain distance apart. Ignition was done with an electric spark on an octane droplet. At 1-g and μg gravity, the maximum distance for a burning coal particle to ignite the adjacent one was determined. The experiments revealed that gravity induced remarkably difference in ignition and flame propagation speed. However, coal particles were actually asymmetrically ignited and the data were not 1-D in nature. In addition, no data of ignition temperature were reported.

In the mid-1990s, Katalambula et al. [15–17] conducted much more deliberated and well design μg experiments on ignition of single coal particles, assessing the ignition temperature, time, volatile matter released with different particle size and coal type at both μg and 1-g. The μg experiments were carried out using a 10 s drop tower located at JAMIC (Japanese Microgravity Center). The experimental system is shown in Fig. 3. During the experiments, a single coal particle with a size around 1.0 mm was ignited by radiant energy from spot heaters at heating rates of 770–900 K/s, and the ignition temperature, time for coal particles with different particle sizes and coal types were measured. It was found the volatile accumulation and gas transport in the vicinity of the coal particle were significantly affected by the natural and/or forced convection, and thereby played an important role for coal particle ignition mechanism and ignition temperature. Under negligible and natural convections, coal particle ignited homogeneously, while under forced convection, coal particle ignited heterogeneously. Ignition temperatures were highest when a particle was surrounded by the largest amount of volatiles and lowest when the volatile amount was the least. Further, the ignition temperature decreased with the volatile matter content of coal under negligible and natural convection. Under forced convection, however, the ignition temperature remained more or less constant. [17]. Obviously, the results were rather qualitative than quantitative, and the boundary to divided the natural convection and forced convection was not very clear.

In the late 1990s, using the same experimental apparatus, Wendt et al. [18] conducted some μg experiments on the variation of ignition delay time with particle shape. It was found the volatile release time and ignition time increase with the decreasing of specific surface area for the homogenous ignition mode. For the heterogeneous ignition mode, the results are opposite.

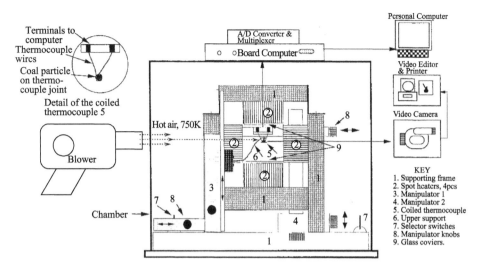

Fig. 3 The schematic μg coal experimental system for single coal particles used in the by Katalambula et al., reprinted from Ref. [17], copyright 1997, with permission from American Chemical Society

However, there are some controversies between existing experimental data and theory analysis. For example, reported by Katalambula et al. [17], the ignition temperature is higher and ignition time is longer at μg than those at 1-g. The authors attributed the differences to the weaker diffusion transport at μg. However, just as discussed in their discussions [15–17], the surrounding volatile cloud is a crucial factor for ignition and its concentration increases at μg rather than decreases as it happens at 1-g due to the sweeping by the nature convection. As a result, the ignition temperature should decrease and ignition time should be shorter at μg.

The experimental setup and measurement method used in previous studies conducted by Katalambula [15–17] were also questionable. In the experiments, the coal particles were heated up by the spot heaters, controlled at desired heating rate. However, the surrounding oxidizer, as well as the evolved volatile cloud was in a relatively low temperature environment. This setting differed from the hot furnace used in conventional boilers. Furthermore, the measurement method for the surface temperature with a set of fine thermocouples coiling around the particle also introduced a few drawbacks. First, the bean of the thermocouples was away from the particle. It was questionable that thermocouples' reading could represent the particle surface temperature since the thermocouples were just emerged in the gas phase environment around the particle. Second, the thermocouples used in experiments were of Pt/Pt–Rh type and could be an effective catalyst for the homogeneous ignition reaction. Third, the thermocouples interfered the flow field, and such interference was indeed buoyancy depended, affecting data obtained at both μg and 1-g.

The results about the gravity effect on the ignition temperature of the single coal particles was challenged by the μg experiments conducted by Zhang and his

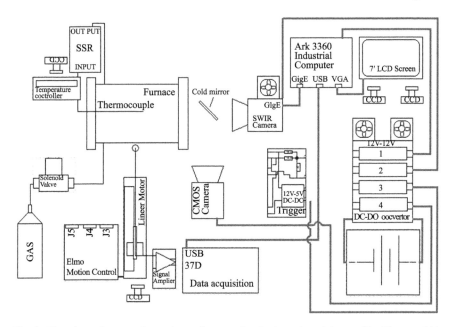

Fig. 4 The schematic μg coal experimental system for single coal particles used by Zhang and his coworkers, reprinted from Ref. [21], copyright 2015, with permission from Springer Nature

coworkers [19–21], using the 3.5 s drop tower at the Chinese National Microgravity Laboratory (NMLC) with μg level of 10^{-2} to 10^{-3} g.

Figure 4 shows the schematic of the experimental system used by Zhang and his coworkers [18, 19]. Figure 5 illustrates a side-view of the experimental apparatus. The main body of the apparatus was a tube furnace with dimensions of ϕ 25 × 300 mm, heated by AC current with the power of 200 W. The experimental apparatus included a coal particle feeding system, a PLC control device and an image and data recording system as well as auxiliary power supply.

During an experiment, a single coal particle was glued at the tip of a fine glass fiber with high-temperature inorganic glue. The glass fiber was mounted to a linear motor which could quickly translate the particle in and out of the furnace. In later experiments, the coal particle was held by a K-type thermocouple through a tiny center hole (Fig. 6). The thermocouple could measure the variation of central temperature during the combustion.

Once the temperature of the furnace reached the desired point which was heated up by the AC power, the particle was sent into the furnace, heated for 1.0–1.5 s, prior to being dropped in the drop-tower. The preheating of the particle before dropping was to ensure the ignition take place at μg. It was checked that the heating rate of the furnace in experiments was about 200–300 K/s and no volatile content in the coal was released under 600 K.

Fig. 5 Experimental apparatus for single coal particle ignition, left: reprinted from Ref. [19], copyright 2008, with permission from Springer Nature; right: reprinted from Ref. [21], copyright 2015, with permission from Springer Nature

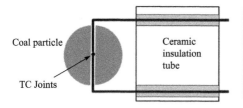
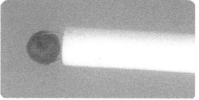

Fig. 6 The measurement of central temperature of single coal particle ($d_p \approx 2$ mm)

Three kinds of high volatile coals with the size of ϕ 1.5 and ϕ 2.0 mm were studied. Some difficulties existed for mounting the small particles and the ignition time of the low volatile lean or anthracite coals with the same size was usually more than 5 s. The tested coals were Datong bituminous (DTB), Sandaocha bituminous (SDB) and Huolinhe lignite (HLL).

The entire burning process of the particle was recorded visually with a CCD camera, with a frequency of 25 frames/s showing the flame shape and its development, flame radius, i.e., flame standoff from the particle surface, and the luminosity of the solid surface. The images recorded by the color CCD camera visually show the flame shape and its development, flame radius, i.e., flame standoff from the solid surface and the burning of the solid phase. To increase the experimental accuracy, flame standoff was determined by pixel-based image processing using the Matlab image processing tools. The flame was defined to be located at the pixel whose energy density was of a maximum value in radial direction.

A non-intrusive method, i.e., RGB colorimetric method, was developed by analyzing the images recorded by the color CCD camera. Here, any color image of a radiative object was assumed to be uniquely composed by three single colors, red (R), green (G) and blue (B). Thus, the two values of RGB can be regarded as the two colors in the twin color pyrometry used in previous studies [19], representing the radiative intensity at corresponding wavelength [20, 21]. In the experiments, the R and G values were used.

Using the calibration with a black body furnace, the linear relationship between $1/T$ and $\ln(R/G)$ can be obtained. Then based to the measured R and G values, the temperature of the pixel can be derived. Though this method was not with high accuracy for low temperature object, and blazingly burning particles, it was convenient and with acceptable accuracy for surface temperature measurement around ignition. The measurement error was estimated within ± 15 K.

Figures 7 and 8 are selected pictures showing the ignition process of ϕ 1.5 mm DTB and ϕ 2.0 mm HLL coal particles at μg and 1-g respectively. It can be seen that the ignition of tested coal particles commenced in homogeneous phase, while the shape, structure, brightness and development of the flames, as well as the volatile matter release during the ignition process are different. At μg, it was observed that the part of volatile was released as a jet, while such a phenomenon was masked by nature convection at 1-g. After ignition, flames were more spherical, thicker, laminated and dimmer at μg than at 1-g. When the particle was preheated for a certain time, a flame appeared before the particle combustion commences. Namely, the coal ignition is of homogeneous type under the testing conditions at both μg and 1-g. At μg, this ignition and process, however is quite different than at 1-g, and some interesting phenomena were first observed. The whole process of ignition and combustion with visual flame can be divided into three phases: the beginning phase, intermediate phase, and the ending phase.

In the beginning phase, e.g., in the first 200 ms after ignition, the flame is irregularly shaped, neither of spherical shape, nor of an upward tail like the one at 1-g. The irregular shape of the flame is attributed to the non-uniform evolution of the volatile content around the particle surface. The gravity effect in this phase is rather weak.

Shortly after the beginning phase is the intermediate phase, in which the flame becomes spherical, enclosing the solid particle. At the same time, its volume increases and then decreases after a certain period. Compared with the 1-g flames at the same time after ignition, it can be seen that μg flames were more spherical, thicker, laminated and dimmer. At 1-g, the flame shape is strongly affected by the buoyancy. Flame center is offset upwardly. The top of the flame is of a pear (Fig. 7b) or a tulip (Fig. 8b) shape, depending on the coal properties.

It can also be seen from Fig. 10a, for the ϕ 1.5 mm DTB, there are some sparks on the outer surface of the flame during the intermediate phase. The results indicate that volatile was partially released as a jet, very probably from the pores of the coal particles. The spark presence in the intermediate phase, rather than the beginning phase, also indicates that the combustion of volatile content in the gas phase around the coal particle will further facilitate the release of volatile content left inside. However, under present experimental condition, sparks were only found for DTB

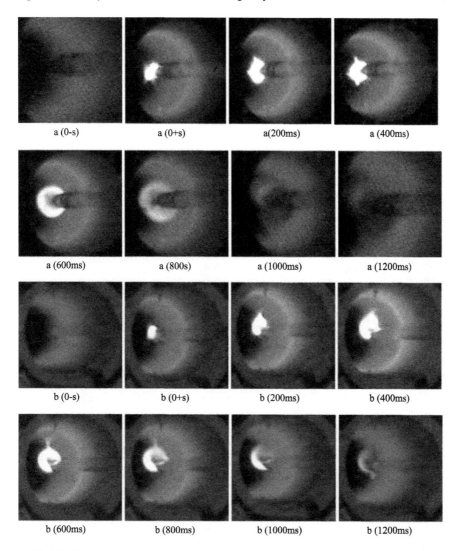

a (0-s)   a (0+s)   a(200ms)   a (400ms)

a (600ms)   a (800s)   a (1000ms)   a (1200ms)

b (0-s)   b (0+s)   b (200ms)   b (400ms)

b (600ms)   b (800ms)   b (1000ms)   b (1200ms)

Fig. 7 Selective photos taken in single coal particle ignition and burning process for Datong bituminous coal with diameter of ϕ 1.5 mm under **a** μg, **b** 1-g, time starts at ignition, reprinted from Ref. [19], copyright 2008, with permission from Springer Nature

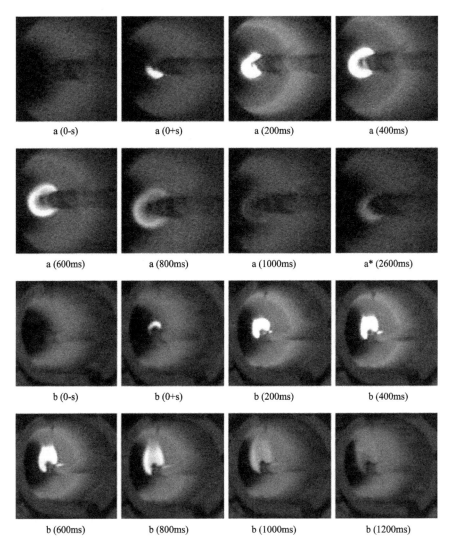

Fig. 8 Selective photos taken in single coal particle ignition and burning process for Huolinhe lignite coal with diameter of ϕ 2.0 mm under **a** μg, **b** 1-g, time starts at ignition, *2nd Ignition when the drop was finished at 1-g, reprinted from Ref. [19], copyright 2008, with permission from Springer Nature

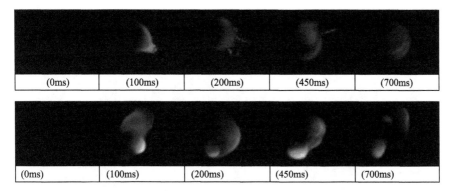

Fig. 9 The burning of a ϕ 1.5 mm bituminous coal particle in the air at μg (lower row) and 1-g (lower row) respectively, reprinted from Ref. [21], copyright 2015, with permission from Springer Nature

at μg. At 1-g, it was expected the jet-type devolatilization still existed for DTB but was covered by the nature convection. While for some coals like HLL, even with high volatile matter, due to their special pore structure, heating value, volatile and moisture content etc., jet-type devolatilization may not exist or just unappreciable.

A certain period after ignition, the flame starts to shrink, and the combustion enters an ending phases. The gravity effect in this phase is significantly strong. At 1-g, the flame keeps in bright for rather long time. Heterogeneous combustion occurs in the intermediate phase and lasts in the ending phases. Flame extinction happens as the char particle is burnt out. When the environment is free of nature convection, at μg, the molecular transports of the oxidizer to the flame front and the products (e.g., CO_2 and H_2O) are weaker. Thus, the burning intensity is lower; thereby the flame brightness is dimmer. Furthermore, close to extinction, in the center of the spherical flame, there is a large dark area, indicating that combustion barely occurs in the heterogeneous phase. The flames are of multiple color layers. Given μg is enough long, a remarkable part of the particle would remain unburnt, even though its surface temperature would be high. The statement is further supported by the 2nd ignition shown in the last picture of Fig. 8. The 2nd ignition occurred right after the rig reached at 1-g environment.

Figure 9 shows the images of ignition process for a ϕ 1.5 mm bituminous coal particle at μg in a furnace of 1123 K. The images were taken with a CMOS camera (IDT NR3-S2). It can be seen more clearly that some volatile content is released as small jets at μg. At 1-g, the volatile is more favor to form a tar droplet at the bottom of the particle due to the gravity force.

Figure 10 further shows the images taken immediately after the single coal particles were ignited at four different O_2 concentrations under both μg and 1-g conditions. It was observed that in all microgravity experiment conditions, homogeneous ignition occurred. The volatile matter flames dominated the early combustion process regardless of the gravity. For low oxygen concentration, i.e., $VO_2 = 21\%$, under

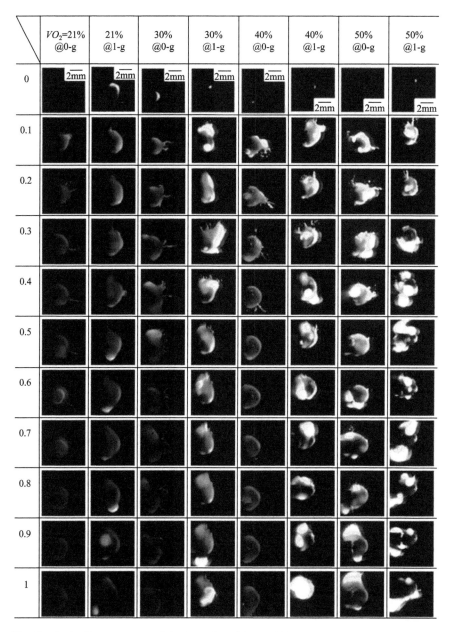

Fig. 10 The ignition processes of single coal particles at different oxygen concentrations under normal- and micro-gravity, reprinted from Ref. [21], copyright 2015, with permission from Springer Nature

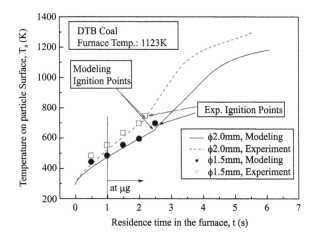

Fig. 11 Temporal variations of the surface temperature of coal particle for DTB coal with size of ϕ 1.5 and ϕ 2.0 mm during ignition, reprinted from Ref. [20], copyright 2009, with permission from Elsevier

both μg and 1-g conditions, coal ignition was observed with the presence of a small yellow spot located close to the surface on the side facing the furnace wall. The heat loss of the ignition was apparently homogeneous. At μg, with negligible convection, a spherical volatile flame appeared. While at natural convection (quiescent air at 1-g), the shape was slightly distorted by the buoyancy. In addition, volatile flames were dimmer in luminosity and smaller in size at μg than at 1-g. This was because that at μg the molecular transport and heat transfer became weaker, inducing a slower heating rate and burning process. However, at a high O_2 concentration, e.g., $VO_2 \geq$ 30%, under μg and 1-g conditions, the particle burnt very intensively with a volatile matter flame of irregular surface. The flame surface burst at some locations where the burning intensity was obviously higher than average or and more jet-like mini flames were ejected from the surface. In some cases, especially as $VO_2 \geq$ 40%, there were some bright spots around the volatile flame surface, with obviously higher luminosity.

The volatile flame could be regarded as the burning of light gases pyrolyzed from intermediate product tar. At low O_2 concentrations, the released tar rather uniformly covered around the surface and the thereafter burning of the particle was also rather uniform. At high O_2 concentrations, the released tar cover was ruptured due to uneven local heating or the ejection of volatile from particle inside, and when O_2 concentration was sufficiently high, no full tar cover could be formed.

Figure 11 shows the temporal variation of the surface temperature of DTB coal particles with ϕ 1.5 and ϕ 2.0 mm. The particle temperature increases with the residence time during ignition. The slope of temperature variation changes from a smaller value to a larger one at a certain time, which is corresponding to the starting point of homogeneous ignition. In the experiments, as introduced before, the ignition temperature was defined as the as particle temperature when the gaseous volatile cloud around it was lit into a visual flame. No temperature was retrieved after ignition since the image brightness was saturated. Neither in low temperature region, as radiation was too weaker. Nevertheless, the ignition temperatures obtained by the modeling and experimental agree very well. The ignition time at 1-g is slightly longer than that at μg.

Figure 12 shows the ignition temperatures for the three testing coals with 1.5 mm diameter at μg and 1-g respectively. It can be seen at both μg and 1-g, the ignition temperature decreases with the increasing of volatile content of the coal. The result agrees with that found by Katalambula et al. [10]. However, present experimental data show that for the same coal, the ignition temperature is lower at μg than that at 1-g. As discussed before, it is suggested that the difference might attributed to their spot heater heating system, and the temperature measurement with the coiled thermocouple around the particle. The temperature obtained in present study is non-intrusive and independent of flow motion around the particle. At μg, the transport intensive is weaker than that at 1-g. On the other hand, the evolved volatile matter is accumulated around the particle rather than be swept away. Based on present results, it is suggested that at μg, the favorable volatile acclamation prevail over the weakening of transport and thus lead an ignition temperature reduction. The reduction amplitude is about 50–80 K, increasing slightly with volatile content.

Figure 13 shows the temporal variations of the surface temperature (T_s), center temperature (T_c) of the DTB particles during ignition at $T_w = 1123$ K, $VO_2 = 21$ and 50%, respectively. For a ϕ 2 mm coal particle, the center temperature of the particle was as much as 150 K lower than the surface temperature when ignition occurred. Therefore, for mm-sized coal particles, the inner temperature gradient should be considered in the ignition model as that may affect the ignition temperature, delay time and ignition mechanism, via lower the heat transfer, the devolatilization rate and the surface reaction rate. It is also seen that the measured temporal variation of T_s and T_c showed a good agreement with the calculated ones.

Figure 14 shows the effect of O_2 concentration on the ignition delay time and ignition temperature under both μg and 1-g conditions. It can be seen that the ignition delay time of the coal particles has no obvious difference between μg and 1-g. On one hand, the evolved volatile matter was accumulated in the boundary layer around

Fig. 12 The ignition temperature of single coal particles at different oxygen concentrations under normal- and micro-gravity, reprinted from Ref. [20], copyright 2009, with permission from Elsevier

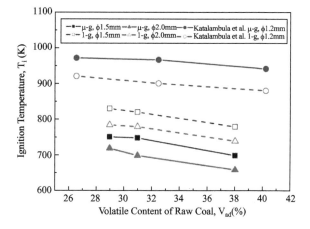

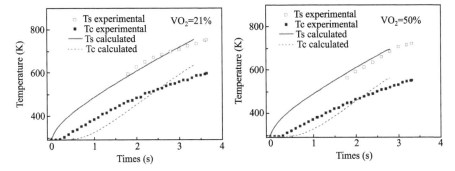

Fig. 13 Temporal variations of the surface temperature (T_s) and center temperature (T_c) measured as compared to the calculated particle surface temperature, reprinted from Ref. [21], copyright 2015, with permission from Springer Nature

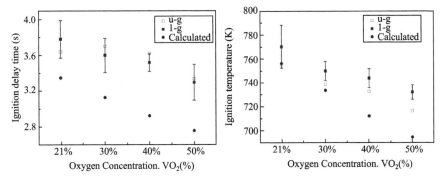

Fig. 14 Variations of the ignition delay time and ignition temperature with oxygen concentration at both μg and 1-g, reprinted from Ref. [21], copyright 2015, with permission from Springer Nature

the particle rather than being swept away under μg, leading to a shorter ignition delay time. On the other hand, the lack of natural convection suppressed the mass and heat transfer between the ambient gas and the particle surface, resulting in a longer ignition delay time. It can be found that the calculated ignition delay time is noticeably smaller that the experimental one. This could be caused by the over-estimation of the volatile release rate.

At μg, the ignition temperature was 20–40 K than that at 1-g, indicating that the weakening effect of the heat and mass transport was prevailed by the volatile accumulation effect. The results were in good agreement with those reported Zhu et al. [20], in which ignition temperatures were 50–80 K lower at μg, but contrast to the experimental results reported by Katalambula et al. [16]. The difference was attributed mainly to the different particle heating methods used in the studies.

It also can be seen that at both μg and 1-g, ignition temperature and ignition delay time slightly decreases with increasing O_2 concentrations. The results confirmed the previous findings obtained by various kinds of experimental methods [22–24]. With a higher O_2 concentration, the mass diffusion of oxygen is stronger and less volatile matter concentration is needed for a mixture to be the flammable. The results also show that the homogeneous criterion applied in the present model can correctly predict the variation trend of ignition temperature and ignition delay time at different O_2 concentrations, against the homogeneous criterion used by Annamalai and Durbetaki [25], Du and Annamalai [11].

Figure 15 shows the ignition regime plotted against the convection intensity under air conditions for single bituminous particles [26]. All three ignition mechanisms were found in the temperature range of the fluidized bed combustion, depending on T_w and convection intensity (Re_d). In general, at a given Re_d, below a critical T_w, heterogeneous ignition dominated; while above the critical T_w, hetero-homogeneous ignition dominated. However, when Re_d was small and T_w was within a certain range, homogeneous dominated. The critical temperature for the transient between heterogeneous and hetero-homogeneous increased with Re_d, while the critical temperature for the transient between homogeneous and hetero-homogeneous in the low Re_d region decreased with Re_d.

It was also found that as O_2 concentration increased, the curve dividing the heterogeneous and hetero-homogeneous zones moved slightly to the right while the curve dividing the homogeneous and hetero-homogeneous zones moved slightly downward. Asymmetric hetero-homogeneous ignition occurred at low temperatures and under medium convection conditions. Hence, the hetero-homogeneous ignition mechanism was more favored at high O_2 concentrations at the same Re_d and the transient to heterogeneous occurred at a lower temperature.

Fig. 15 Ignition regimes for single particles of DT bituminous coal with ~ϕ 1.5 mm in size under different convection intensities, reprinted from Ref. [26], copyright 2014, with permission from Elsevier

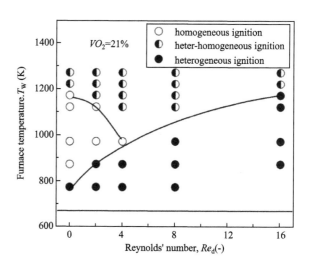

2.2 The Experiments on Ignition of Coal Particle Clouds

In the 1990s, Kiga et al. [27] studied the pulverized coal combustion in different gaseous environments using JAMIC 10 s drop tower. The experimental system is shown in Fig. 16. The fine coal particles were loaded on a cathode in a cylindrical combustion chamber with an inside diameter of 230 mm and height of 245 mm. The particles were levitated by the electrostatic force before the rig was dropped down. The flame propagation was recorded by a high speed CCD and more accurate flame speeds were obtained against the 1-g experiments.

After that, Suda et al. [28] modified the experimental apparatus and pre-placed the fine coal particles into an inner cave located in the bottom of the combustion chamber. Shown in Fig. 17, the fine particles were blown upwards into the combustion chamber ~1 s before the rig was dropped and ignited at ~7 s during the drop. It was found that the coal cloud flames were of rather spherical shape, indicating the dispersion of the fine particles was sufficient. Also, as the coal particle concentration increases, the flame speed first increases and then decreases when it reaches a peak value. The flame speed strongly depends on the coal types. However, only high volatile coals (with $V_{daf} = 32.9$–46%) were studied and no burnt-out data were reported.

The results show that at O_2/CO_2 environment, flame propagates much slowly than that at O_2/N_2 and O_2/Ar environment, and the addition of O_2 promotes flame propagation, as shown Fig. 18. Similar studies were also reported by Fujita et al. [29] and Ito et al. [30].

Fig. 16 The schematic μg coal experimental system for pulverized coal particles or coal dust clouds used by Kiga et al., reprinted from Ref. [27], copyright 1997, with permission from Elsevier

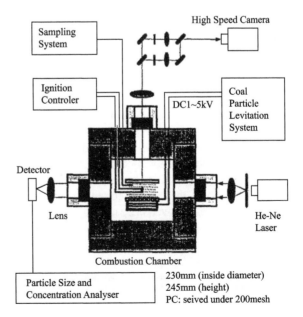

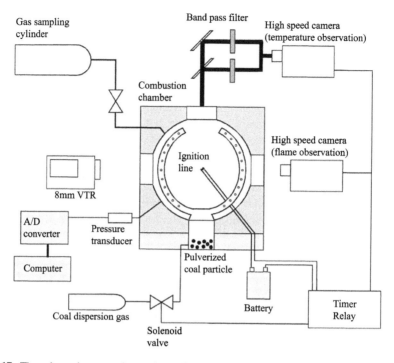

Fig. 17 The schematic μg coal experimental system for pulverized coal particles or coal dust clouds used by Suda et al., reprinted from Ref. [28], copyright 2007, with permission from Elsevier

3 SJ-10-Based Study of Coal Combustion at Microgravity

3.1 The Project Description

Based on the previous introduction, μg coal combustion research can reveal some combustion phenomena which are difficult or even impossible to be observed in 1-g. It can also obtain more accurate data to describe the combustion processes and validate the combustion models. However, due to the limited facilities and μg time, the existing μg studies are very scarce and many aspects, including the entire combustion process, the ignition/combustion of low volatile coals, and the pollutant formation, have not be covered. Space program is prevailing in providing much longer μg time than any ground-based facilities. With the development of Chinese space technology, the SJ-10 satellite space program was proposed for scientific research in microgravity [31]. In this program, the project of investigation of the coal combustion and pollutant formation characteristics under microgravity was selected and led by Tsinghua University and Huazhong University of Science and Technology.

Using the sufficiently long μg time provided by the first dedicated scientific research satellite, a set of experiments on the entire combustion process of single

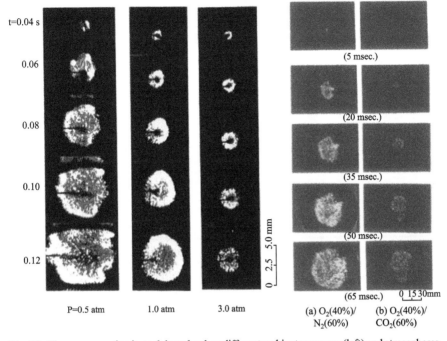

t=0.04 s

0.06

0.08

0.10

0.12

5.0 mm
2.5
0

P=0.5 atm   1.0 atm   3.0 atm

(5 msec.)

(20 msec.)

(35 msec.)

(50 msec.)

(65 msec.)   0 15 30mm

(a) O_2(40%)/   (b) O_2(40%)/
N_2(60%) CO_2(60%)

Fig. 18 Flame propagation in coal dust clouds at different ambient pressures (left) and atmospheres (right) [30]

particles and pulverized clouds of a few kinds of typical Chinse coals were conducted. The experiments were in the first time to study the entire coal combustion process in an ideal buoyancy-free environment where surrounding mass and heat transfer was isotropic. Through the experiments, we can observe some fundamental coal combustion phenomena which are covered by the buoyancy in the ground or limited μg time in drop tower experiments. Besides, some more accuracy data can be obtained to improve the modelling and theory development for coal combustion and emission control on the Earth [32].

3.2 The Experimental System

The experimental system (coal combustion box, CCB) is consisted of 6 units: furnace unit (FU), coal feeder unit (CFU), gas (supply and exhaust) unit (GU), measurement unit (MU) and control unit (CU). The functional link is schematically shown in Fig. 19 and the 3-D layout of the system is given in Fig. 20. The satellite controls the power supply and data/image storage and transmission. The CCB functions to realize multiple single coal particle combustion experiments and pulverized coal combustion experiments under different preset conditions. It can control the furnace

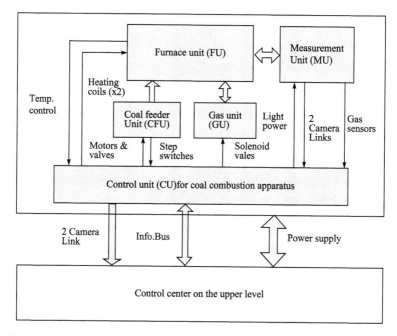

Fig. 19 Functional layout of the experimental system for coal combustion study at microgravity using SJ-10 satellite

temperature, gas supply and exhaust, as well as the coal feeding. It also can acquire two sets of images and some pressure/temperature data. The overall dimension of the coal combustion box is $400 \times 450 \times 500$ mm with the weight of ~55 kg. Figure 21 shows the front and side views of the experimental system.

3.2.1 The Mini-Furnace Unit (MFU)

Similar to the previous studies [19, 20], coal combustion is arranged in an enclosed mini-furnace rather than open space. The furnace is the key unit in the system. The structure of the furnace is shown in Fig. 22. The dimension of the cylindrical combustor is of ϕ 60 $\times$ 90 mm. This mini-furnace is heated up with electric coils. Redundant design is adopted to have two identical sets of coils. The power of the heating coil can be adjustable through extremal programming and the maximal power is 150 W. The most inner layer of the furnace is made of stainless steel to have an enclosed environment whose atmosphere is easy to control. Next to the inner layer is a ceramic layer on which the coils are placed around. Outside of that is a layer with high quality isolated material. And a vacuum jacket is right outside of the isolation layer. Then, the most outside of the furnace is wrapped with isolation fabrics. The outer diameter of the furnace is about ϕ 200 mm and the surface temperature of furnace is kept below 60 °C.

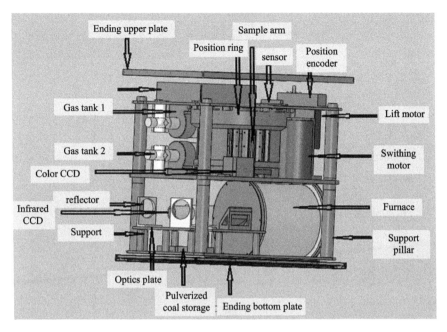

Fig. 20 The 3-D layout of the experimental system for coal combustion study at microgravity using SJ-10 satellite

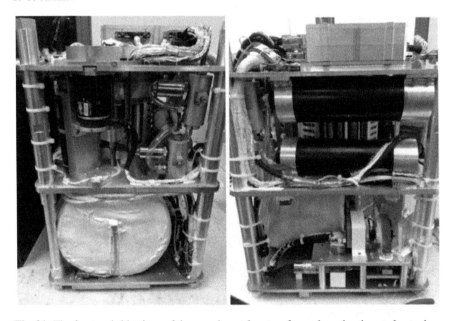

Fig. 21 The front and side views of the experimental system for coal combustion study at microgravity using SJ-10 satellite

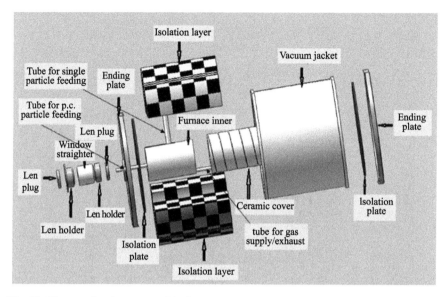

Fig. 22 The exposional structure of the furnace

One end of the furnace is blind and the other end has an optical winder with two lens separated with a distance of 40 mm. The window section is ~50 mm. The two lens are parallelly aligned.

In the middle of the body furnace, there is tube perturbing for the single coal particle feeding. On the ends of the furnace, there are two tube connected, one for the pulverized coal (p.c.) coal supply and the other for the gas supply/exhaust.

For high reliability, four sets of thermocouples are used for the temperature measurement and control, two for measurement and two for control using redundant design. Two thermocouples are mounted in the middle section of the furnace, and the pother two are mounted close to the each end of the furnace. All of them are mounted on the ceramic layer. The readings of the each thermocouple are precalibrated. The temperature inside the furnace is measured by a calibrated thermocouple.

As shown in Fig. 23, numerical simulation shows the furnace can reach 1073 K, the maximal temperature for the experiments in about 3 h at μg under the heating power of 140 W.

3.2.2 The Coal Feeding Unit (CCU)

Single coal particle is fixed to the end of a test probe, as shown in Fig. 24. To minimize the conduction heat loss, the front section of the probe could be made of ceramic. The probes with coal particles are stored in a cage with multiple slots (Fig. 25). Inside the slots, the coal particle is well protected from external impact and vibration (Fig. 26), which has been validated by the mechanic experiments.

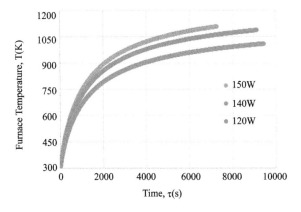

Fig. 23 Simulation results of heating history at different power

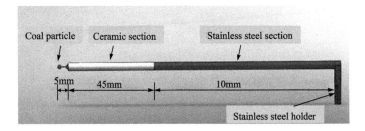

Fig. 24 The probe to fix the single coal particle

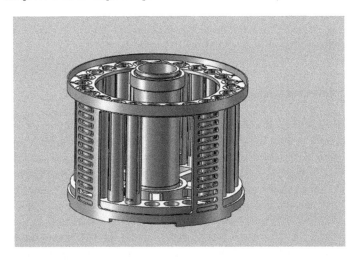

Fig. 25 The cage used to store the probes for single coal particle experiments

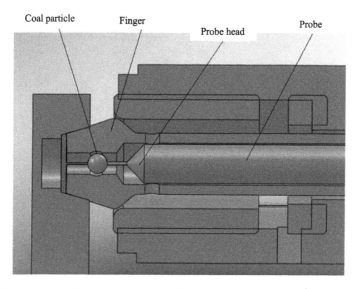

Fig. 26 The assemble of the probe and storage chamber

The cage is assembled in a rotary device (as shown in Fig. 28) for the single coal particle feeding. The device is mounted next to the top surface of the furnace. The linear motor precisely drives the cage to a designed angle and the lift motor controls the insert of the probe to the preset axial location. With such a design, single coal particles can be placed into the center of the furnace in a preset order. After each run, the problem is extracted from the furnace to the cage. To control the temperature rise of coal particle, the inserting speed of the lift motor is preset.

3.2.3 The Gas Unit (GU)

Gas unit can be further divided into gas supply unit (Fig. 27) and gas exhaust unit (Fig. 28). Shown in Fig. 27, the gases stored in the tanks are of 10 MPa and are reduced to ~0.2 MPa for the tests. During the experiments the gas can be directly supplied to the furnace or directed to the pulverized coal chamber to carry out the fine particles. For example, when valve S_2 is closed, with valve S_1 and the S_{N2} are opened, the reactant gas can carry the pulverized coal particles stored in the second tank.

Shown in Fig. 28, after each run of experiments, the flue gas is cooled down through the gas cooling coil. Before it is sent out to the space, part of the flue gas is released into a gas sampling tank to avoid the sensors to operate in a vacuum atmosphere. The measured data are stored and transmitted to the ground with the images.

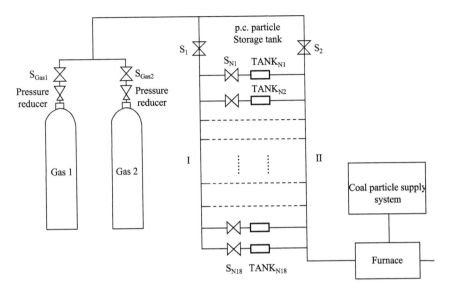

Fig. 27 The schematic of gas supply unit

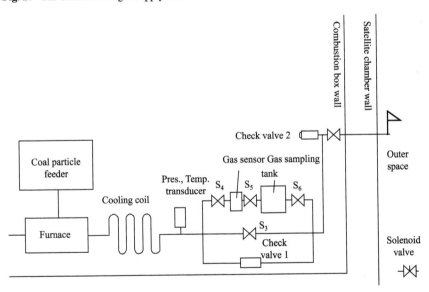

Fig. 28 The schematic of gas exhaust unit

3.2.4   The Measurement Unit (MU)

The MU includes a color CCD camera, a SWIR camera and a beam splitter. The combined use of a visible and short wavelength infrared imaging system is to be achieved by adopting a beam splitter (CLDM) to split the visible light and infrared light emitted from the coal particle. 90% of the visible light emitted from the coal particle is to be redirected to the CCD camera while 80% of the infrared light is to be transmitted through the beam splitter and detected by the SWIR camera. Thus, the entire combustion process of single particle and pulverized clouds in a furnace can be recorded in both visible (380–700 nm) and short-wavelength infrared (900–1700 nm) range simultaneously.

For single coal particle study, a monochromatic imaging technique is applied to determine the particle surface temperature during ignition using a SWIR camera. The measurement accuracy is verified by applying the system to measure the blackbody temperature from 673 to 913 K with an interval of 40 K. It was found that the maximum error of 5 K occurred at the blackbody temperature of 673 K, equivalent to a relative error of 0.7%.

For pulverized coal clouds study, the color CCD camera and the SWIR camera record the entire ignition and combustion process. The ignition temperature is determined as the furnace temperature just at the moment the visible volatile flame observed.

3.3   Some Experimental Results

Figure 29 shows the selected image representing the entire process of combustion for the particles of Datong bituminous (DTB) coal with size of ϕ 2 mm burning at air in a furnace with preset temperature $T_w = 913$ K and 983 K respectively at microgravity.

In both cases, due to the high volatile content in the coal, the ignition is homogenous. When the environment is of relatively low temperature ($T_w = 913$ K), the volatile burning around the coal particle and the flame is rather spherical during the burning process, remarkably different from the asymmetric flame at 1-g. The maximum flame size can be 4–5 times of that of initial coal particle. From the infrared images, we can see some dense matter surrounding the solid particle, with a diameter about 2–3 times of the solid particle. This matter is believed to the tar formed by the volatile released from the coal. When most volatile is burnt, the flame becomes less bright. A portion of tar surrounding the solid particle breaks into small pieces and some eject from tar layer as small jets.

When the environment is of rather high temperature ($T_w = 983$ K), the flame is brighter. Moreover, the burning of the volatile matter is much more intensive and the flame becomes asymmetric. However, this kind of symmetricity is totally different from the one occurring at 1-g. This can be seen more clearly in Fig. 30, with the comparison between the coal particle burning at µg and at 1-g. In the both case,

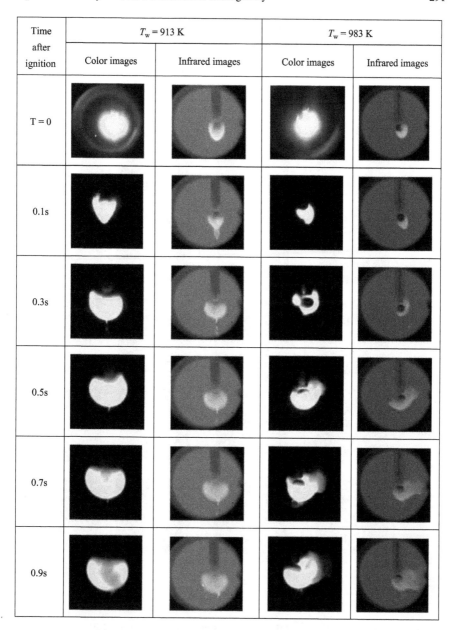

Fig. 29 The combustion process of ϕ 2 mm DTB particles in the air at μg

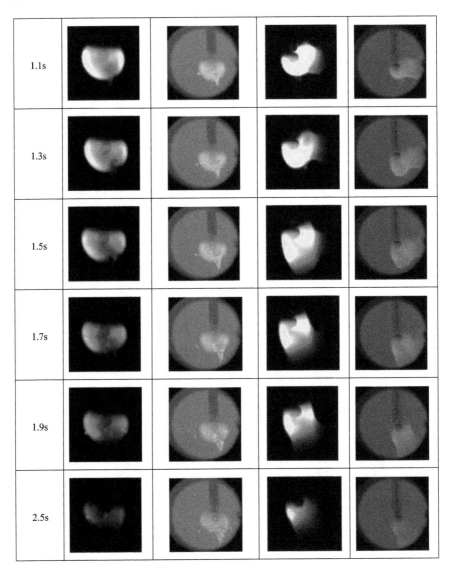

Fig. 29 (continued)

the environment gas is air, and the furnace temperature remains at $T_w = 913$ K. At μg, the symmetricity of the flames is basically caused by the ejection of the volatile matter and or orientation of volatile jet depends on the pore structure of the coal particles. As discussed before, most volatile is released as tar, forming a tar layer on the outside surface of the coal particle. Then, tar is vapored and burnt homogenous around the particle. However, at a high temperature, volatile released amount and rate increases, not all volatile will remain in the tar layer. Instead, some could behave as high pressure liquid jet ejecting from the pores inside the coal or as gas jets breaking through the tar layer. Thus, asymmetric flames are formed even at µg. In the later phase of the combustion, no obvious debris of tar is found. The reason could be at such a high T_w, tar layer is thinner and more quickly burnt. At 1-g, the flame is much brighter, indicating buoyancy enhances the volatile combustion. The burnout time is shorter. Right after ignition, a liquid drop of tar hanging in the bottom of the coal particle. After ignition of volatile, a tulip shaped flame is formed. Indeed, when τ = 0.2–0.4 s, the flame is not really a symmetric tulip. This flames are synergically affected by the volatile jets and buoyancy. However, as the combustion proceeds, the buoyancy effect becomes dominant.

Figure 31 compares the burning process of Jingxi (JX) anthracite coal particle with ϕ 2 mm size, when T_w = 983 K in the air at µg and 1-g. The results showed due to low volatile content, at both µg and 1-g, the ignition of anthracite coal is heterogeneous, and the flame is much smaller than those of high volatile bituminous coal, with maximum size of the twice of size of the coal particle. However, some remarkable differences still exist between µg and 1-g combustion. At µg, the particle is ignited uniformly, and the ignition process is slower. The brightness of the flame increases first and then decreases as the char is depleted. At 1-g, ignition commences at the bottom of the coal particle, and then happens to the entire coal. The bottom of the coal particle burns more intensively than the top part. Also, the flame is brighter. The results indicate that the natural convection still plays important role on the burning of the anthracite coal particles. Due to the enhancement of mass and heat transfer, the ignition delay time and burnt out time are shorter.

Figure 32 shows the ignition and combustion process of Datong pulverized coal (with average size of ~50 µm) at µg. The pulverized coal particles were sent into the combustion chamber more than 30 s and disturbed before the furnace was heated up to the ignition temperature to make uniform distribution of particles in the chamber. It can be seen that a few of particles are ignited first and more and more particles are ignited gradually. The particles are moving fast after the ignition. About ~700 ms of the first ignition, the majority of coal particles are ignited with a bright cloud flame. The flame induces turbulent front quickly. The major combustion completes in ~200 ms.

4 Concluding Remarks

Coal is the primary energy supply worldwide and will remain dominant in energy consumption structure in China for next a few decades. The fundamental studies on coal combustion are still of great demand due to the increasing pursuit of the higher efficiency and lower emission. Microgravity (µg) environment minimizes the buoyancy effect that the coal particles suffer during the combustion under normal gravity, and thus provides an ideal environment to discover new phenomena and obtain more accurate data.

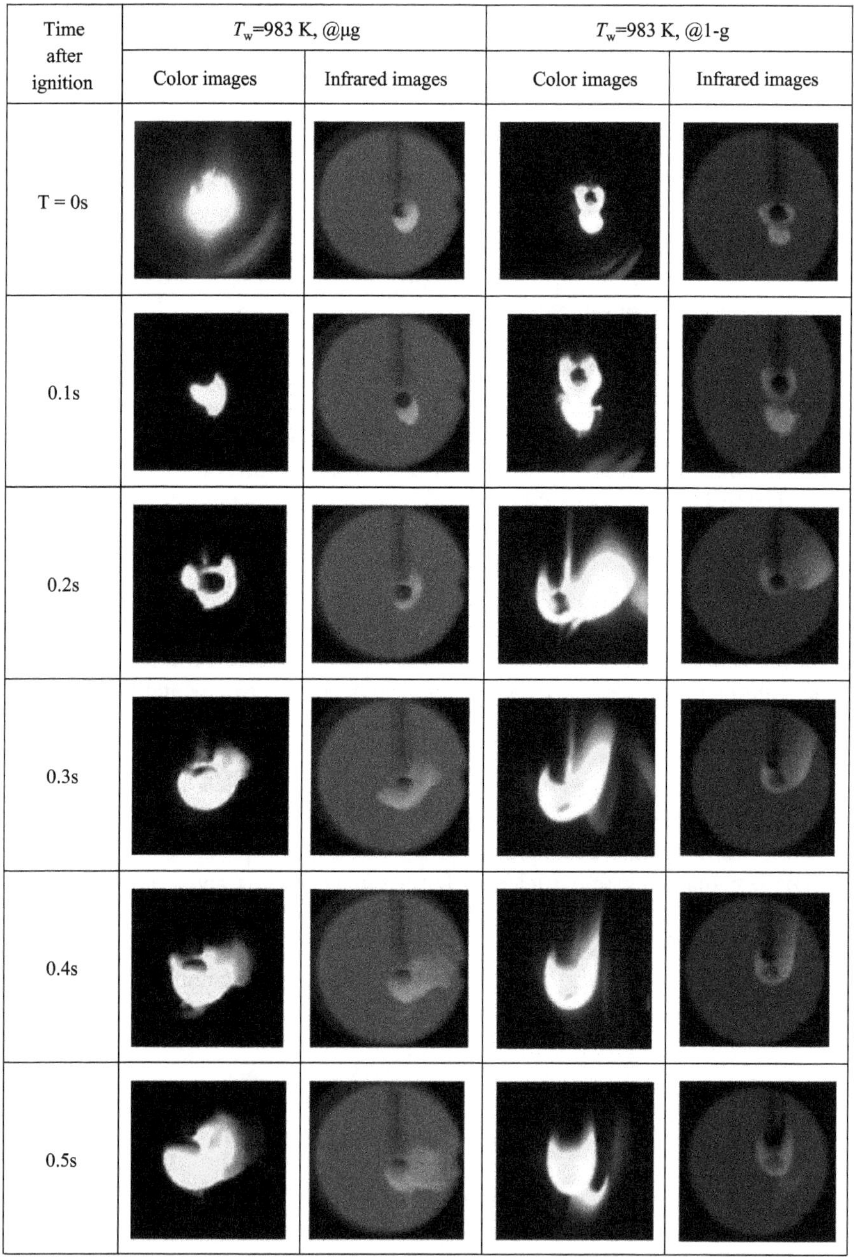

Time after ignition	T_w=983 K, @μg		T_w=983 K, @1-g	
	Color images	Infrared images	Color images	Infrared images
T = 0s				
0.1s				
0.2s				
0.3s				
0.4s				
0.5s				

Fig. 30 The combustion process of ϕ 2 mm DTB particles in the air at μg and 1-g

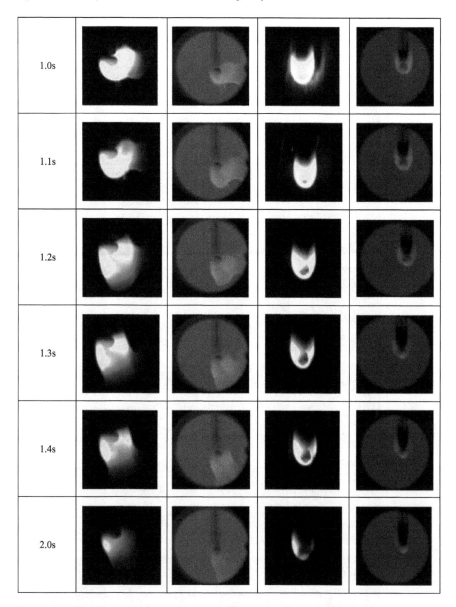

Fig. 30 (continued)

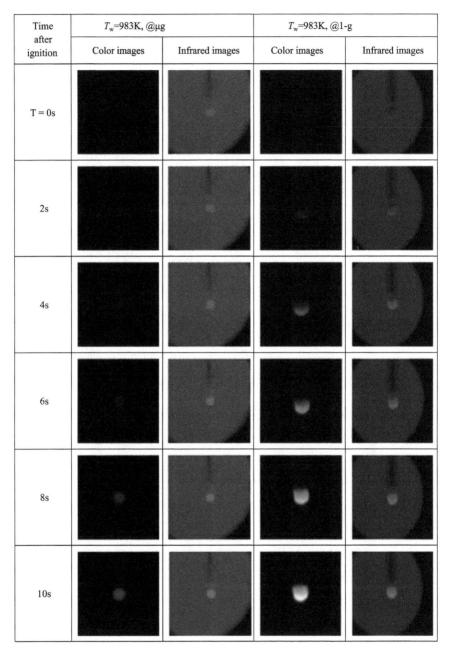

Fig. 31 The combustion process of ϕ 2 mm JX anthracite in the air at μg and 1-g

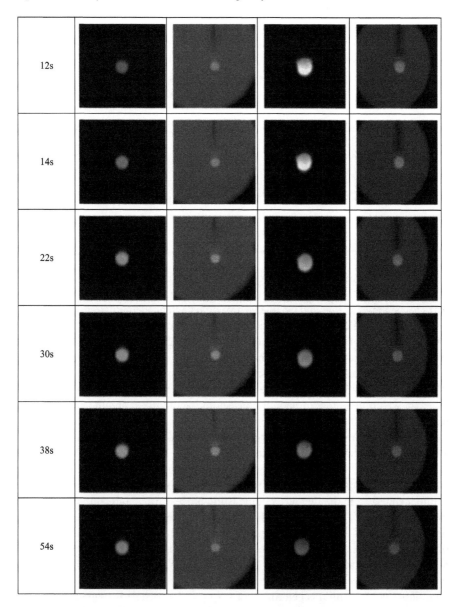

Fig. 31 (continued)

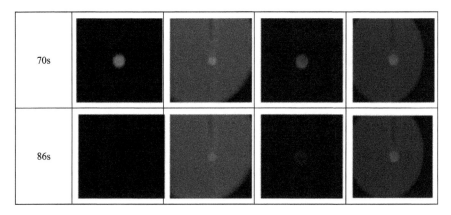

Fig. 31 (continued)

Microgravity coal combustion experiments using drop tower revealed that coal ignition is remarkably influenced by the gravity. At µg, the transport around the coal particles becomes slower and the released volatile content accumulation is enhanced. For the high volatile coals, the ignition delay time becomes longer and the ignition temperature could be 50 K lower that that at 1-g. The gravity also changes the flame shape, brightness and induces asymmetric ignition.

Satellite provides a long-time microgravity environment to study the fundamentals of coal combustion, including the entire combustion process and low volatile coal types, in which the effect of natural convection is minimized and the experimental condition is simplified. With the careful design of the µg experimental system, space experiments on both single coal particles and pulverized coals were conducted using Chinese SJ-10 satellite in the first time worldwide. The results revealed that the combustion process becomes noticeably longer at µg. Moreover, for high volatile coals, in a low temperature environment, the ignition is homogeneous and the flame is rather spherical in the thereafter combustion process. In a high temperature environment, much of the volatile ejects in jet flow, severely distorting the shape of the flame. The size of the flame surrounding the coal particle could be 5 times of that of coal particle. At 1-g, the flame behavior in the beginning stage is synergically affected by the volatile jets and buoyancy. However, as the combustion proceeds, the buoyancy effect becomes dominant and the flame is of a tulip shape. For low volatile bituminous coal, ignition is heterogeneous. At µg, the particle is ignited uniformly, and the ignition process is slower. At 1-g, ignition commences at the bottom of the coal particle, and then happens to the entire coal. The bottom of the coal particle burns more intensively than the top part. Before the cloud flame of pulverized coal is formed, a small amount of coal particles are ignited and the coal particles move randomly.

Through µg coal combustion studies, new phenomena and data could be obtained to get better understanding of the insights about the coal ignition/combustion mechanisms. And more accurate models could be developed for the ground coal utilization.

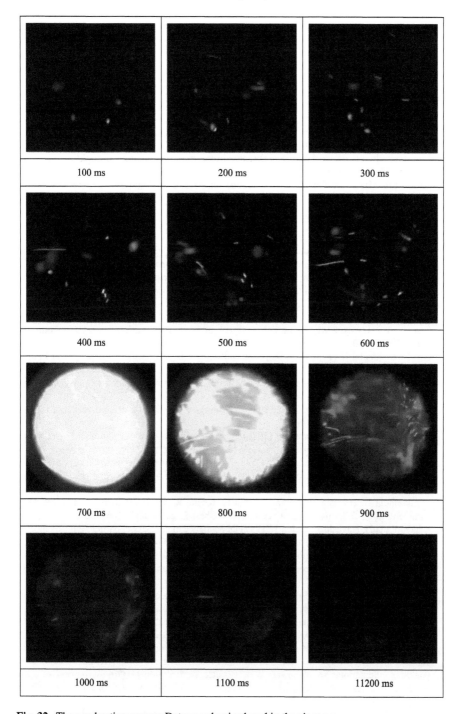

Fig. 32 The combustion process Datong pulverized coal in the air at μg

Acknowledgements The support by the CAS Project (XDA04020202-09 and XDA04020409) and NSFC project (11872231) for this study is acknowledged.

References

1. World Energy Balances (2018) IEA report
2. Coal Information Review (2018) IEA report
3. Bhandyopadyay S, Bhaduri BD (1972) Prediction of ignition temperature of single coal particle. Combust Flame 18(3):411–415
4. Tognotti L, Malotti A, Petarca L et al (1985) Measurement of ignition temperature of coal particles using a thermogravimetric technique. Combust Sci Technol 44(1):15–28
5. Zhang D, Wall TF (1994) Ignition of coal particles: the influence of experimental technique. Fuel 73(7):1114–1119
6. Su S, Pohl J H, Holcombe D, et al (2001) Techniques to determine ignition, flame stability and burnout of blended coals in p.f. power station boilers. Prog Energy Combust Sci 27(1):75–98
7. Ronney PD (1998) Understanding combustion processed through microgravity research. In: 27th international symposium on combustion, Combustion Institute
8. Zhang X (2004) Research advances on microgravity combustion. Adv Mech 34(4):507–528
9. Zhu MM, Zhang H, Zhang DK (2009) Combustion of coal under microgravity conditions: a review. In: 7th Asia-pacific conference on combustion, Taipei, May 2009
10. Essenhigh RH, Misra MK, Shaw DW (1989) Ignition of coal particles: a review. Combust Flame 77(1):3–30
11. Du X, Annamalai K (1994) The transient ignition of isolated coal particle. Combust Flame 97:339–436
12. Gururajan VS, Wall TF, Gupta RP et al (1990) Mechanisms for the ignition of pulverized coal particles. Combust Flame 81(2):119–132
13. Gieras M, Klemens R, Wójcicki S (1985) Ignition and combustion of coal particles at zero gravity. Acta Astronaut 12(7/8):573–579
14. Gieras M, Klemens R, Wolanski P (1986) Experimental and theoretical study of ignition of single coal particles at zero gravity. Acta Astronautica 13(5):231–239
15. Katalambula H, Kitano K, Ikeda K, Chiba T (1996) Mechanism of ignition of single coal particle: effect of heating rate on particle-size dependence of ignition temperature. J Chem Eng Jpn 29(3):523–530
16. Katalambula H, Hayashi JI, Chiba T, Ikeda K, Kitano K (1997) Mechanism of single coal particle ignition under microgravity condition. J Chem Eng Jpn 30(1):146–153
17. Katalambula H, Hayashi JI, Chiba T (1997) Dependence of single coal particle ignition mechanism on the surrounding volatile matter cloud. Energy and Fuels 11:1033–1039
18. Wendt C, Ikeda IM, Katalambula H, Kitano K, Eigenbrod C, Rath HJ (1999) Dependence of single coal particle homogeneous ignition on particle shape under microgravity. Microgravity Sci Technol 12(2):51–55
19. Tang GT, Zhang H, Zhu MM et al (2010) Experimental study on the ignition process of single coal particles at microgravity. Microgravity Sci Technol 22:27–35
20. Zhu MM, Zhang H, Tang G et al (2009) Ignition of single coal particle in a hot furnace under normal and micro-gravity condition. Proc Combust Inst 32:2029–2035
21. Liu B, Zhang Z, Zhang H, et al (2015) Volatile release and ignition behaviors of single coal particles at different oxygen concentrations under microgravity. Microgravity Sci Technol, 1–8
22. Ponzio A, Senthoorselvan S, Yang WH, Blasiak W, Eriksson O (2008) Ignition of single coal particles in high-temperature oxidizers with various oxygen concentrations. Fuel 87(6):974–987
23. Howard JB, Essenhigh RH (1965) The mechanism of ignition of pulverized coal. Combust Flame 9(3):337–339

24. Shaddix CR, Molina A (2009) Particle imaging of ignition and devolatilization of pulverized coal during oxy-fuel combustion. Proc Combust Inst 32(2):2091–2098
25. Annamalai K, Durbetaki P (1977) A theory on transition of ignition phase of coal particles. Combust Flame 29:193–208
26. Liu B, Zhang Z, Zhang H, Yang H, Zhang D (2014) An experimental investigation on the effect of convection on the ignition behavior of single coal particles under various O_2 concentrations. Fuel 116:77–83
27. Kiga T, Takano S, Kimura N, Omata K, Okawa M, Mori T, Kato M (1997) Characteristics of pulverized-coal combustion in the system of oxygen/recycled flue gas combustion. Energy Conversion Manag 38:S129–S134
28. Suda T, Masuko K, Sato J et al (2007) Effect of carbon dioxide on flame propagation of pulverized coal clouds in CO_2/O_2 combustion. Fuel 86(12):2008–2015
29. Fujita O, Ito K, Tagashira T, Sato J (1993) Flame propagation in coal dust cloud under microgravity environment. Proc Int Symp Aerosp Fluid Sci, 274–281
30. Ito K, Fujita O, Tagashira T, Sakamoto A, Sato J (1994) Utilization of 10 second microgravity environment for the measurement of coal dust flame propagation speeds. In: Proceedings of drop tower days workshops, pp 115–119
31. Hu WR, Zhao JF, Long M et al (2014) Space program SJ-10 of microgravity research. Microgravity Sci Technol 26(3):159–169
32. Zuo C, Liu B, Chen J, Zhang P, et al (2015) Space program SJ-10 on coal combustion research at microgravity. In: The 8th international symposium on coal combustion, Beijing, China, 19–22 July 2015

Material Processing Facilities in Space

Xiuhong Pan, Fei Ai and Yan Liu

Abstract Many researchers around the world have lots of interests in discovering what and how may happen when materials processing is conducted in space. Usually, for the investigation of space material sciences, experimental facilities are important and necessary. Up to now, more than one hundred of material experimental facilities have been developed in the world. They include many high-temperature heating furnaces, in situ observation and diagnosis equipments, as well as facilities for crystal growth from water solution in microgravity. Some of them are prepared by international cooperation among two or even more countries. In this chapter, we will give a brief summary for the material processing facilities in the world. The facilities developed in the recent ten years serving on International Space Station (ISS) are particularly focused. Furthermore, some material experimental devices built by China which have served in Chinese recoverable satellites and man-made space crafts are also discussed emphatically.

Keywords Microgravity material science · Facility · Heating furnace · In situ observation equipment

1 High-Temperature Heating Furnaces in Microgravity

For many kinds of materials, such as metal alloy, semiconductor, oxide crystal, ceramic, their experiments are generally performed under high temperature. So heating furnace is one important kind of experimental facilities for materials processing in microgravity. For such a furnace, its temperature can arrive up to

X. Pan (✉) · F. Ai · Y. Liu
CAS Key Laboratory of Space Manufacturing Technology, Beijing, China
e-mail: xhpan@mail.sic.ac.cn

State Key Laboratory of High Performance Ceramics and Superfine Microstructure,
Shanghai Institute of Ceramics, Chinese Academy of Sciences, Shanghai 200050, China

© Science Press and Springer Nature Singapore Pte Ltd. 2019
W. R. Hu and Q. Kang (eds.), *Physical Science Under Microgravity: Experiments on Board the SJ-10 Recoverable Satellite*, Research for Development,
https://doi.org/10.1007/978-981-13-1340-0_12

several hundreds of centigrade or even higher than one thousand centigrade. Almost all the countries aiming at materials science research in space have developed their own heating furnace. The follows are some represents.

1.1 Russian POLIZON Facility

Russian/Soviet Union is one of the earliest countries to perform material experiments in microgravity [1]. More than ten material processing facilities have been built and launched on the Soyuz space capsule, Salyut space station, the MIR space station, or on the Photon satellite. Among these, POLIZON is the latest one. The POLIZON facility can perform material experiment using the PROGRESS transport spacecraft as a flying platform, which is aimed to deliver the payload on the international space station (ISS) [2].

The POLIZON facility includes an electrovacuum furnace with one 8-sectional heater, in which up to 12 different experiments on crystal growth can be performed in an automatic mode (Fig. 1). The furnace is equipped with the loading–unloading device in which 12 capsules with material samples are installed. The capsules are loaded by turn into the furnace, the heaters are switched on, and the melting process begins. Due to adjusting temperature on the separate sections of the heater, it is possible to obtain the optimal temperature profile along the capsule axis for each experiment. Crystal growth by Bridgman method is realized by transition of the capsule from the furnace at a given velocity.

Fig. 1 The photo of POLIZON furnace, reprinted from reference [2]. Copyright 2018, with permission from Acta Astronautica

The furnace is equipped with a system of magnetic stirring of the melt that allows to heat control and mass transfer during crystal growth. The maximum temperature of POLIZON is 1200 °C.

1.2   Heating Furnace in Microgravity Developed by German

Some other European countries have also developed their heating furnace for material experiment in microgravity, among those German contributed much. In 1996, a multizone furnace including a magnetic damping array consisting of two radially magnetized permanent magnet rings was developed in Germany [3]. In this furnace, a three-zone heater has been designed which is suited to the utilization in the zone melting facility (ZMF). It is designed for crystal growth of semiconductors and provides standard subsystems like the heater translation mechanism, gas supply, and process chamber for the implementation of a specific multizone heater. The heater has been laid out for a maximum achievable temperature of the central heating zone of 1300 °C with a maximum total power of 1200 W. It has three heating zones, two isothermal heaters made of Kanthal with a length of 80 mm, and the central heater with a length of 22 mm. In order to compensate for the magnetic field arising from the dc current all heater wires are winded bifilarly.

Another heating furnace for solidification investigation of aluminum alloys had been prepared by German [4]. This facility named TITUS furnace has a 6-zone heater. It served onboard the Russian MIR Space Station. Al–Ti alloy experiments were performed in this furnace in 1999. The pulling rate of solidification can arrive at 1.9 μm/s. This furnace is used to study the transition from equiaxed to columnar morphologies.

1.3   Materials Heating Furnaces on ISS

NASA has sponsored the flight investigation on growing crystals with baffle under microgravity condition [5]. This project named SUBSA (Solidification Using a Baffle in Sealed Ampoules) was first conducted in the Microgravity Science Glovebox (MSG) on ISS.

The flight furnace located in the MSG had a transparent gradient section, and a video camera, sending images to the earth. The SUBSA furnace which designed for growing InSb is shown in Fig. 2. The axial temperature gradient in the transparent section was approximately 80 K/cm. The seed was placed about 10 mm from the furnace, so that seeding and the initial 10 mm of growth could be observed. The heater temperature is lower than 820 °C. Crystal growth was performed in this furnace by lowering the furnace temperature.

Several other heating furnaces can serve for materials experiments on ISS [6]. For example, NASA developed the Quench Module Insert (QMI) and the Diffusion

(a)

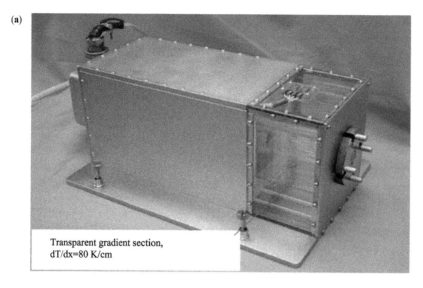

Transparent gradient section,
dT/dx=80 K/cm

(b)

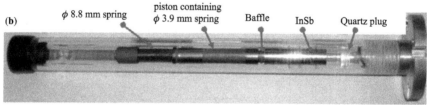

Fig. 2 SUBSA furnace with a transparent gradient region **a** and the ampoule with the baffle **b**, reprinted from reference [5]. Copyright 2018, with permission from Journal of Crystal Growth

Module Insert (DMI). ESA prepared a Low Gradient Furnace Module Insert (LGFMI) and the Solidification Quench Furnace Module Insert (SQFMI), which are both for material processing on ISS. JAXA built the Gradient Heating Furnace (GHF) which is in the Kobairo Rack (Kobairo) of ISS. Some of them are fixed in the Materials Science Research Facility (MSRF) of ISS, which is designed to accommodate materials science investigations selected to conduct research in the microgravity environment of ISS. The MSRF consists of modular autonomous Materials Science Research Racks (MSRRs).

The Quench Module Insert (QMI) operates inside the Materials Science Laboratory (MSL), which was launched at 2009 and installed in NASA's first Materials Science Research Rack (MSRR-1) [7]. This unique material processing furnace has been designed to create an extremely high temperature gradient for the directional solidification processing of metals and alloys. But at the same time it is flexible enough to process samples in either low-gradient or isothermal environments. QMI can heat samples up to 1400 °C, suitable for the processing of aluminum samples and its alloys. The QMI has also been designed with "quench capability." This allows

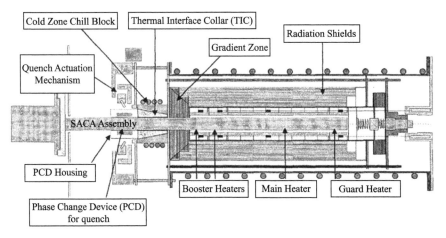

Fig. 3 Schematic view of the QMI furnace on ISS, reprinted from reference [7]. Copyright 2018, with permission from IEEE Aerospace Conference Proceedings

the furnace to rapidly freeze a sample at the solid-liquid interface. The design of the QMI can be broken down into three main parts: the hot zone responsible for melting the sample, the cold zone responsible for freezing the sample, and the quench zone which rapidly freezes the samples when desired to preserve the solid-liquid interface. A diagram of the furnace is shown in Fig. 3. The hot zone is constructed from four individually controlled heaters: the two booster heaters, the main heater and the guard heater. The main heater sets the overall sample temperature. The guard heater prevents the temperature at the end of the sample from falling off. The booster heaters keep the temperature of the sample from cooling off too far ahead of the chill block. The sample is quenched, or rapidly cooled, with the "Phase Change Device" located at the end of the "Thermal Interface Collar" (TIC). The samples are contained in "Sample Ampoule Cartridge Assemblies" (SACAs). The QMI works in vacuum.

The Diffusion Module Insert (DMI) is the second NASA-built insert for flight on the MSRR-1 [8]. DMI is a Bridgman type furnace, which is designed to accommodate processing temperatures up to 1600 °C. The requirements are for this Module Insert to have appropriate isothermality for Fickian diffusion measurements, but to support adequate temperature gradients for Soret experiments and Bridgman growth. Fickian and Soret diffusion investigations levy substantially different requirements on the furnace. Both processes focus on the study of diffusion. The Fickian experiment design calls for an isothermal furnace chamber at least 10 cm in length with temperature capabilities up to 1600 °C. Soret diffusion calls for a Bridgman-Stockbarger furnace that can provide an axial gradient of approximately 100 °C/cm. NASA's Advanced Pattern Formation and Coarsening Research Module (APFCRM) is an on-orbit replacement for an experiment module sponsored under NASA's Space Product Development program. It consists of a low temperature facility with a precisely controlled both for in situ observation of the solidification and growth of transparent model materials that simulate the behavior of metals and alloys.

With the ELIPS (European Program for Life and Physical Science) Program of the ESA, the MLS including furnace inserts have been developed to allow directional solidification and crystal growth processing on the ISS. ESA's Low Gradient Furnace Module Insert is a furnace for crystal growth capable of reaching 1400 °C. Samples can be translated at slow and precise rates within a temperature controlled environment. Magnetic field capabilities, both static and rotating are available to influence the liquid flow and improve the properties of the crystalline product. ESA's Solidification Quench Furnace Module Insert is a furnace designed primarily for metallurgical experiments capable of reaching 1400 °C, and including a quench capability. While initially designed to be used for ESA experiments, these latter two insert modules may be made available for NASA experimenters. By these facilities, samples of 8 mm diameter and 245 mm length have been processed and other sizes are possible. Up to 12 thermocouples can be accommodated in the crucible along the sample for temperature measurements.

The Gradient Heating Furnace (GHF) by JAXA is in the Kobairo Rack (Kobairo) ISS [9]. The GHF is a vacuum furnace that contains three heating blocks, and is used mainly to conduct high quality crystal growth experiments using unidirectional solidification. The three heater-units can generate the high temperature gradients needed to produce large scale pure crystals. Some preliminary experiments of $In_xGa_{1-x}Sb$ crystal growth in GHF have been performed recently [10, 11].

1.4 Materials Heating Furnaces for Microgravity Investigation Developed in China

The first heating furnace aimed at materials science research in space was designed and successfully carried out into orbit at the end of 1980s in China. Since then, a series of different materials processing experiments have been performed more than 10 times by virtue of piggyback opportunities onboard both the recoverable satellites and the manmade spacecrafts in China. Most of these furnaces had the ability of heating for material experiments at high temperature.

The DGW facility on Chinese Manmade Spacecraft. The temperature of Chinese early materials processing furnaces in microgravity could not be programmably controlled, until the occurrence of Duo Gong Wei (DGW) facility. The project of DGW materials processing facility began in 1993. This facility was designed mainly for crystal growth, solidification of metals and alloys, and the process of glasses, as the payloads of Chinese "SZ (ShenZhou)" series spacecrafts in space. DGW facility has been carried into orbits 3 times, as the payloads of SZ-1, SZ-2, SZ-3 Chinese unmanned spacecrafts from 1999 to 2002 [12, 13].

The DGW facility is a one-zone heater furnace. It consists of 3 main parts: the heating furnace, the sample management system and the controller. Both the heating furnace and the sample management system are enclosed inside a vacuum-tight housing. The heating furnace is constructed of an alumina tubular chamber of 24 mm

Fig. 4 The DGW facility for the Chinese SZ-3 flight mission

in diameter and 40 mm in length, a Ni–Cr filament resistance heating unit, multi-layer radiation reflectors, compound insulation structures and five S-type thermocouples. The Ni–Cr heating filament is enwound tightly adjacent to the inner bore of the alumina chamber to enable easy and direct heating to the sample cartridge with minimum heat loss. The furnace is able to maintain at 1000 °C. Totally six samples cylindrical cartridges can be installed in this furnace for microgravity experiments.

The controller is a programmable microprocessor system, which has the duel functions of process control and data acquisition. It controls the heating process of the furnace and the movements, including axial temperature-time profiles, the exact position of each cartridge, the rotating speed of the magazine and the translating speed of the cartridges. The PID mode controller of the facility has the ability of obtaining temperature stability better than $\pm 1.0°$C/h. Figure 4 is the photo of DGW facility served in SZ-3 Chinese unmanned spacecraft.

The MMP facility on Chinese SJ-10 Recoverable Satellite. Recently, much more requirements proposed by the materials scientists greatly accelerate the developments in the field of space materials science in China. In other words, differing from before, more and more different materials including semiconductors, metal and alloys, single crystals, etc., are required to be processed in microgravity by taking advantage of very valuable opportunities given by recoverable satellite. Considering the varieties of materials and their experimental requirements, and also, the constraints of size, mass, energy given by the satellite, the Multi-functional Materials Processing (MMP) facility has being developed [14]. This heater furnace is designed for materials experiments planned to carry out in space by Chinese "SJ (Shijian)-10" recoverable satellite.

The design of MMP facility can be broken into four main parts as shown in Fig. 5: the vacuum house 1, the heating furnace 2, the sample arrangement assembly 3, and the controller 4. The later three parts are all in the vacuum house and fixed tightly with each other.

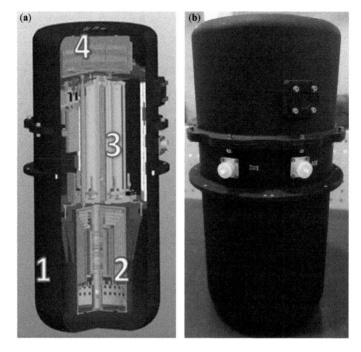

Fig. 5 The structure of MMP facility **a** and its photo **b** served on Chinese SJ-10 recoverable satellite

The vacuum house is a cylindrical house prepared with Al alloys. It can supply the vacuum environment when MMP facility works. This is benefit to the heat energy conservation as well as to preventing the sample cartridges being polluted at high temperature. Furthermore, all the mechanical and electrical interfaces of MMP facility with the satellite are designed on this vacuum house, including a flange ring and two electric connectors.

The heating furnace is designed with two adjacent heating zones, which can adjust the temperature file along the axis. It is a style of Bridgman method. The heater core of the furnace is constructed by two individual heaters, main brackets, four S-type thermocouples and high purity alumina pans. These two heaters are fabricated by resistance coils which enwound into a tube. The surface of the tube is a helical resistance coil. These tubes are encased along the circular grooves in pans whose symmetrical axes are in the same line. The cartridge of φ16 mm is structured by these pans. These two heaters can be controlled separately to achieve the various temperature profiles along the axes of the cartridge in accordance with each special experimental requirement. These two half main brackets shaping into a cylinder are able to fix these pans tightly through several grooves in themselves. The feature of the structure is efficient to focus the energy in the region needed and at the same time, enough long operation life. In fact, the average life of it is more than 1000 h under the maximum temperature of 950 °C validated by experimental tests on ground.

The sample arrangement assembly consists of two kinds of drive gears, two motors, several hall locators, and 6 sample magazines etc. Totally 6 sample cartridges with 231 mm in length are mounted in the 6 sample magazines individually which can be automatically exchanged by these drive gears. One group of the drive gears are used for displacing the cartridge in or out the heating region along the axes of the heating core. And the other group of drive gears can make the 6 sample magazines around the axes of the center and move them to the right place to do the next experiments located accurately. The maximum processing length of the sample is 178 mm with the translation rate ranging from 0.5 to 120 mm/h.

The sample translation and sample movement, as well as the heater process, are controlled by the controller. The controller is a programmable microprocessor system. The PID mode is used for temperature control with temperature stability of better than ± 0.5 °C/h. The controller can communicate with the satellite by the two electric connectors on the vacuum house. It should be noted that, as for all of other Chinese materials processing facilities with high temperature, their electric controllers are fixed separately from the heating furnace. So there are at least two machines for each of these facilities. However, the MMP facility is only one machine since its controller is integrated into the vacuum house. This design makes the temperature controlling being more accurate due to the shorter cables between the heater and the controller.

In Table 1, the main technical specifications of MMP facility are summarized.

MMP facility was designed mainly for the project of "Research of Melt Growth in Space" (RMGS) on Chinese "SJ-10" recoverable satellite, which is supported by the Strategic Priority Research Program on Space Science from the Chinese Academy of Sciences. In this project totally 8 material sample experiments have been arranged in space. There are:

- Crystal growth of GaMnSb magnetic semiconductor in space;
- Growth and study of fluoride laser crystal with tunable ultraviolet wavelength;
- Growth and property study of new infrared detector material;
- Wetting of metallic molten and formation of composite;
- Solidification and defect controlling of high temperature alloy single crystal;

Table 1 The technical specifications of the MMP facility in Chinese SJ-10 recoverable satellite

Specifications	Values
Number of heating zones	2
Maximum temperature	950 °C
Maximum power consumption	Lower than 140 W
Translation rate range	0.5–120 mm/h
Max of axial temperature gradient	75 °C/cm
Number of magazines	6
Cartridge length	ϕ 16 mm × 231 mm
Temperature stability	Better than ± 0.5 °C/h

Fig. 6 The returned sample
cartridges in MMP facility
served on the Chinese SJ-10
recoverable satellite

- Synthesis of metallic matrix composite by self-propagating high-temperature technique;
- Interface phenomena of Ti-based alloy during melting;
- Growth, simulation and property study of ternary InGaSb phototransistor crystal.

This MMP facility was carried into orbits by the "SJ-10" recoverable satellite launched at April 4, 2016, and automatically performed 8 material experiments in space lasting more than 210 h. It should be noted that, the 8 material samples were fixed in 6 cartridges, with the fourth and the fifth cartridges containing two samples, respectively. Figure 6 shows the sample cartridges returned to the ground after the experiments in space. Although two heating zones have been designed in the MMP facility, they were not allowed to work simultaneously in space due to the electrical current limitation. As a result, several samples in MMP had not got to their programmed maximum temperature during the microgravity experiments.

The Heating Furnace in Chinese TG-2 Space Lab. Another two-zone heaters furnace has been laughed in Chinese "TG (Tiangong)"-2 space laboratory in September 15, 2016. The structure of this heater furnace is familiar to that of MMP facility on Chinese "SJ-10" recoverable satellite. It also has two adjacent heating zones, which are fabricated by resistance coils enwound into a tube and the temperature of each heating zone can be controlled separately. The maximal temperature of this furnace can also arrive at about 950 °C. Furthermore, sample cartridges in this facility can be manned exchanged batchedly by the cosmonaut. Firstly, six sample cartridges with ϕ 16 mm × 260 mm were mounted in the magazines and finished their heating experiments in space automatically. Then these six sample cartridges were drawn out of the heating furnace by the cosmonaut and another group of six sample cartridges were put into the furnace for space experiments. At last, the third group of six sample cartridges were put into the furnace replacing the second group of six sample cartridges when the later finished their experiments. This is the first time that cosmonaut can participate in the material experiments in space in Chinese history. The heater furnace has performed 18 sample experiments in "TG-2" space laboratory in space.

In the future, Chinese Space Station (CSS) will be built, onboard which a High Temperature Experimental Rack (HTER) will be developed. This rack is a comprehensive equipment for materials science investigation in space. It will take on the functions of high temperature processing, in situ observation by X-ray or optical technique, and measurement of melting physical properties. Furthermore, this rack will be designed with "quench capability" as well as with a system of magnetic stirring of the melt during crystal growth.

2 In Situ Observation Facilities for Materials Processing in Space

2.1 In Situ Observation Facility by Optical Techniques

Optical observation with visible light is an important method to visualize the experimental process under microgravity conditions. Most of the optical observation systems are applied for the material experiment at room temperature or water solutions, so their hardware facilities can be much more comprehensive and sophisticated. They can be used in the investigation of fluid phenomenon as well as in the study of transparent materials, such as ice, succinonitrile, and protein crystal growth. Here, we will give several examples.

The PFMI on ISS. The Pore Formation and Mobility Investigation (PFMI) was conducted in the Microgravity Science Glovebox (MSG) onboard the ISS [15]. It was selected by the NASA Glovebox Investigation Panel and assigned to the Glovebox Program Office at the Marshall Space Flight Center (MSFC) on December 3, 1997. The PFMI hardware and samples were launched onboard Space Shuttle STS-111/UF-2 on June 5, 2002 and the first experiment was initiated on September 19, 2002.

The PFMI apparatus is a Bridgman type translation furnace with a main zone, a booster zone, and a cold zone. It is designed to process transparent samples such that the translating solid/liquid interface can be directly observed and recorded. Figure 7 shows a photograph of the PFMI ground test unit and the schematic of the PFMI sample installed in the MSG aboard the ISS. The maximum continuous operating temperature on the main and booster heaters is 130°C and the minimum temperature of the cold zone is 0 °C. The unique design allows the sample to be completely visible during all phases of processing. To view and record the sample processing, two cameras are mounted 90° apart on a translation system that is separate from the cold zone and electrode ring translation system.

The PFMI investigation utilizes an innovative approach for heating the sample. A thin, transparent layer of Indium Tin Oxide (ITO) is deposited on the exterior surface of the ampoule. The ITO coating is electrically conductive and acts as a resistance heater, effectively melting the sample, when current is applied. The PFMI is applied mainly to investigate the bubble formation from low-temperature melt materials, such as succinonitrile, a transparent material that solidifies in a manner analogous to metals [16].

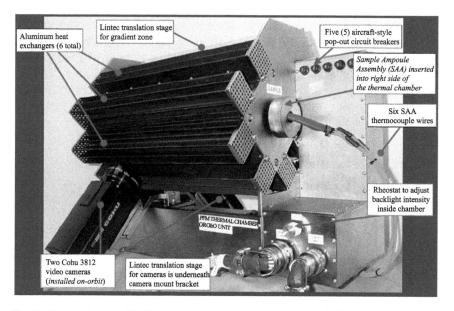

Fig. 7 Photograph of the PFMI ground unit, reprinted from reference [15]. Copyright 2018, with permission from Acta Astronautica

SCOF and EU on ISS. The Solution Crystallization Observation Facility (SCOF) was already set up to be used onboard ISS KIBO [17]. The facility was used for an ice crystal experiment and a phenyl salicylate crystallization experiment. For each experiment, the Experiment Unit (EU) was developed according to specific scientific requirements. The EU was connected to the SCOF, which provided electric power and communication lines to the EU. Cooling water (16–23 °C) was supplied to a cold plate on which the EU was installed.

The SCOF was equipped with an amplitude modulation microscope (AMT), a bright field microscope, a Mach-Zehnder type interference microscope with two wavelength light sources (532 and 780 nm), and two cameras (×2 and ×4). The field views were 2.4 × 3.2 mm and 1.2 ×1.6 mm. It enabled the in situ observation of the crystal growth process by using the Mach-Zehnder interference microscope. The EU that was developed specifically for this experiment was named "NanoStep".

DECLIC on ISS. DECLIC is a joint CNES/NASA research program implemented in the ISS. The facility can be used to study of materials science and the processing operations were performed in an ISS EXPRESS rack from 2006 to 2008 [18].

DECLIC comes in the form of two lockers, one electronic (ELL) and one experimental locker (EXL), accommodated in an EXPRESS rack inside the ISS. The EXL contains all the optics of DECLIC, which is divided in two sets: an optical emitter bench (OEB) and an optical receiver bench (ORB). In the OEB are all the light source, together with their optics and various devices. The OEB and ORB are mounted on a central titanium structure which gives the EXL mechanical and thermal stability.

The experiment is being led inside an insert located in the EXL. This insert contains the sample under study and all the associate scientific instrumentation in a given environment.

Two kinds of light source are available in the OEB: a He–Ne laser at 633 nm, and 6 red led (665 nm) with various beam shapes. Three cameras and one photodiode are available in the ORB. Cells containing the materials are placed together with their specific instrumentation (heaters, actuators, and scientific probes) in inserts that can be easily put in and removed from an experiment locker. This facility allows to study the directional solidification of transparent materials such as succinonitrile, too.

Chinese High-temperature In Situ Observation Instrument (HISOI). A High-temperature In Situ Observation Instrument (HISOI) is dedicated to visualize and record the whole oxide crystal growth with high temperature solution in space. This facility is developed by Jin et al. and served on the 17th Chinese recoverable satellite launched in the year 1996 [19] as well as served on "SZ-2" Chinese man-made spacecraft in the year 2001.

The HISOI is composed of three individual parts: (1) the space high-temperature microscope system, (2) the command and control system and (3) the video recorder system. The heart of it is a space optical microscope, by which both crystal surface and environmental melt flows can be observed. The Photo of HISOI and the experimental unit in it are shown in Fig. 8a. The test section is made by a heater chamber and a loop-like Pt wire heater. As shown in Fig. 8b, the Pt wire (ϕ 0.2 mm) is usually employed to heat and suspend the solution during the in situ observation experiment. A S-type thermocouple (ϕ 0.08 mm) is used to measure the temperature of the loop. The diameter of the loop is about 2 mm. Two V-typed electrodes are used to prevent the loop-like heater from deformation under high temperature. The maximal temperature of HISOI is up to above 1000 °C.

The video recorder system is a differential interference microscope coupled with Schlieren technique to visualize the crystal growth processes from the melt/solution. Figure 9 shows the schematic of the Schlieren optical system including two lens. The fore lens L1 is used to form parallel rays, and a knife edge is placed at the rear focal point of lens L2. Similarly, these parallel lights are also used to pass through the objective of the optical system of differential interference microscope. If a knife edge is installed at the rear focal point of the objective of the microscope, part of the light which has passed through the ununiform region of the object will be refracted and shielded, and the Schlieren effect can be obtained, i.e., the mass flow can be observed. With this method, the growth pattern and the mass transportation phenomenon can be visualized simultaneously. The video from the microscope is recorded and visualized by the video recorder system. Figure 10 shows the convection flow patterns observed by this facility in microgravity with temperature of 923 °C [20].

Chinese Optical Facilities for Materials Research at Low Temperature. To study the growth kinetics and structural transformation of colloidal crystals under micro-gravity condition, an experimental device with three crystallization cells, each with two working positions was designed [21]. It uses direct-space imaging with white light to monitor morphology of the crystals and reciprocal-space laser diffraction

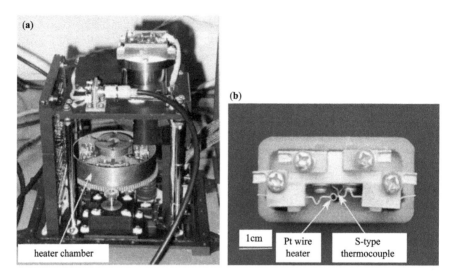

Fig. 8 Photo of Chinese HISOI **a** and the experimental unit in it **b**

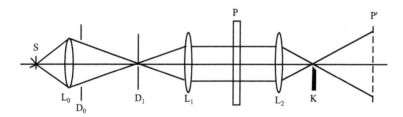

Fig. 9 Schematic figure of the Schlieren optical system for Chinese HISOI. S: light source; L0, L1, L2: lens; D0, D1: diagrams; P: specimen; K: knife edge; P′: image of P

Fig. 10 The convection photograph of the surface tension observed by Chinese HISOI in space

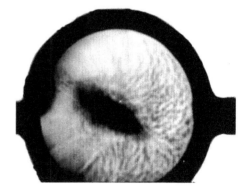

(Kossel lines) to reveal the lattice structure. The device has served for colloidal crystal growth on Chinese "TG-1" manmade spacecraft launched at September 29, 2011. Hundreds of images and diffraction patterns were collected via the on-ground data receiving station. The data showed that single crystalline samples were successfully grown on the orbit.

An optical in situ observation facility has been built to study the mass transfer process of water droplet in EAFP protein solution under microgravity condition provided by the Chinese "SJ-8" satellite [22]. This facility is based on the Mach-Zehnder Interferometer for imaging of liquid at room temperature. It can be applied to investigate the convective flows as well as the crystallization from water solution. So it is a convenient method for both fluid science and materials science.

2.2 In Situ Observation Facility by Other Techniques

Optical in situ observation technique can only applied to the investigation of transparent melt, which has much limitation. Some other diagnose techniques have been developed in microgravity for materials sciences investigation, such as X-ray radiography, infrared ray radiography, scanning probe microscopy, and diagnoses by Seebeck thermoelectric effect, and so forth. They are mostly designed for research of material processing especially for the non-transparent melt. Infrared ray radiography has been used in microgravity condition to obtain the temperature distribution of liquid and has not been applied to investigation of melt solidification. Scanning probe microscopy is used to observe the surface morphology of materials in nano scale. In addition, X-ray radiography and diagnoses by Seebeck effect have been used in space for investigation of alloy processing with high temperature.

In Situ Diagnose Technique by X-ray Penetration Method. X-ray penetration and image technique has been a very popular method on the ground for solidification research of metal or alloy. The solidi-liquid interface as well as the dendritic growth may be observed by this technique. However, in space it is much more difficult to observe the solidifying process of metal by X-ray image. In the frame of the XRMON (In situ X-ray monitoring of advanced metallurgical processes under microgravity and terrestrial conditions) project, which is a Microgravity Application Promotion (MAP) program of the European Space Agency (ESA), a facility dedicated to the study of directional solidification of aluminum-based alloys, with in situ X-ray radiography is developed [23, 24]. This facility, named XRMON-GF(GF for Gradient Furnace) device consists of a gradient furnace system for solidification of alloys and an attached high-resolution X-ray diagnostic system. The metallic samples had a sheet-like geometry with a length of 50 mm, a width of 5 mm and a constant thickness of less than 0.2 mm along the sample. Each sample was first mechanically polished down to the desired thickness (with a surface rugosity of 1 μm), and then spray coated with boron nitride (BN). Then it was sandwiched between two rectangular glass plates, welded together. The field of view

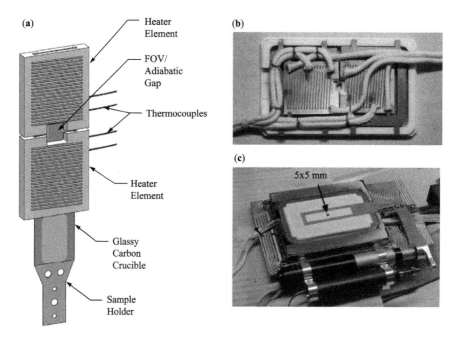

Fig. 11 a Schematic layout of the XRMON-GF furnace, **b** Pictures of the open furnace with the two heaters(top), **c** assembled furnace, with the hole for the X-ray radiation transmission, crucible frame placed on top to show sample placement, reprinted from reference [23]. Copyright 2018, with permission from Journal of Crystal Growth

for the X-ray diagnostics was about 5×5 mm^2 with a spatial resolution of 3–5 μm and a temporal resolution of 2–3 Hz. The X-ray source was based on a microfocus transmission target in molybdenum using polychromatic radiation with a peak at 17.5 keV. Figure 11 is the schematic layout of the XRMON-GF furnace.

This facility was successfully validated during the MASER-12 sounding rocket campaign, in spring 2012. Radiographs were successfully recorded during the entire experiment including the melting and solidification phases of the sample. The heaters are independently regulated by a PID-regulator implemented in software. The temperature of the facility can arrive above 500 °C.

In Situ Diagnose Technique by Seebeck Thermoelectric Effect. The interfacial morphology of alloys can be in situ diagnosed by the Seebeck thermoelectric effect [25]. This named Mephisto furnace is built under the auspices of the French nuclear energy commission CEA and the French space agency CNES. Mephisto furnace is a sophisticated Bridgman furnace with two heating and cooling sub-systems, where three separate samples are solidified in parallel. Sample 1 is dedicated to the measurement of Seebeck electric signal, which is used to obtain most of the experimental data presented here. A measurement of the electrical resistance is performed on sample 2. When the different resistivities of the base material between solid and liquid states are taken into account, this measurement can be translated in an estimation of the solidification velocity. In addition, sample 2 contains thermocouples for the measurements of the temperature gradients within the alloy. Sample 3 is for Peltier pulse marking and post-mortem analysis of the shape of the quenched growth interface. Sample 3 also contains thermocouples for temperature gradient measurements. Such experiments have been done on USMP1 and USMP3 flights.

Such Seebeck and resistance measurement device have also worked on board the ISS in 2004/2005 [26].

Scanning Tunneling Microscopy Technique. Scanning tunneling microscopy allows for atomic resolution imaging of surfaces and the observation of individual adsorbed molecules, as well as in situ studies of chemical reactions on an atomic scale. Tanja et al. built a scanning probe microscopy setup and it was carried out on a parabolic flight campaign in November 2001. This scanning probe microscopy setup is small, lightweight and do not require vacuum or high voltage supply. In addition, samples can be investigated directly without further preparation. By this facility, surface morphologies of graphite sample in nano scale were obtained in microgravity [27].

Facility for Electrical Resistivity Measurements of Melt. Some facilities are designed for the measurement of melt properties, such as diffusion coefficient, electrical resistivity. Here a facility for electrical resistivity measurement is introduced [28]. The experimental apparatus was developed for electrical resistivity measurements of liquid metals and alloys under microgravity due to the launch of rocket. Whole assembly, contained in the small space of 360 mm (length) × 360 mm (width) × 397 mm (height), was composed of the cell box, the electronic circuit box, and the battery box. The cell for electrical resistivity measurements due to the dc four probe method was made of Ti metal because of its strength and no reaction to corrosive liquid alloys at high temperatures. The application of this method was made successfully to the measurement of electrical resistivity for liquid Bi–Ga alloys with critical mixing both under gravity condition and under microgravity condition due to the launch of the rocket.

3 Facilities for Special Material Processing in Microgravity

Besides the high-temperature heating furnaces and the in situ observation facilities, which are the most popular two kinds of facilities for materials processing in microgravity, another kind of material experimental facilities are also typical. Such facilities are usually designed for crystal growth from low temperature solutions. For example, the α-LiIO$_3$ crystals were grown by the method of free evaporation of water solutions and evaporation with controlled condensation at a constant temperature. Experiments of α-LiIO$_3$ crystal growth were performed several times using the Chinese Scientific Recoverable-Satellites [29]. The zeolite crystal growth furnace unit is an integrated payload designed for low temperature crystal growth in solution on ISS [30]. These experiments are carried out mainly in solutions with temperature lower than 300 °C.

In addition, there are several other facilities that are developed for some special investigation of material experiments. Here are several examples in the following.

3.1 Containerless Processing Facility for Materials

To support several materials science experiments in microgravity, it is sometimes necessary to counter g-jitters and external forces imparted to bodies under study and to position accurately a sample in space. This requirement triggered the development of positioners or levitators that make use of different forces (e.g., electrostatic, aerodynamic, electromagnetic, acoustic).

Applying electromagnetic levitation under micro-gravity conditions the undercooled regime of electrically conductive materials becomes accessible which allows unique investigations of nucleation phenomena as well as the measurement of a range of thermo-physical properties both above the melting temperature and in the undercooled regime. Hence, based on a long and successful evolution of electromagnetic levitation facilities for microgravity applications (parabolic flights, sounding rocket missions and spacelab missions) the Electro Magnetic Levitator (EML) is developed by Astrium Space Transportation under contracts to ESA and DLR. EML is accommodated in the European Drawer Rack (EDR) within the European Module COLUMBUS on ISS. The design of the payload allows flexible experiment scenarios under ultra-high vacuum or ultra clean noble gas atmosphere individually targeted towards specific experimental needs and samples including live video control of the running experiments and automatic or interactive process control. EML can perform processing of samples with 5–8 mm diameter with temperature up to 2000 °C. The experiments by EML can be done under inert gas or ultrahigh vacuum possible.

However, the latest Electrostatic levitation furnace (ELF) for materials containerless processing is developed by JAXA, which is serving on ISS today. Before this ELF, a similar furnace was prepared hand worked under reduced gravity on-board a sounding rocket in 2000 by JAXA. The ELF on ISS by JAXA [31] consisted of a stainless steel chamber which was surrounded by diagnostics instruments (camera, pyrometers, etc.) and heating lasers. This furnace can handle evaporative materials over 3500 K. In JAXA's studies, samples with diameters ranging from 1 to 3 mm can be levitated. The ELF facility has launched onboard ISS in 2015 and is installed into the Work Volume (WV) of the Multi-purpose Small Payload Rack (MSPR) in the Japanese Experiment Module (JEM). And some experiments have been carried out [32]. Due to the limitation of space and electrical power, vacuum pumps necessary to get high vacuum environment have been omitted. This is one of the reasons why ISS-ELF will mainly process oxide sample.

3.2   Plasma Research Facility Under Microgravity

Complex plasma research under microgravity condition is one of the present key research topics in fundamental physics and materials science on ISS.

In 2001 the so-called PKE-Nefedov facility performed its first experiments in radiofrequency induced complex plasmas on ISS [33]. This bilateral German–Russian research facility operated successfully in over 13 missions nearly five years until its internal resources were consumed. The next generation experiment apparatus PK-3 Plus with refined and more powerful instruments and diagnostics had its on-orbit commissioning in January 2006. It is performing flawlessly since then and has already been successfully employed during various ISS increments including Russian cosmonauts and ESA astronaut Thomas Reiter.

Here PK-3 Plus is introduced (Fig. 12): the RF plasma chamber (1) is thermally improved to have a more evenly distributed temperature background in order to reduce thermo-phoretic effects; the RF generator (2) and the control possibilities for physical parameters are extended; the gas supply (3) was increased to two gases types: neon and argon; the number of cameras (4) was increased; three of the cameras (CCD, analog output) can be moved along a 2-axis translation table (5) allowing a much larger 3D analysis of the plasma; the particle injectors (6) are now mounted outside the electrode and their number was increased from 2 to 6, therefore allowing a larger variety of particle sizes; a small turbo molecular pump (7) is now integrated inside the experiment container.

Fig. 12 PK-3 Plus experiment block integrated in experiment container, reprinted from reference [33]. Copyright 2018, with permission from Acta Astronautica

3.3   In-space Soldering Investigation (ISSI)

In-space Soldering Investigation (ISSI) was conducted in the microgravity environment aboard the ISS. The ISSI took place over four increments aboard the ISS in the maintenance work area (MWA). The investigation utilized commercially available soldering tools and materials to conduct a series of experiments and was conducted as "Saturday Science." The intent of the experiments was to look at joining techniques, shape equilibrium, wetting phenomena, and microstructural development in a microgravity environment. The desired zero up mass requirement a quick search of equipment and tools aboard the ISS found that a well-equipped soldering and pin kit was available. The components are all off-the-shelf items which facilitated ground testing. The ISS soldering iron was adapted for battery usage, reaching a maximum temperature of 600 °F [15].

3.4   Facilities for Thin-film Material Fabrication in Space

Materials development can also occur in space based on the presence of vacuum. The utilization of the vacuum of space for thin-film materials development has been pioneered by the Wake Shield Facility (WSF) [34]. The WSF is a 3.7 m diameter disc-shaped free-flying platform designed to generate an ultra-vacuum in low earth orbit space, and to utilize that ultra-vacuum for the fabrication of thin-film materials by epitaxial growth. The WSF was designed, built, deployed and operated by the Space Vacuum Epitaxy Center and its industry and government. The WSF is transported to orbit by the Shuttle in its payload bay mounted on a specifically designed cross-bay carrier. The carrier not only restrains the WSF in the payload bay, but also protects the central epitaxial growth region of the wake side from pre-launch payload processing environments, and payload bay launch environments through the incorporation of an integral vacuum chamber.

   The thin-film growth apparatus on the WSF includes eight source cells (small furnaces) for the evaporation of elemental atoms for MBE thin-film growth, and two gas nozzles for the effusion of organometallic gaseous species for CBE thin-film growth. The WSF has flown several missions, STS-60 (February 3, 1994), STS-69 (October 11, 1995), and STS-80 (November 4, 1996). In the three flights of WSF, high-quality GaAs-based epitaxial thin-films were grown.

   In references [35, 36], one experimental reactor cell designed and developed for the growth of films and crystals by closed cell physical vapor transport on the Space Shuttle Orbiter is discussed. The cell design has been proven by over 60 experiments, including 18 on two space flights, to be very satisfactory from the standpoints of minimal power use at operating temperatures of 400 °C, mechanical vibration resistance and approximation to ideal steady-state growth conditions.

4 Summary

Experimental facilities are very important for the research of materials science in space. Up to now several kinds of experimental facilities have been prepared, including many high temperature heating furnaces, in situ observation facilities and some specially designed facilities for materials science research.

Heating furnace is one presentative kind of facility for materials science research in space since many material experiments require high temperature environment. Almost all the countries or groups in the world aiming at microgravity materials science study have built their own heating furnaces serving in space, such as Russia, America, Japan, ESA, and China. Most of these furnaces work through heat generation from resistance when applied electricity due to its comparatively high stability. Generally, such heater furnaces are designed as Bridgman styles which allow the investigation of single crystal growth or directional solidification of metals possible for samples with large size. The maximum of working temperature is the most important parameter characterizing the ability of heating furnace. Most of the recent heating furnaces can heat up to more than 1000 °C. The largest value can even arrive at 1600 °C. The sample diameter usually is larger than 15 mm.

As for the in situ observation facilities in space for materials science research, optical imaging method by visible light is the most popular technique owing to its convenience. Several facilities for process visualizing by other techniques are also developed and used in space, such as X-ray penetration method, and diagnoses method by Seebeck thermoelectric effect. Most of the in situ observation facilities can only be used at low temperature, applying to the investigation of solidification or crystal growth from water solution. Only a little can work at temperature up to several hundred centigrade. So newly kind of in situ observation facilities for more higher temperature should be designed in the future.

The following features can be obtained for the materials science processing experiment furnace in microgravity developed recently: highly integrated synthesis experiment ability such as preparing materials, measuring, analyzing, with help of external force such as magnetic field; high precision for temperature control and sample movement control; ability of arriving higher temperature; stability and long lifetime; modularization, standardization and assembled, which allows the facility to be exchanged more conveniently.

References

1. Barta C, Triska A, Trnka J et al (1984) Experimental facility for materials research in space CSK-1. Adv Space Res 4:95–98
2. Barmin I, Bryukhanov N, Egorov A et al (2002) Using the progress transport spacecraft in structure of the international space station for realization of scientific experiments under microgravity conditions. Acta Astronaut 51:255–259

3. Stenzel Ch, Dold P, Benz KW (1996) Magnetic damping array implemented into a space-compatible multizone furnace for semiconductor crystal growth. Rev Sci Instrum 67:1985–1988
4. Bernard Z, Rainer K, Judith R (2003) Cooperative research in microgravity sciences during the French manned MIR missions. Acta Astronaut 53:963–970
5. Ostrogorsky AG, Marin C, Churilov A et al (2008) Reproducible Te-doped InSb experiments in microgravity science glovebox at the International Space Station. J Cryst Growth 310:364–371
6. Cobb SD, SzofranF R, Schaefer DA (1999) Preliminary concepts for the materials science research facility on the International Space Station. AIP Conf Proc 458:459–464
7. Carswell W, Kroeger F, Hammod M (2003) QMI: a furnace for metals and alloys processing on the International Space Station. IEEE Aerosp Conf Proc 1:65–74
8. Myscha RC, William EC, Jeff F et al (2000) Quench Module Insert (QMI) and the Diffusion Module Insert (DMI). AIP Conf Proc 504:493–498
9. Yoda S, Shibukawa K, Murakami K et al (1992) The development of isothermal furnace (IF) and gradient heating furnace (GHF). In: International symposium of space technology and science, 18th, Kagoshima, Japan, 17–22 May 1992, vol 1 & 2, pp 2211–2216
10. Kaoruho S, Midori M, Govindasamy R et al (2014) Thermal properties of molten InSb, GaSb, and $In_xGa_{1-x}Sb$ alloy semiconductor materials in preparation for crystal growth experiments on the international space station. Adv Space Res 53:689–695
11. Kinoshita K, Arai Y, Inatomi Y et al (2014) Growth of a $Si_{0.50}Ge_{0.50}$ crystal by the traveling liquidus-zone (TLZ) method in microgravity. J Cryst Growth 388:12–16
12. Liu Y, Ai F, Feng CD et al (2006) A new kind of multi-task materials processing facility for space applications. Space Technol 26:87–90
13. Wang JC, Liu Y, Ai F et al (2002) The DGW-I furnace a materials processing facility in "SZ-2" spacecraft. Chin J Space Sci 22(supplement):154–158
14. Zhang XW, Yin ZG, Pan XH (2016) Crystal growth form melts: materials science program in the SJ10-recoverable scientific experiment satellite. Physics 45:213–218
15. Richard NG, Paul L, Guy S et al (2008) Materials research conducted aboard the International Space Station: facilities overview, operational procedures, and experimental outcomes. Acta Astronaut 62:491–498
16. Richard NG, Lucien NB, Amrutur VA (2012) Disruption of an aligned dendritic network by bubbles during re-melting in a microgravity environment. Microgravity Sci Technol 24:93–101
17. Izumi Y, Katsuo T, Tomoya Y et al (2013) Growth rate measurements of lysozyme crystals under microgravity conditions by laser interferometry. Rev Sci Instrum 84:103707 (8 p)
18. Laubier D, Martin B, Durieux A (2004) The optical diagnostics of DECLIC. In: Proceedings of the 5th international conference on space optics (ICSO 2004), 30 March–2 April 2004, Toulouse, France, pp 441–446
19. Jin WQ, Chen JY, Li WS et al (1993) Development of optical system for high temperature in situ observation of oxides crystal growth. Ferroelectrics 142:13–18
20. Jin WQ, Pan ZL, Cai LX et al (1999) The studies of $KNbO_3$ cellular growth in high-temperature solution at microgravity. J Cryst Growth 206:81–87
21. Hu SX, Li XL, Sun ZB et al (2014) Study on structure transformation of colloidal crystal growth in space. Manned Spacefl 20:261–266
22. Duan L, Kang Q, Sun ZW et al (2008) The real-time Mach-Zehnder interferometer used in space experiment. Microgravity Sci Technol 20:91–98
23. Nguyen-Thi H, Reinhart G, Salloum Abou Jaoude G et al (2013) XRMON-GF: a novel facility for solidification of metallic alloys with in situ and time-resolved X-ray radiographic characterization in microgravity conditions. J Cryst Growth 374:23–30
24. Nguyen-Thi H, Bogno A, Reinhart G et al (2011) Investigation of gravity effects on solidification of binary alloys with in situ X-ray radiography on earth and in microgravity environment. J Phys: Conf Ser 327:012012 (11 p)
25. Garandet JP, Boutet G, Lehmann P et al (2005) Morphological stability of a solid–liquid interface and cellular growth: Insights from thermoelectric measurements in microgravity experiments. J Cryst Growth 279:195–205

26. Pissard JP, Le GG, Salvi C et al (2002) See beck and resistance diagnostics in the ESA MSL facility for the ISS. Acra Astronautica 51:1–9
27. Tanja D, Michael R, Wolfgang MH et al (2004) Scanning probe microscopy experiments in microgravity. Appl Surf Sci 238:3–8
28. Masaki T, Itami T, Kuribayashi K et al (1996) The experimental apparatus for electrical resistivity measurements of liquid metals and alloys under microgravity due to the launch of a rocket. Rev Sci Instrum 67:2325–2331
29. Chen WC, Li CR, Liu DD (2003) Post-research on α-LiIO$_3$ crystal growth in space. J Cryst Growth 254:169–175
30. Nurcan B, Joseph H, Robert A M et al (2000) A low temperature furnace for solution crystal growth on the International Space Station. In: Genk MS (ed) Space technology and applications international forum, pp 499–504
31. Paul-Francois P, Takehiko I, Shinichi Y (2008) Experiments in materials science on the ground and in reduced gravity using electrostatic levitators. Adv Space Res 41:2118–2125
32. Takehiko I, Junpei T O, Yuki W et al (2015) Thermophysical property measurements of oxide melts at high temperature by electrostatic levitation furnace on ISS. Int J Microgravity Sci Appl 32: 320410 (4 p)
33. Hofmann P, Seurig R, Stettner A et al (2008) Complex plasma research on ISS: PK-3 Plus, PK-4 and impact/plasma lab. Acta Astronaut 63:53–60
34. Alex I (2001) Advanced thin-film materials processing in the ultra-vacuum of space. Acta Astronaut 48:115–120
35. Debe MK, Poirier RJ, Schroder FS et al (1990) Design and performance of a vapor transport cell for operation onboard the Space Shuttle Orbiter. Rev Sci Instrum 61:865–870
36. Debe MK (1986) Industrial materials processing experiments on board the Space Shuttle Orbiter. J Vac Sci Technol A4:273–280

Melt Growth of Semiconductor Crystals Under Microgravity

Zhigang Yin, Xingwang Zhang, Wei Wang, Xiaoya Li and Jianding Yu

Abstract The low-gravity environment aboard the space offers an advanced platform to prepare materials with improved properties as compared with their terrestrial counterparts, and allows an in-depth understanding of crystal-growth-related phenomena that are masked by gravity on the earth. The main achievements in the melt growth of semiconductors under microgravity are listed below: (i) on the way towards high-quality semiconductors with chemical homogeneities on both the macroscopic and microscopic scales, the dependence of solute transport on the buoyancy-driven convection and Marangoni convection was better understood; (ii) the underlying mechanism of detached Bridgman growth was clarified based on the microgravity experiments, which in turn guided the ground-based crystal growth processes; (iii) new crystal growth techniques including the traveling liquidus-zone method and the vertical gradient freezing method were proposed and developed, and chemically homogeneous semiconductor alloys such as $Ge_{1-x}Si_x$ and $In_xGa_{1-x}Sb$ were successfully grown using these methods under microgravity. In this part, the main progresses in these areas are reviewed and summarized.

Keywords Microgravity growth · Semiconductor · Detached growth ·
Buoyancy-driven convection

Z. Yin (✉) · X. Zhang (✉)
Key Lab of Semiconductor Materials Science, Institute of Semiconductors,
Chinese Academy of Sciences, Beijing 100083, China

College of Materials Science and Opto-Electronic Technology,
University of Chinese Academy of Sciences, Beijing 100049, China

Z. Yin
e-mail: yzhg@semi.ac.cn

X. Zhang
e-mail: xwzhang@semi.ac.cn

W. Wang · X. Li · J. Yu
State Key Lab of High Performance Ceramics and Superfine Microstructure,
Shanghai Institute of Ceramics, Chinese Academy of Sciences, Shanghai 200050, China

© Science Press and Springer Nature Singapore Pte Ltd. 2019
W. R. Hu and Q. Kang (eds.), *Physical Science Under Microgravity: Experiments
on Board the SJ-10 Recoverable Satellite,* Research for Development,
https://doi.org/10.1007/978-981-13-1340-0_13

1 Introduction

The progress in space technology provides new opportunities for materials research. The first crystal growth experiment performed during the Apollo emissions in the late 1960s announced the birth of space materials science [1]. The unique microgravity environment in space offers a new degree of freedom for materials processing, allowing in-depth insights into the fundamental issues in crystal growth. Investigations in this field mainly aim at the following three goals:

(i)   Manufacturing perfect crystals with properties that cannot be obtained on the earth by eliminating the influence of gravity. An exemplified example is that the space-grown GaAs semi-insulating crystals are featured with improved stoichiometry and lower defect concentration, and the field effect transistors fabricated based on these crystals have performances much better than the terrestrial samples [2, 3].
(ii) Understanding fundamental phenomena that are masked by gravity and therefore are difficult to study quantitatively on the earth. For instance, fluids are affected by buoyancy-driven convection, and microgravity helps isolating such effect from the targeted systems [4].
(iii) Improving the ground-based crystal growth. An excellent example is the understanding of the underlying mechanism of detached directional solidification and its guidance to the terrestrial preparations. The detached growth on the earth greatly reduces the dislocation densities of the crystals and provides new opportunities for their device applications [5].

Space materials science covers the whole materials family, and the preparations of metals, semiconductors, and biological materials, are all intensively studied under microgravity. In this review, we only limit our scope in semiconductors. Different growth techniques, including melt-related methods (floating zone method and Bridgman solidification), solution-based routes (liquid epitaxy and traveling heater method with solution starting materials), and chemical or physical vapor deposition were employed in experiments aboard spacecraft. Melt growth is the most widely exploited technique to produce high-quality, single-crystalline semiconductor chips for commercial applications. Among the melt growth methods, Bridgman directional solidification is the main-stream technique for microgravity semiconductor crystal growth. The main advantage of Bridgman growth as compared with other crystal growth techniques is its simplicity. Typically, the ampoule structure used in Bridgman method is easily modified to fix the charges along both the axial and radial directions. Such a characteristic is highly attractive for the microgravity experiments, taking into account the harsh mechanical environment during the spacecraft emission. The Bridgman method involves heating a polycrystalline material in a crucible above its melting point and then slowly cooling it from one end where the seed crystal locates. During this process, single crystal is progressively formed along the axial direction of the initial ingot. Bridgman solidification can be performed under horizontal or vertical geometries, and the latter is dominant in the reported experiments.

We mainly focus on the progresses of the melt growth of semiconductor crystals under microgravity, and special attention is paid to the Bridgman solidification and related issues. The review is divided into four parts: in part I, the studies towards chemically homogeneous semiconductors both macroscopically and microscopically are reviewed, and the difficulties and recent progresses are introduced; part II introduces the new understandings on detached solidification, a phenomenon found 40 years ago, and its guidance to the terrestrial materials research are discussed; in part III we introduce the emerging crystal growth concepts, including the traveling-liquidus zone method and the vertical gradient freezing method, and the efforts towards high-quality binary or ternary semiconductor alloys by using these methods are also discussed; part IV is a summary of this review.

2 Towards Diffusion-Controlled Growth

2.1 Segregation and Solute Transport Under Microgravity

According to the classical Tiller's theory, steady-state growth can be achieved on condition that the convective mixing is negligible and the solute transport is dominated by diffusion [6]. The initial transient that follows the Tiller's theory can be expressed as [7]

$$C_s(x)/C_0 = k + (1-k)\left[1 - \exp\left(-kRx/D\right)\right] \tag{1}$$

where $C_s(x)$ is the dopant concentration in the solid at a distance x from beginning of the sample, C_0 is the uniform concentration of the initial melt, k is the equilibrium distribution coefficient, D is the diffusion coefficient of the solute in the melt, and R is the growth rate. Steady growth can occur only if the growth rate R is fixed and the impurity concentration adjacent to the interface remains constant. These conditions are not satisfied during the ground-based normal freezing because of the buoyancy-driven convection, while space provides an ideal platform to achieve the steady-state crystal growth. Without the perturbation by buoyancy-induced convection, diffusive transport can be obtained in the melt and therefore, after a short initial time, the crystals should have a uniform chemical composition.

In an early work performed by Witt et al., a steady-state growth was really observed in the Te-doped InSb crystals grown by Bridgman method on Skylab spacecraft [8]. A uniform axial distribution of Te was obtained across the entire grown crystal, as evidenced by both Hall-effect measurements and ion-microprobe scanning. Thermoelectric semiconductor crystal of $Bi_2Te_{2.79}Se_{0.21}$ was grown by Zhou et al. on the Russian Foton-M3 spacecraft, and EPMA analysis show that the axial composition is more homogeneous than that of the ground-based crystal [9]. Diffusion-controlled growth was also found in Ge crystals grown by Bridgman technique and homogeneous composition distribution along the axial direction was

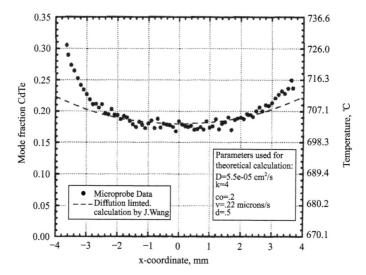

Fig. 1 Radial composition profile of HgCdTe crystal; the dashed line indicates the calculated results according to the diffusion-controlled scheme. Reprinted with permission from reference [12]

achieved [10]. However, notable radial segregation was observed in the sample. Bly et al. also found that the axial Se distribution for the space-grown Se-doped GaAs is diffusion-limited, whereas an evident radial segregation is present in the crystal [11]. Similarly, the axial compositional profile of space-based HgCdTe crystal agrees well with the diffusion-controlled scenario while the radial distribution not (see Fig. 1) [12]. Although uniform axial composition distribution can be achieved, researchers always fail in producing diffusion-limited radial composition profiles.

All these studies inform that the crystal growth performed in space cannot be simply treated as a process in which convection is absent. Although the gravity level in spacecraft is significantly lower than that on the earth, it is not zero. To explain the observed experimental results, residual accelerations should be considered. Residual accelerations on spacecraft are mainly composed of two categories—quasi-static and vibrational accelerations [13]. The magnitude and direction of the quasi-static accelerations are determined by the orbit altitude and the orientation of spacecraft axes with respect to the earth. In most cases, the residual quasi-static accelerations in the spacecraft are on the order of 10^{-4} to 10^{-6} g_0 (g_0: the gravitational acceleration on the earth) with frequencies below 0.01 Hz. The vibrational accelerations (also called g-jitter) are mainly originated from the elastic vibrations of the spacecraft, thruster firings for orbital correction, daily crew activities, and active machinery and so on. In general, the vibrational component frequencies range from 0.01 to 300 Hz and the amplitude can be as high as 10^{-2} g_0 (e.g., during orbital adjustment). In order to fabricate homogeneous semiconductor crystals, particular attention should be paid on the residual accelerations. For example, Alexander et al. reported that the vibrational accelerations recorded by spacecraft accelerometers have a strong influence on the solidifying system, and a close correlation between the vibrational accelerations and the disturbances in the transport conditions were clearly resolved [14].

We now discuss the different segregation behaviors along the axial and radial directions for crystals grown under microgravity. The convective flows induced by residual gravity mix the melt and cause fluctuations in the axial and radial segregation of solute. As discussed by Brown et al., the degree of melt mixing can be categorized into three cases: no mixing, poor mixing and strong mixing [13]. For the axial case, the effective segregation coefficient (k_{eff}) is nearly unity, suggesting a negligible segregation at the no mixing condition; k_{eff} gradually deviates from unity with the enhancing flow intensity, and eventually approaches the equilibrium distribution coefficient (k) as a result the full mixing of the diffusion layer with the melt (strong mixing). Similarly, a nearly uniform radial composition distribution is predicted in both the no mixing and strong mixing conditions. A uniqueness of the radial segregation is that it does not vary monotonically with the flow intensity. As nontrivial cellular convection appears in the melt, it interferes with the diffusion layer adjacent to the growth interface and causes a drastic variation of the radial composition [13, 15]. This corresponds to the poor mixing condition. For many of the space crystal growth experiments, the residual accelerations induce non-negligible convective flows which produce a poor mixing condition. These discussions explain why the space growth experiments usually yield an improvement of axial segregation and a deterioration of radial composition distribution.

Intensive numerical modeling works focused on the microgravity growth of semiconductors from melts were performed by various groups, in order to achieve better understandings of the experimental results and offer guides for the modified experimental designs. Griffin et al. investigated the influence of temporal variations in the gravity magnitude on the natural convection during the Bridgman solidification of semiconductors [16]. The results show that for low Rayleigh numbers the response time to step changes in g is controlled by the momentum diffusive time scale, while for higher Rayleigh numbers the response time is reduced due to the inertial effects. Alexander and coworkers investigated the effect of quasi-static and vibrational accelerations on the dopant distributions, and found that the composition non-uniformity is strongly dependent on the orientation of quasi-static acceleration. Moreover, they pointed that the influence of vibrational acceleration can last for thousands of seconds after the termination of the pulse [17]. Naumann developed an analytical model for the two-dimensional flow field to examine the effects of small residual accelerations [18]. He found that the quasi-static and vibrational accelerations on the order of $10^{-4}g_0$ can induce severe solute redistribution, and orientating the furnace axis along the residual gravity vector can ameliorate this effect.

Derby and coworkers developed a three-dimensional (3D) method to model the vertical Bridgman growth which incorporates melt convection, heat transfer, segregation, and solidification related phenomena in the crucible [19]. By implementing the deforming-grid, front-tracking method, they built a framework in which the 3D shape and movement of the crystal-melt interface is represented. They developed a Galerkin mixed-order formulation which employs a higher-order, quadratic representation of the velocity field with a linear, discontinuous basis to meet the challenges to produce robust, accurate solutions for incompressible 3D flows. The boundary conditions can be supplied in a self-consistent way by coupling another furnace-scale model which deals with heat transfer by radiation and conduction. These methods have been successfully used to compute the growth of CdZnTe [20, 21].

Most of the experimental works performed the Bridgman technique is based on diluted systems. The Duffar group focused on the growth of concentrated semiconductors and used GaInSb as a model system [22, 23]. The ground-based experiments show that the considerable chemical segregation which is coupled with thermosolutal convection results in huge deformations of the crystal-melt interface, and the resultant interface curvature can exceed the crystal diameter [23]. A semi-empirical analytical expression was developed to account for the evolution of the interface deflection during the growth, and the results show that low pulling rate and low thermal deflection favors for achieving flat growth interface [24]. The axial indium component distribution for the space-grown crystal exhibits a Scheil-like solute profile apart from the beginning of the solidification. To understand this puzzling result, theoretical simulations were performed by Stelian et al. and based on the results they proposed an experimental scheme to improve the chemical homogeneity [25]. Moreover, the influence of g-jitter on the chemical segregation was also simulated by varying its amplitude and frequency [26]. The results show that low-frequency ($<10^{-2}$ Hz) residual gravity have a noticeable influence whereas the crystal growth can tolerate the high-frequency g-jitters.

2.2 Influence of Marangoni Convection

Another source that contributes to the considerable segregation under microgravity is Marangoni convection, which is usually masked on ground-based crystal growth due to the strong buoyancy-driven flow. According to its surface-related nature, it is difficult to study the Marangoni convection and related phenomena in a quantitative way. Nakamura et al. developed an in situ visualization system to investigate the Marangoni convection in a molten silicon bridge, and comparative studies on ground- and space-based experiments were performed [27]. Such a method is also applicable to other semiconductor molten systems. Two types of temperature oscillation were observed in the microgravity experiment in the molten silicon, in which one is featured with a frequency of about 0.1 Hz and the other has no remarkable frequency. The former oscillation has an antiphase correlation in thermocouples with 90° separation in azimuthal angle, while the latter is characterized by an antiphase correlation with 180° azimuthal angle separation. The Marangoni number (Ma) was extracted by carefully measuring the temperature fluctuations, and a value of 1900 was determined for the microgravity condition [28]. In a recent study, the critical temperature difference and the critical Marangoni number (Ma_{c2}) which are needed for the occurrence of oscillate flow are determined in a rather wide range of aspect ratio [29].

Due to its connection with the occurrence of free surface, Marangoni convection is prevalent in the floating zone growth experiments [13]. For the Bridgman growth case, Marangoni convection can also play an important role in the heat and solute transportation, when free surface appears. For example, considerable chemical segregation of Te in Te-doped GaSb has been observed along the axial direction for the space-grown samples, demonstrating a strong mixing of the melt during the whole growth process [30]. The influence of residual gravity during the experiments is effectively controlled and it cannot account for the observed macroscopic segregation. This conclusion is corroborated by numerical studies in which the extracted velocity field is in good agreement with the strong disturbance of solutal boundary layer close to the solid-liquid interface [30]. Mathematical simulations confirm that as the buoyancy-induced convection is reduced, the intensity of convective flow is completely determined by the Marangoni convection [13]. Under microgravity, the velocity of convective flow near the solid-liquid interface can reach a value as high as 0.5 cm/s even at a small radial temperature gradient of 1.2 K/cm [13]. Such a strong convection flow is comparable with the conventional level of the pure-buoyancy-dominated case on ground. As a result, it is hard to achieve chemical homogeneity under microgravity when free surface is present.

Marangoni convection not only results in macrosegregation, but also leads to microsegregation—microscopic variations of the dopant concentration (dopant striations). It is known that the Rayleigh number, which is a measure of the driving force for buoyancy convection is reduced by a factor of 10^4 to 10^6 under microgravity, depending on the residual gravity level [31]. Moreover, it was revealed that, the magnitude of vibrational accelerations only slightly influences the convective flows near the growth interface in the absence of Marangoni convection. A close examination of the ground-based Ge:Ga growth shows that on a closed melt surface, the vibrational effects do not yield noticeable striations even at amplitudes of $2 \times 10^{-1} \, g_0$ [13]. An extrapolation to the microgravity condition yields an upper limit (at which striations are absent) of vibrational acceleration of $\sim 2 \times 10^{-3} g_0$ [13]. Therefore, the residual gravity and the resultant buoyancy-driven convection do no play a role in the striation formation as important as Marangoni convection. Obviously, the reported striations by various groups for the space-based crystals are mainly caused by unsteady (or oscillatory) Marangoni convections. The unsteady Marangoni convection is characterized by the temperature variations in the melt near the growth interface, which gives rise to the variations in the growth rate and the resultant local inhomogeneity. For example, Schweizer and coworkers found that under temperature fluctuations of 0.5–0.7 °C with a frequency range <0.5 Hz, the microscopic growth rate fluctuates significantly around the average growth rate of 1 mm/min [32]. Growth rates up to 4, ~0 mm/min, and negative values were all observed in the experiments. Dopant striations are clearly visible from the obtained crystals, as shown in Fig. 2.

It is the low Prandtl number (Pr, defined as the ratio between the fluid kinematic viscosity and the thermal diffusivity) of semiconductor melt that results in the easy formation of striations during the semiconductor crystal growth [33, 34]. As revealed by 3D numerical simulations, Ma_{c2}, which determines the onset of time-dependent convection is strongly dependent on Pr [34]. For low-Pr melts, the

Fig. 2 Micrograph of the etched μg crystal of Si, in which dopant striations induced by Marangoni convection are clearly visible. Reprinted with permission from reference [32]

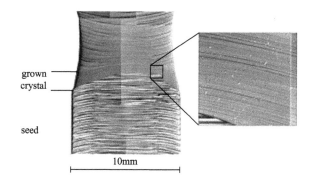

grown
crystal

seed

10mm

steady-to-unsteady transition is mainly caused by the inertial effect [35]. Microgravity provides an ideal platform to accurately determine the parameter of Ma_{c2} for various semiconductor melts. For example, Cröll et al. have carefully investigated the critical Marangoni number of GaSb and obtained a Ma_{c2} of 375 ± 125 from the striation to non-striation transition [37]. Such a value is in good agreement with 3D numerical simulations which yields $Ma_{c2} = 355 \pm 90$. Their experiments were performed during the German-Canadian Spacelab-4 mission with a GaSb ingot diameter up to 15 mm. For Si and GaAs melts, the Ma_{c2} values are 150 ± 50 and ~400, respectively [36]. Based on these data obtained by space experiments, the values of Ma_{c2} for semiconductor melts can be expressed empirically as

$$Ma_{c2} = 2.2 \times 10^4 Pr^{1.32} \qquad (2)$$

Taking into account the difficulties in measuring Ma_{c2} for all the semiconductor melts, this emprical formula extracted by data fitting is very useful for predicting the onset of steady-unsteady transition of a given melt.

The driving force for Marangoni convection is proportional to the local thermal gradient or component gradient at the free surface. Therefore, reducing these gradients at the free surface seems to be a powerful way to eliminate or ameliorate the influence of microsegregation. However, for a true crystal growth process, there leaves no enough room to tune these parameters. For instance, a smaller thermal gradient results in a more deformed interface shape, although it can considerably reduce the convection level. On the other hand, according to classical theory, the steady diffusion-controlled growth yields a non-zero solute gradient near the growth interface. Here we note that the occurrence of free surface on microgravity condition is not necessarily associated with the formation of dopant striations, even when the Marangoni number exceeds the critical value of Ma_{c2}. GaSb crystal growth was conducted on the scientific platform of Chinese recoverable satellite and a notable diameter shrinkage of the crystal was observed [37]. The longitudinal cross-section characterizations of the crystal do not resolve any growth striations, indicating no unsteady Marangoni convection was involved in the growth. The GaSb growth free of Marangoni convection is, in a rather possibility, associated with the formation of a thin oxide layer surrounding the crystal [37]. Another work also highlights the role of thin oxide layer in suppressing the Marangoni convection [38]. These observations provide a possible means to suppress the occurrence of impurity striation and improve the quality of the grown crystals.

2.3 Damping the Convections Under Microgravity

Both the experimental observations and numerical simulations indicate that the growth of semiconductors under the microgravity environment is still influenced by the buoyancy-induced convection. It has long been anticipated that the application of a static magnetic field during the growth can dampen the convective flow in the melt. This idea works in principle for the semiconductor systems since most of the semiconductors are good conductors in their melt states. However, to suppress the convection to the diffusion-controlled level on the earth, the applied magnetic field should be as high as a few Tesla or even larger [39–41]. Such a large magnetic field results in a rather high energy consumption and is not economically acceptable. Furthermore, a strong magnetic field is commonly associated with non-negligible, additional effects like thermoelectric convection [41, 42]. Combining the microgravity with the magnetic field seems to be a more effective way to dampen the buoyancy-driven convection to the desired level and produce the pure diffusion-controlled growth.

Many groups have explored the possibility of applying magnetic field during the semiconductor crystal growth. Fripp et al. performed numerical simulations for the microgravity growth of PbSnTe using the vertical Bridgman configuration, in which an axial magnetic field is applied to interfere with radial convection component [43]. Their results show that the applied magnetic field is very effective in controlling the convective flow, and a moderate axial magnetic field on the order of kilo-Gauss can dampen the buoyancy-induced convection. The computed effective segregation coefficient approaches to 1 when the magnetic field exceeds 3 kilo-Gauss, suggesting the achievement of the near-diffusion-controlled growth. Cröll et al. reported that a complete suppression of dopant striations caused by time-dependent Marangoni convection has been achieved with fields of several Tesla [44]. Baumgartl et al. and Ma et al. investigated the effects of static magnetic field in suppressing the g-jitter and acceleration sparks induced convections under microgravity by numerical modellings [45, 46].

Herrmann et al. have employed static magnetic field to suppress the Marangoni convection in the GaAs growth by floating zone method, and striations are still clearly visible in the crystals due to the limited magnetic field strength [46]. It revealed that the applying of a rotating magnetic field may help addressing this issue [47]. Unlike static magnetic field, rotating magnetic field induces an electromagnetically driven flow to counteract the existing one in the melt. The merit of rotating magnetic field is that the field strength can be considerably decreased as compared with the static field case, and therefore it is more technically and financially acceptable. Ge:Ga crystals were grown via the Bridgman method under a rotating magnetic field on Foton-M3, and a considerably weakened striation pattern as compared to the crystal without applying the magnetic field [48]. Feonychev et al. pointed out that for the floating zone method, the use of rotating magnetic field can result in a reduction of azimuth velocity which is responsible for the striations in the crystals under a proper frequency and field intensity [49]. Dold and coworkers found that by applying a rotating magnetic field to the Bridgman growth of Ga-doped Ge, the impurity striations are decreased in intensity and increased in frequency [50]. However, Feonychev et al. showed that, for the Bridgman method, the rotating magnetic field cannot reduce the segregation or even worse, cause side effects during the crystal growth [49].

On the other hand, Lan et al. performed three-dimensional simulations to investigate the possibility of using ampoule rotation to control segregation and suppress the g-jitter effects during the Bridgman growth [51]. They have shown that under terrestrial conditions, the idea works only when the rotation speed is on the order of 100 rounds per minutes (RPM) [52, 53]. Their work revealed that under microgravity, the rotation speed can be lowered to less than 20 RPM [51]. The modellings reveal that the unsteady nature of the g-jitter is greatly suppressed by rotation and more axisymmetric flow and dopant fields can be obtained. The results demonstrate that the axial segregation is close to the diffusion-controlled limit if the growth distance is much longer than the diffusion length, and the radial segregation is also significantly improved. Moreover, the authors also considered the effect of eccentric rotation and revealed that considerable improvement in the composition homogeneity can still be achieved once the rotation axis does not deviate much from the growth axis.

Besides the uses of external forces, the convective flow can also be modulated through careful designs of the growth apparatus and ampoule. In general, the buoyancy-driven convection level in the melt is determined by the dimensionless Rayleigh number Ra [54]

$$\mathrm{Ra} = g\beta\Delta T L^3 / \alpha\gamma \qquad (3)$$

where g is the gravitational acceleration, β is the thermal expansion coefficient, ΔT is the destabilizing temperature difference, L is the melt height, α is the thermal diffusivity, and γ is the kinematic viscosity. Besides reducing g, the regulation of the parameters ΔT and L is also a powerful way to ameliorate the convection flow. For instance, the buoyancy-driven convection level can be reduced by a factor 10^3 when L decreases from 10 to 1 cm. On the other hand, reducing ΔT is also an effective way to suppress the buoyancy-driven convection.

In the common Bridgman furnace design, the heat is supplied radially inward which causes a radial temperature gradient. This design results in a concave growth interface and convective mixing in the melt and, therefore, gives rise to a notable macroscopic segregation. The numerical simulation performed by Ostrogorsky reveals that this effect can be ameliorated by supplying the heat axially instead of radially by using a submerged heater coupled with a guarding heater surrounding the enclosed melt [55]. The typical curve of temperature versus radial position in the melt and crystal regions shows that a significantly reduced radial temperature gradient is achieved. Ostrogorsky also shows that by using a guarding heating source around the enclosed melt, both the macroscopic and microscopic homogeneity of the crystals are greatly improved. Moreover, the insertion of the submerged heater greatly reduces the parameter L, which can also contribute to the alleviation of convection. Using this method, uniform doping of Te in InSb has been realized on the earth in both the axial and radial directions, suggesting the occurrence of near-diffusion-controlled growth [56]. Replacing the submerged powered heater by an unpowered plate-like baffle, a uniform, steady-state Te segregation was still observed in Te-doped GaSb crystals [57]. According to analytical model and scaling analysis, this scheme applies to the system with equilibrium distribution coefficient in range from 0.3 to 1.0 and low diffusion coefficient [58]. Recently, Te- and Zn-doped InSb crystals were grown under microgravity by using the submerged baffle structure at the International Space Station [54, 59, 60].

3 Detached Solidification

3.1 Observations and Merits of Detached Solidification

In ground-based vertical Bridgman solidification, the outer-surface of the resulting crystal replicates the inner-surface of the crucible. By comparison, directional solidification under microgravity usually results in the so-called detached growth [5]. During detached growth the melt contacts with the crucible wall whereas the crystal grows without touching the crucible, as schematically shown in Fig. 3 [61]. As a result, a narrow gap, typically on the order of 10^{-5} m, appears between the crystal and the crucible wall where a vapor resides. The melt above the gap is supported by the formation of a liquid meniscus between the out edge of the growing crystal and the container wall. Detached growth was firstly observed in the Te-doped InSb crystal grown in Skylab by Witt et al., in which the diameter of the crystal was a little smaller than that of the container [8, 62]. Similar results were found in the subsequent microgravity experiments performed on various semiconductor materials, including Ge, GaSb, GaInSb, InAs, CdTe, CdZnTe, HgCdTe, and so on [63]. In these experiments, a wide spectrum of detached configurations including bubble, necking, shrinkage and wavy surface were observed. A comprehensive review of these observations has been conducted by Cröll et al., and Regel et al., in references [62, 64].

During the ground-based normal Bridgeman growth, the crystal adheres to the crucible wall which leads to considerable thermal mismatch due to the difference in the thermal expansion coefficients of the crystal and the crucible material. For instance, the thermal expansion coefficient of quartz is $\sim 5 \times 10^{-7}$/K, an order of magnitude lower than those of the commonly used semiconductors. Such a thermal mismatch results in an increased defect (such as dislocation) density or even worse, macroscopic cracks in the grown crystals. As a sharp contract, the lack of direct

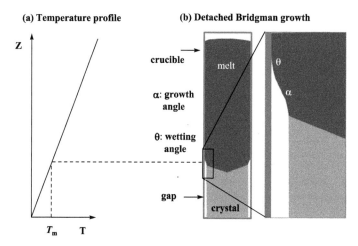

Fig. 3 **a** Temperature profile together with **b** the schematic illustration of the detached Bridgman growth configuration

contact between the grown crystal and the container crucible in the detached growth provides an effective way to reduce the defect density and therefor to improve the total crystal quality [61]. Experiments demonstrated it is true that detached growth can result in a decline of the etched (dislocation) pit density (EPD) and an improved crystallographic perfection. For example, Dold et al. and Schweizer et al. found that Ge crystals grown by detached growth exhibit an EPD reduction of ~2 orders of magnitudes [65, 66]. Similar results were also observed for CdZnTe crystals prepared under μg condition in the detached part [67]. The most striking report is that the EPD of the space InSb crystal can be reduced to $2.5 \times 10^1/\text{cm}^2$, far lower than that ($10^4$ to $10^5/\text{cm}^2$) prepared under terrestrial conditions [63]. The greatly reduced EPD has important implications for the device applications of the crystals. For example, GaAs wafers obtained from the earth and space were used to fabricate analogy switch integrated circuits, and the tests revealed that the threshold voltage of backgating, the photosensivity, and the average turn-on time for the space-grown samples are all superior to those of the terrestrial counterparts [2].

Here we stress that although detached growth seems to be a typical feature of the flight experiments, the microgravity condition itself is not necessarily associated with detached growth of semiconductors. It was found that the grown materials, the growth processes, the designed ampoule structures, and the chosen crucible materials, all together determine a growth detached or attached. Despite the consensus that hydrostatic pressure plays an important role in the detachment, the underlying origins of detached growth were not fully understood in a rather long period of time. Moreover, it is highly desirable to reproduce detached experiments on the earth due to the on-going demands on high-quality semiconductor crystals. On the other hand, it is noteworthy that although detached growth was frequently observed, the obtained crystals are not uniform in diameters. Elucidating the underlying mechanism of detached growth helps providing a possible way towards controllable growth of semiconductor crystals with uniform diameters on the earth, which may open up a new way for cost-effective practical applications.

3.2   The Underlying Mechanism

It was shown geometrically that detachment would occur if $\theta + \alpha \geq 180°$, provided that no pressure appears [61]. The two angles, θ and α, denote to the wetting angle of the melt on the crucible wall and the growth angle, respectively (Fig. 3). Generally, the growth angle α mainly depends on the chosen materials and the crystal orientation. For most of the semiconductors, the growth angle ranges from 10° to 30°. Therefore, to achieve detached growth the wetting angle should be 150–170° or even larger, which provides a strict constraint on the selection of crucible materials. Deviations to the $\theta + \alpha \geq 180°$ criterion also appeared in literature. For instance, Wang et al. observed detached growth in the InSb system in which the sum of growth angle and wetting angle is far less than 180° [68]. Their results show that besides the material properties, detached solidification also depends on the operating parameters. The reduced value of $\theta + \alpha$ can be ascribed to the introduction of a flux of gas into the gap through the meniscus, which leads to an increase of the pressure in the bottom gas reservoir.

The introduction of additional pressures in the system makes it more complex to derive the conditions needed for the detached growth. The hydrostatic pressure of melt and the vapor pressure difference in the volumes below and above the melt have strong influences on the crystal growth process. Figure 4 schematically illustrates the key factors that influence the contact-free growth. According to the differential Yang-Laplace equation, the pressure difference ΔP across the meniscus is [68, 69]

$$\Delta P = \sigma(1/R_1 + 1/R_2) \tag{4}$$

where σ is the melt surface tension and R_1 and R_2 are the main radii of the curvature (R_1 in the plane of the figure whereas R_2 in the plane perpendicular to R_1), respectively. From Fig. 4, it follows

$$R_1 = -ds/d\beta \tag{5}$$

$$R_2 = r/\cos\beta \tag{6}$$

where s is the curvilinear length along the meniscus, and r is the radial coordinate. The pressure difference across the meniscus can be expressed as

$$\Delta P = \Delta P_o + z\rho g \tag{7}$$

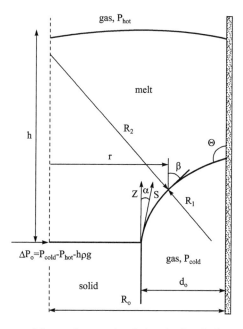

Fig. 4 Schematic diagram of the meniscus region during the detached growth. P_{cold} and P_{hot} are gas pressures below and above the melt, h is the height of the melt, R_0 the radius of the crucible, and R_1 and R_2 are the meniscus radii (R_1 in the plane of the figure, R_2 in the plane perpendicular to R_1). Reprinted with permission from reference [68]

where $\Delta P_o (P_{cold} - P_{hot} - h\rho g)$ is the pressure difference at the three phase junction, ρ the melt density, and g the gravitational acceleration. From Eq. (7) it is clear that the pressure difference is determined by the vapor pressure above the melt (P_{hot}), the vapor pressure below the melt (P_{cold}), and the hydrostatic pressure. Based on Eqs. (4)–(7), it can de deduced that

$$d\beta/ds - \cos\beta/r = -(P_{cold} - P_{hot})/\sigma + \rho g(h - z)/\sigma \qquad (8)$$

A very important implication of Eq. (8) is that the meniscus shape is dependent on the vapor pressures below and above the melt, the hydrostatic pressure, and the melt surface tension. Equation (8) indicates that if the pressure difference varies linearly with the solidification length, the solidification itself does not affect the meniscus shape, i.e., it can be treated as steady-state problem. Under boundary conditions that $\beta = 180° - \theta$ at the crucible wall and $\beta = \alpha$ at the gap distance of d_0, the meniscus shape can be computed numerically [68, 69]. It is noted that a growth parallel to the crucible wall is anticipated at a gap width of d_0, given that $\beta = \alpha$ at the three phase line. Any deviation of β from α would result in the crystal growth towards or away from the crucible wall, i.e., the change of the gap width.

In general, the numerical solutions can be divided into two categories: $\theta + \alpha <$ 180° and $\theta + \alpha >$ 180° [61, 68, 69]. Figure 5 shows the dependence of gap width on the pressure difference across the meniscus for $\theta + \alpha <$ 180°. The calculation was based on Ge, and the growth angle, surface tension and the inner crucible diameter was selected as 11°, 0.5969 N/m and 12 mm, respectively [68]. It can be seen that to achieve a growth parallel to the crucible wall, a small negative or positive pressure difference is needed. The maximum gap width, on the order of several hundred microns, occurs at $\Delta P_m \approx -2$ mbar. According to the Lyapunov stability criterion, the gap width is stable if $\partial \beta / \partial r < 0$, i.e., the meniscus curvature at the crystal/melt interface is convex [68]. When the meniscus at the crystal/melt interface is concave with respect to the melt, the growth is dynamically unstable. For the concave meniscus case, any perturbation of the meniscus would lead to a sudden change of the growth direction. If the solidification direction is swung to the crucible wall, stable attached growth occurs. Calculations show that $\partial \beta / \partial r < 0$ is tenable when ΔP_m is less than -1 mbar. However, the gap width drops drastically to the value that is comparable with the roughness of the crucible wall, as ΔP_m decreases from -2 mbar. Under this situation, the gap tends to disappear and the complete detachment is not available. Considering these facts, the stable detached growth is possible only in the ΔP_m range of -2 mbar $< \Delta P_m < -1$ mbar. Figure 6 shows the gap width as a function of pressure difference for the $\theta + \alpha >$ 180° case [69]. Under negative pressure difference, the calculations reveal that the meniscus is convex with respect to the melt. That is, a stable gap width can be maintained in a rather broad ΔP_m range of $\Delta P_m < 0$ mbar. However, for ΔP_m very close to 0 mbar, the gap width is so large that the melt run-down occurs under terrestrial conditions. Moreover, for very high pressure difference, $\Delta P_m < -30$ mbar, the gap width is too small to guarantee a true detached solidification.

Figures 5 and 6 also reveal that, the larger the sum of θ and α, the easier the realization of the contact-free growth. Since the wetting angle of Ge on BN is much larger than that on fused silica, the use of BN crucible is beneficial for the realization of detached solidification of Ge. On the other hand, the growth angle of InSb (20–30°) is higher than that of Ge (7–13°), and therefore detachment is more frequently observed during the Bridgman growth of InSb. To achieve detached growth for a given semiconductor material, the chosen crucible should have a wetting angle as large as possible with the semiconductor melt. The contact angles of various semiconductor melts on the commonly used crucible materials are listed in Table 1. Because the surface condition has a close correlation with the wetting angle, any surface contamination should be eliminated before the ampoule preparation. Moreover, since rough surface favors for an increased wetting angle, utilization of crucibles with coarse side surface is beneficial for detached growth. However, rough crucible walls usually enhance heterogeneous nucleation and therefore may cause polycrystalline growth.

Fig. 5 Gap width as a
function of pressure
differential across the
meniscus for θ + α < 180°.
Reprinted with permission
from reference [69]

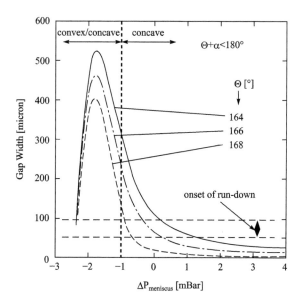

Fig. 6 Gap width as a
function of pressure
differential across the
meniscus for θ + α > 180°.
Reprinted with permission
from reference [69]

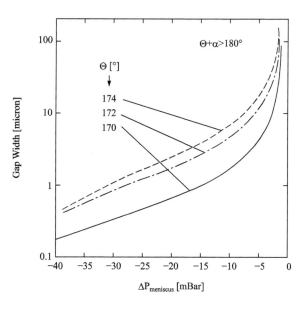

Table 1 Contact angles of various semiconductors on commonly used crucibles

Crucibles	Semicond.				
	Si	Ge	InSb	GaSb	CdTe
Fuzed silica	87 [70]	117 [71]	165–175 [72]	130 [73]	83–108 [74]
BN	145 [75]	173 [71]	134 [76]	132 [77]	132 [74]
Graphite	10–30 [78]	166 [71]	124 [76]	129 [77]	

3.3 Guides to Ground-Based Solidifications

As discussed in Sect. 2, the pressure difference across the meniscus has two sources—the pressure difference between vapors in the gap and above the melt, and the hydrostatic pressure. Provided that the density of the semiconductor melt is 5 g/cm^3 and the melt height is 10 cm, the hydrostatic contribution to the pressure difference across the meniscus under 1-g_0 condition is −50 mbar. Such a value is one order of magnitude larger than what is needed by the detached solidification, under both $\theta + \alpha <$ 180° and $\theta + \alpha >$ 180°. This explains why detachment has scarcely been observed in the Bridgman growth on the earth. To meet the criterions for detachment illustrated in Sect. 2, an overpressure in the gap is of vital importance for the solidification processes performed on the earth.

There are various methods to generate the overpressure in the gap: (i) active control of the pressure difference by using an external gas source, (ii) inducing overpressure via temperature gradient variation, and (iii) the creation of overpressure by introducing internal gas source. The latter one is applicable for the compounds in which one element has a higher saturated vapor pressure than the others, and has been successfully utilized to the CdTe system [79]. The drawback of this method is that the adding of volatile element into the ampoule can induce the non-stoichiometry of the grown crystals.

The former two methods are not restricted to the volatile compounds like CdTe, but also apply to other material systems. Duffar et al. have proposed the active pressure control scheme and used it to grow InSb crystals [69, 80]. They designed an ampoule architecture in which the top part and the bottom part are connected to the gas systems through openings, and therefore the pressures in these two parts can be controlled separately. Using this method the overpressure in the bottom part can be easily tuned to close to the hydrostatic pressure of the melt and, as a result, contact-free growth was achieved over the entire length of the sample under 1-g_0 condition. As the growth proceeds, the gap in the bottom part of the ampoule increases in volume and thus the static pressure decreases. To insure stable detached solidification and control the crystal radius, the vapor pressure has to be tuned continuously to compensate this gap increase. Yeckel et al. scheduled a scheme to incorporate the proportional-integral (PI) feedback controller to stabilize the detached vertical Bridgman growth [81]. This is very challengeable due to the fact that detached growth is a strong nonlinear process. Very recently, they also investigated the possibility to manipulate the gas pressure through nonlinear feedback control [82, 83]. Their results show that both the solution multiplicity and the strict operability limits encountered by using the linear PI controller are well addressed.

Another efficient way to achieve detached growth is inducing an axial temperature gradient variation, as proposed by Szofran et al. [84]. In this approach, the sealed crucible is filled with inert gas and then the feed material is molten using the normal Bridgman temperature profile. The melting of the feed material isolates the two gas reservoirs located at the top and bottom of the crucible. After that, the temperature in the top crucible region is lowered which results in a notable vapor pressure difference. Owing to the overpressure in the bottom reservoir, a meniscus is formed and the crystal starts to grow without contact to the crucible. To achieve the desired axial temperature gradient change, a furnace with multiple heating zones and separate temperature controls is needed. As the growth proceeds, the crystal moves out of the modified temperature profile, resulting in the breakdown of detached growth. To address this issue, the temperature in the top crucible region should be tuned successively during the growth. Duffar et al. [85] modified this method by introducing a gas reservoir below the seed and separately controlling its temperature through an additional heating zone. The experimental results revealed that this improved approach is self-controlling in terms of pressure adjustment [85].

Experimental results demonstrate that the detached growth is highly beneficial for improving the quality of ground-based crystals. Figure 7 presents the surface images of detached and attached Ge crystals grown on the earth [65]. For the detached growth of Ge, the typical gap width is ~10 μm. The samples were etched for 10 min at 85 °C with the Billig etchant which is prepared by dissolving 12 g KOH and 8 g $K_3[Fe(CN)_6]$ into 100 ml H_2O. As can be clearly seen in Fig. 7, detached growth leads to an EPD reduction as high as two orders of magnitude in the edge regions of the crystals. It is known that the EPD of the ground-grown crystals via vertical Bridgman

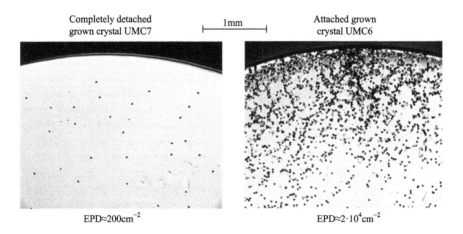

Fig. 7 Surface photographs of the detached and attached Ge crystals. Reprinted with permission from reference [65]

method is on the orders of 10^3 to $10^8/cm^2$. The edge region of the detached Ge crystal has an EPD as low as $2 \times 10^2/cm^2$. In the middle of the crystals, detached growth also results in a notable reduction of EPD from $7 \times 10^3/cm^2$ to $5 \times 10^2/cm^2$. The results are further confirmed by X-ray topography using synchrotron radiation which show that the EPD decreases from $10^4/cm^2$ in the seed to 10^2 to $10^3/cm^2$ in the detached crystal. Similar observations were also observed in other semiconductor materials. Obviously, the understanding of the mechanism of detached growth and the guide to terrestrial experiments are one of the greatest achievements in the microgravity materials research.

3.4   Implications of Detached Growth Mechanism for Microgravity Experiments

As discussed in Sect. 3.2, semiconductor melts are fluids with low Prandtl number, which are characterized by high thermal conductivity and low viscosity. Due to this fact, stabilized natural convection cannot be observed in semiconductor melts on the earth. However, it is hard to fully understand and account for the natural convection featured with unsteady and random flow patterns using the analytical tools. Under microgravity conditions, the natural buoyancy-driven convection is greatly reduced which makes it possible to realize the diffusion-controlled segregation. The diffusion-controlled segregation simplifies the growth process and the growth parameters such as diffusion coefficient can be derived analytically. However, due to the residual acceleration and the appearance of free surface, true diffusion-limited process is not available in most cases. We have shown in Sect. 3.2 that the influence of residual acceleration can be dampened by applying a magnetic field or using a submerged baffle. On the other hand, free surfaces are frequently observed in semiconductor melts due to detached growth during the flight experiments. The in-depth understanding of detached growth provides powerful measures to eliminate free surfaces.

Ostrogorsky et al. have performed series of Bridgman solidification experiments on Te-doped InSb in the Microgravity Science Glovebox (MSG) at the International Space Station (ISS) [60]. To preclude the influence of Marangoni convection, all the charges were pressurized by specially designed piston and spring devices to prevent the occurrence of de-wetting. The furnace in MSG is equipped with a viewing port, and the growth interface can be visualized via the in situ observation system. Real-time observations revealed that de-wetting did not occur on the side of the crystal. The dopant distribution was measured by secondary ion mass spectrometry, and a typical result along the central axis of the crystal is shown in Fig. 8. A diffusion coefficient of 1×10^{-5} cm^2/s, several times lower than previously reported values, provides a reasonably good fit to the data using the one-dimensional diffusion-controlled segregation relationship. Reproducible initial transients in dopant concentration were observed in other Te-doped InSb growths in the series of experiments.

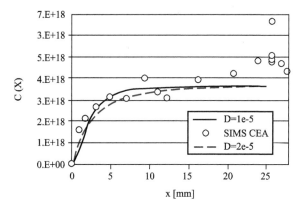

Fig. 8 Dopant segregation profile of the Te-doped InSb crystal grown at ISS; the solid and dashed lines are based on finite element calculations using diffusion coefficients of 1×10^{-5} cm^2/s and 2×10^{-5} cm^2/s, respectively. Reprinted with permission from reference [59]

Zn-doped InSb solidifications were also performed at ISS using the similar ampoule configuration [61]. To prevent the formation of free surfaces, a static pressure of ~40 mbar was applied on the melt through a piston controlled by the carbon spring. A diffusion-controlled initial transient was also observed, with a fitted Zn diffusion coefficient of 1.2×10^{-4} cm^2/s. Zn-doped InSb has an equilibrium distribution coefficient of 2.9 and materials systems with $k > 1$ were scarcely studied in the microgravity experiments. The work performed on Zn-doped InSb is the first one that reports diffusion-controlled growth for systems with $k > 1$. All these results show that by damping the formation of free surface based on the knowledge of detached growth, new experimental schemes were designed and an in-depth understanding of the microgravity solidification process has been achieved.

4   New Melt Growth Techniques

4.1   Difficulties in Preparing Semiconductor Alloys

Semiconductor alloys (or concentrated semiconductors), such as $Ge_{1-x}Si_x$, $In_xGa_{1-x}Sb$ and $In_xGa_{1-x}As$, are of considerable interest for technological applications since their physical properties, including lattice constants, electronic performance and optical bandgaps can be tuned simply by controlling the alloy composition. For example, $In_xGa_{1-x}Sb$ is an important opto-electric material which exhibits an enhanced thermoelectric performance than InSb due to its proper bandgap [85]. On the other hand, it is well known that $Ge_{1-x}Si_x$ is very promising for seamlessly connecting the optoelectronics and microelectronics on the Si-based platforms [86]. Due to these facts, there has been attracted considerable interest in semiconductor alloys in recent years. However, it is difficult to grow large-size, chemically homogeneous Semiconductor alloy crystals because the large separation between the solidus and liquidus lines in the pseudo-binary phase diagrams. The attempts to grow semiconductor alloys via traditional techniques, e.g., Bridgman method, fail to produce homogeneous crystals, even under microgravity conditions [22, 23]. The obtained crystals have compositions vary considerably along the axial and radial directions, and cannot meet the needs for actual applications.

Under the diffusion-controlled steady-state growth, there occurs a boundary layer adjacent to the growth interface where a solute concentration gradient (solute concentration varies from C_0/k to C_0) is built. Any perturbation near the boundary would break the delicate balance and produce a variation of the solute concentration gradient and therefore, leading to a notable segregation in the crystal. This effect is particularly strong for a system which has a large separation between the solidus and liquidus lines in the pseudo-binary phase diagrams near the targeted composition. To address this issue, researchers proposed new schemes which yield more stable solute profiles in the melt. These new schemes are the traveling liquidus-zone (TLZ) method and the vertical gradient freezing (VGF) method. The key of these methods relies in the degree of saturation in the melt—the melt is saturated and therefore the solute concentration is equivalent to that on the liquidus line. This feature distinguishes these two methods from other techniques. On condition that the slope of liquidus line is a constant, the solute concentration gradient in the melt is proportional to the temperature gradient. Based on these two methods, chemically homogeneous $Ge_{1-x}Si_x$ and $In_xGa_{1-x}Sb$ single crystals were grown under microgravity.

4.2 TLZ Method and Related Microgravity Experiments

The TLZ method is a zone-melting method proposed and developed by Kinoshita et al. [89] As discussed in Sect. 4.1, the most important feature of TLZ is that it has a saturated liquidus-zone. The study on this method starts from $In_xGa_{1-x}As$, and latterly applies to the $Ge_{1-x}Si_x$ system [89].

Figure 9 schematically shows the principle of the TLZ method (taking the InAs-GaAs system as an example). According to the InAs-GaAs phase diagram, the InGaAs ($In_{0.3}Ga_{0.7}As$) crystal growth needs a freezing interface temperature of ~1027 °C, a linear temperature gradient of ~15 °C/cm, and a zone width of ~20 mm. In this case, the solute concentration gradient can be controlled by the imposed temperature gradient because the solute concentration is determined by the temperature on the liquidus-line

$$\left(\frac{\partial C_L}{\partial Z}\right)_{Z=0} = \left(\frac{\partial C_L}{\partial T}\right)_{Z=0}\left(\frac{\partial T}{\partial Z}\right)_{Z=0} \qquad (9)$$

where C_L is the solute concentration at the liquidus line, T is the temperature, Z is the distance measured from the freezing interface, and $\frac{\partial T}{\partial Z}$ is the temperature gradient.

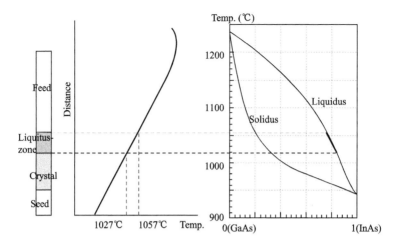

Fig. 9 Sample configuration and the corresponding temperature profile of the TLZ method for the InAs-GaAs system. Reprinted with permission from reference [87]

Therefore, $\frac{\partial C_L}{\partial T}$ represents a reciprocal of the slope of the liquidus-line at around the interface temperature. Furthermore, the freezing rate, R, can be written as

$$-R(C_{Lo} - C_{so}) = D(\partial C_L / \partial Z)_Z = 0 \tag{10}$$

where C_{L0} and C_{S0} are the liquidus and solidus solute concentrations at the freezing interface, and D is the inter diffusion coefficient between solute and solvent. According to the Eq. (10), the freezing rate can be expressed as

$$-R = -\frac{D}{(C_{l0} - C_{s0})} \left(\frac{\partial C_L}{\partial Z} \right)_{Z=0} = -\frac{D}{(C_{l0} - C_{s0})} \left(\frac{\partial C_L}{\partial T} \right) \left(\frac{\partial T}{\partial Z} \right)_{z=0} \tag{11}$$

A very important point obtained from Eq. (11) is that the freezing rate is determined by the temperature gradient ($\partial T/\partial Z$). As a result, it is easy to maintain the freezing interface temperature at a constant value by translating the sample at a speed of R. The relationship between R and ($\partial T/\partial Z$) in Eq. (11) is determined by measuring the growth rate R and the temperature gradient near the interface, and by comparing the measured growth rate with the calculated one. Kinoshita et al. have successfully grown homogeneous InGaAs and SiGe single crystals using the TLZ method on ground, and the chemical inhomogeneity is lower than 1% within the whole growth areas, both along the axial and radial directions [88–90].

Although the TLZ method has been used for growing semiconductor alloys on ground [88, 89], it is still a great challenge to fabricate large-diameter crystals with homogeneously distributed chemical compositions on the earth due to the influence of buoyancy-driven convection. To address this issue, microgravity crystal growth experiments performed by the TLZ method are highly desirable. Kinoshita et al.

reported the growth of SiGe alloy crystal with a diameter of 10 mm in and a growth length of 17.2 mm by the TLZ method under microgravity [91]. The concentration of Ge is about 48.5 ± 1.5 at % along the center line of the grown crystal, and the peripheral concentration profile reveals a similar Ge concentration. That is, both the axial and radial concentration variations are small [92, 93]. They also found that the growth length was short for the crystal grown at higher temperature gradient, due to the strong influence of constitutional supercooling. Numerical simulations of the SiGe crystal growth via the TLZ method were also carried out by the same gr.

4.3 The VGF Method

Conventional methods such as Czochralski and Bridgman techniques are difficult to grow $In_xGa_{1-x}Sb$ single crystals with enough size and high quality due to the buoyancy-driven convection on ground. Therefore, many scientists have done their best to overcome this problem through various methods. Tsuruta et al. developed the ultrasonic-vibrations-introduced Czochralski method to reduce the degree of supercooling. When the $In_xGa_{1-x}Sb$ mixed crystal is grown by introducing ultrasonic vibrations into the melt, the facet growth appear firstly, and the facet growth regions has a high concentration of in matrix component and Te impurity compared to the normal growth region [94–96]. Czochralski method with multi-step pulling was applied to Ga-In-Sb ternary solution for growing GaInSb bulk crystals [97]. Dutta et al. demonstrated that forced convection or mixing in the melt during directional solidification of bulk $In_xGa_{1-x}Sb$ ternary alloys significantly decreases cracks in the crystals [98]. Floating crucible Czochralski method was used to prepare InGaSb polycrystalline crystals, and the results show that although axial segregation can be ameliorated, the reproducibility is still a challenge [99]. As for GaInSb alloys prepared by Bridgman method, the growth at a variable pulling rate can improve the axial chemical homogeneity [100]. Numerical simulations certified that an alternating magnetic field applied near the solid-liquid interface increases the melt flow intensity, and has a notable effect on the radial segregation and the interface deflection [101, 102]. Although considerable efforts have been devoted to the InGaSb growth, it is still a great challenge to achieve chemical homogeneity in this system.

The VGF method is very promising for addressing this issue. The VGF method is different from TLZ in that the sample does not shift during the whole growth process. Instead, homogeneous semiconductor alloys can be grown using VGF by cooling the sample at a fixed cooling rate, on condition that the liquidus temperature at the growth interface keeps constant (Fig. 10). Figure 10a, b show the ampoule configuration and the temperature profile of the furnace. The feed and the seed were placed at the high and low temperature regions, respectively. The sample is composed of three parts: the seed, the low temperature zone and the feed, as shown in Fig. 10c, d. Before growth, the system is heated to a temperature that turns the low temperature material into molten melt. Parts of the seed and feed are dissolved into the liquid zone and thus a solution forms. After that, the crystal growth begins by slowly decreasing the

350   Z. Yin et al.

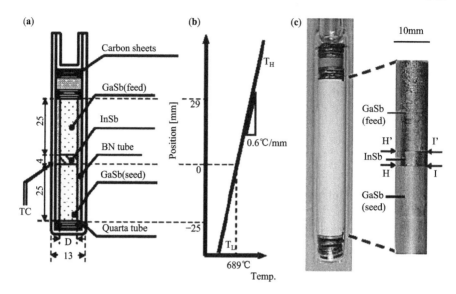

Fig. 10 The vertical gradient freezing method (taking the growth of InGaSb as an example): **a** Ampoule configuration, **b** the temperature profile in the furnace, **c** the ampoule before growth, and **d** the sample before growth. Reprinted with permission from reference [103]

furnace temperature. During the growth, the solid-liquid interface moves towards feed. As a result, homogeneous alloy crystals can be grown by the VGF method.

We now take the growth of $In_{0.03}Ga_{0.97}Sb$ as an example (Fig. 11). The ampoule which locates in the furnace is heated slowly up to the growth temperature T_H, and then the furnace is kept at this temperature for 20 h. During the holding time, InSb melts at its melting point 525 °C and GaSb seed and feed crystals dissolve in the InSb melt to form InGaSb solution. The temperature gradient is kept at 0.6 °C/mm, and the temperature of the feed interface is higher than that of the seed interface. Since the dissolution of GaSb feed is faster than the seed, a compositional gradient between the seed and feed interfaces is formed. When the low temperature seed interface becomes supersaturated, the crystal growth starts and interface moves gradually towards the feed. The key issue for the VGF growth is setting the cooling rate—at a proper cooling rate, the interface moving speed equals to R (Eq. 11) and the liquidus temperature near the growth interface maintains at a constant value. At such a cooling rate, homogeneous semiconductor alloys can be successfully grown. The optimum cooling rate for $In_{0.03}Ga_{0.97}Sb$ growth is 0.15 °C/h, under the temperature gradient of 0.6 °C/mm.

Murakami and coworkers have measured the growth rate using a GaSb(seed)/InSb/GaSb(feed) sandwich sample by characterizing the distances of the Te striations which were intentionally introduced into the grown crystal through thermal pulses [104]. The composition profiles reflect the change of growth rate, which show that the growth rate gradually increases to the saturated value and then rapidly decreases. The same group also explored the possibility of growing $In_xGa_{1-x}Sb$

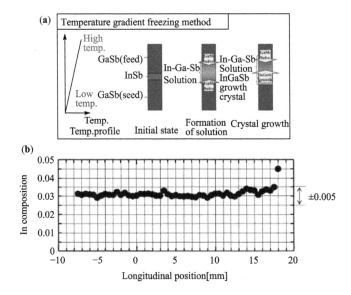

Fig. 11 a The schematic of vertical gradient freezing method. **b** Indium composition profile of InGaSb crystal with a cooling rate of 0.15 °C/h. Reprinted with permission from reference [103]

bulk crystals through a InSb seed by using a InSb(seed)/Te-doped InSb/GaSb(feed) sandwich structure [105]. Homogeneous $In_{0.8}Ga_{0.2}Sb$ and $In_{0.6}Ga_{0.4}Sb$ crystals were grown at cooling rates of 0.77 and 0.33 °C/h, corresponding to the growth rates of 0.45 and 0.19 mm/h, respectively. Hayakawa et al. developed an in situ X-ray penetration method to study the composition profiles of the In-Ga-Sb solution [106]. Rajesh et al. observed the detailed dissolution process of GaSb into InSb by using this method with a CdTe detector [107]. The gallium (Ga) composition profile of the sample was successfully calculated as a function of time by making the calibration line with the penetrated X-ray intensities of GaSb and InSb standard samples.

4.4 Microgravity Experiments Based on VGF

The viscosity, wetting property and evaporation rate of InSb, GaSb and $In_xGa_{1-x}Sb$ were studied for the crystal growth experiments at the International Space Station. The results indicate that the viscosity of $In_xGa_{1-x}Sb$ decreases with the increase of Indium, and the determined value is between those of InSb and GaSb. No noticeable wetting reaction between the molten $In_xGa_{1-x}Sb$ and the quartz, BN and graphite substrates was observed, revealing that these substrate materials can be used to fabricate crucibles for the $In_xGa_{1-x}Sb$ microgravity experiments [33]. Based on these results, an optimized ampoule structure was designed and used in the microgravity experiment by Japan Aerospace Exploration Agency. Inatomi et al. have grown the

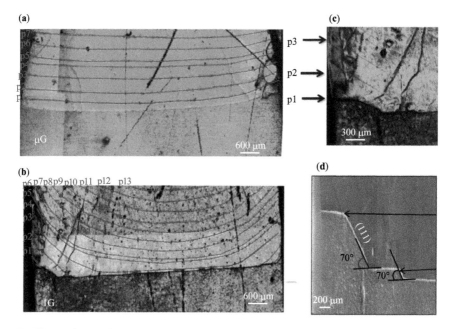

Fig. 12 Polarized optical microscope images of the etched surfaces of **a** μg and **b** 1 g samples. **c** Optical micrograph and **d** field emission scanning electron microscopy image at the periphery of the 1 g sample. The dopant striations are marked as p1, p2 and p3 in (**c**). Reprinted with permission from reference [108]

$In_xGa_{1-x}Sb$ crystals via the VGF method mentioned above under μg conditions at the International Space Station, and the dissolution and growth processes were carefully studied [28]. The optical micrographs of the un-dissolved seed interface in the space- and ground-based samples are shown in Fig. 12. By applying heat pulses, striations were introduced into the crystal and the growth rate was determined. It can be clearly seen in Fig. 12 that the space-based crystal has a rather flat growth interface. By sharp comparison, the crystal grown on earth has a concave growth interface. Moreover, the etch pit density of the μg_0 sample was lower than that of the $1\text{-}g_0$ samples, indicating the μg_0 crystal has a higher quality. The above results clearly show that the buoyancy-driven convection affects both the crystal quality and the solid-liquid interface shape.

The measurements on the composition profiles of the crystals along the growth direction show that the $In_xGa_{1-x}Sb$ crystal growth under μg_0 starts from 20.2 mm and ends at 35.1 mm; therefore, the growth length was 14.9 mm. By comparison, the $1\text{-}g_0$ crystal growth starts from 18.0 and ends at 30.6 mm, and the total growth length is 12.6 mm. The amount of indium gradually decreases along the growth direction and fluctuates severely at the regions x > 35.1 and > 30.6 mm for the μg_0 and $1\text{-}g_0$ samples, respectively. The remaining solution rapidly solidifies due to the crystal cooling after the growth period, leading to the observed compositional fluctuations at the final crystal growth regions. The indium composition of the μg_0 sample at the seed interface was 0.028, much lower than that of the 1-g_0 sample, due to the large amount of seed dissolution on ground. From the InSb–GaSb binary phase diagram, the anticipated growth temperature and temperature gradient can be calculated. However, the obtained temperature gradient for the 1-g_0 growth (0.58 °C/mm) is slightly lower than that of the μg_0 sample (0.64 °C/mm). The crystal growth rate is determined by measuring the grown length between the striations, and the growth rate variations along the growth direction of the μg_0 and 1-g_0 samples are shown in Fig. 12 in reference [108]. The obtained results show that the growth rate of the 1-g_0 crystal is lower than that of the μg_0 sample. Under microgravity, the initial growth rate is about one order of magnitude higher than that under 1-g_0 condition as a result of the absence of buoyancy-induced convection.

InGaSb ternary alloys were also grown by using GaSb (111)$_A$ and (111)$_B$ (Ga-terminated and Sb-terminated faces) seeds and feeds on board the International Space Station by the VGF method [109]. The experiments aim to study the orientation-dependent dissolution and growth of InGaSb along the GaSb(111)$_B$ and (111)$_A$ faces. Assuming the growth of InGaSb crystal in space by VGF is a steady state process dominated by diffusion and the solute was saturated in the solution, the mixed InGaSb crystals growth rate should be directly proportional to the temperature gradient. Therefore, the ratio of growth rates of the (111)$_A$ and (111)$_B$ samples should be same under the same temperature gradients. However, the experimental results show that the (111)$_A$ and (111)$_B$ samples have quite different growth rates (Fig. 13), and the growth rate of the (111)$_B$ face is 15.4% higher than that of (111)$_A$. Apparently, the growth rate is also severely influenced by the concentration gradient $\frac{\partial C_L}{\partial T}$, and the lower growth rate of (111)$_A$ sample has a close correlation with the lower dissolution of the GaSb (111)$_A$ [109]. The growth rates (Fig. 13) at different positions along the growth direction show that the (111)$_A$ face had a lower growth rate than the (111)$_B$ face.

Recently, a Chinese recovery satellite SJ-10 was launched and InGaSb crystals were grown on this scientific platform under gravity level of 10^{-4} to 10^{-5} g_0, by using the VGF method. The grown crystal was cut longitudinally into two pieces along the growth direction and the surfaces were polished by alumina abrasives. The EPMA analysis shows that the high quality $In_{0.11}Ga_{0.89}Sb$ crystal was successfully grown at the Chinese recovery Satellite SJ-10 with a uniform composition distribution.

All these results show that high-quality InGaSb crystals with homogeneous composition distribution can be obtained under microgravity by the VGF method. By sharp comparison, the preparations of InGaSb by other techniques on space did not yield results as good as those obtained by the VGF method [22, 23, 110–112]. Moreover, we note that although the VGF method is mainly used to grow InGaSb, it can also apply to other semiconductor alloy systems.

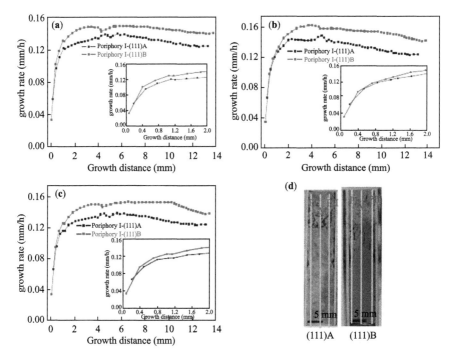

Fig. 13 Growth rates of the InGaSb (111)$_A$ and (111)$_B$ crystals along **a** periphery-I, **b** middle, and **c** periphery-II, **d** crystal surface in which the measured positions are marked by solid lines. Reprinted with permission from reference [109]

5 Conclusion

Melt growth is the most widely used method to fabricate high-quality semiconductor single crystals towards optoelectronic and microelectronic applications. Although the microgravity condition does not always yield the diffusion-controlled growth, the homogeneity of the obtained semiconductor crystals is improved, typically along the axial direction. Based on the microgravity experiments, an in-depth understanding of the interactions among heat transfer, convection, solute transport and chemical segregation, has been achieved. Moreover, with the help of these achievements, active control on the convective flow and the chemical homogeneity is realized via various methods, e.g., by applying a magnetic field during the growth or inserting a submerged baffle in the ampoule.

One of the remarkable breakthroughs in the field of melt growth is the understanding of the detached growth. Detached growth is typically associated with directional solidifications performed on space and is scarcely observed on ground-based experiments. By disclosing the underlying mechanism of detached growth, this magic growth technique can now be realized in the laboratory experiments through improved designs of the ampoules. The most amazing feature of detached growth is that it can reduce the dislocation density by orders of magnitude, which offers new opportunities for low-cost, high-performance device applications.

There emerge new melt growth techniques, including the traveling liquidus-zone method and the vertical gradient freezing method. These two methods distinguish them from the traditional ones in that they provide a saturated liquid region between the growth interface and the feed material. Under this scheme, a constant composition distribution across the whole growth area is achieved by fixing the temperature at the growth interface. The traveling liquidus-zone method and the vertical gradient freezing method are successfully used to grow chemically homogeneous semiconductor alloys, including $Ge_{1-x}Si_x$ and $In_xGa_{1-x}Sb$, in both the space-based and ground-based experiments. These methods also have great potentials for the growth of other semiconductor alloy systems.

References

1. Benz KW, Dold P (2002) Crystal growth under microgravity: present resultsand future prospects towards the International Space Station. J Cryst Growth 237:1638
2. Chen NF, Zhong XR, Lin LY, Zhang M, Wang YS, Bai XW, Zhao J (2001) Comparison of field effect transistor characteristics between space-grown and earth-grown gallium arsenide single crystal substrates. Appl Phys Lett 78:479
3. Chen NF, Zhong XR, Lin LY, Xie X, Zhang M (2000) Semi-insulating GaAs grown in outer space. Mater Sci Eng B 75:134
4. Kamotani Y, Matsumoto S, Yoda S (2007) Recent developments in oscillatory marangoni convection. EDMP 3:147
5. Ebnalwaled AA, Duffar T, Sylla L (2013) Dewetting and transport property enhancement: antimonide crystals for high performance electronic devices. Cryst Res Technol 48:236
6. Tiller WA, Jackson KA, Rutter JW, Chalmers B (1953) The redistribution of solute atoms during the solidification of metals. Acta Met 1:428
7. Smith VG, Tiller WA, Rutter JW (1955) A mathematical analysis of solute redistribution during solidification. Can J Phys 33:723
8. Witt AF, Gatos HC, Lichtensteiger M, Lavine MC, Herman CJ (1975) Crystal growth and steady-state segregation under zero gravity: InSb. J Electrochem Soc 122:27
9. Zhou YF, Li XY, Bai SQ, Chen LD (2010) Comparison of space- and ground-grown $Bi_2Se_{0.21}Te_{2.79}$ thermoelectric crystals. J Cryst Growth 312:775
10. Witt AF, Gatos HC, Lichtensteiger M, Herman CJ (1978) Crystal growth and segregation under zero gravity: Ge. J Electrochem Soc 125:1832
11. Bly JM, Kaforey ML, Matthiesen DH, Chait A (1997) Interface shape and growth rate analysis of Se/GaAs bulk crystals grown in the NASA crystal growth furnace (CGF). J Cryst Growth 174:220
12. Gillies DC, Lehoczky SL, Szofran FR, Watring DA, Alexander HA, Jerman GA (1997) Effect of residual accelerations during microgravity directional solidification of mercury cadmium telluride on the USMP-2 mission. J Cryst Growth 174:101
13. Strelov VI, Kuranova IP, Zakharov BG, Voloshin AE (2014) Crystallization in space: results and prospects. Crystallogr Rep 59:781
14. Alexander JID, Garandet J-P, Favier J-J, Lizee A (1997) g-Jitter effects on segregation during directional solidification of tin-bismuth in the MEPHISTO furnace facility. J Cryst Growth 178:657
15. Coriell SR, Boisvert RF, Rehm RG, Sekerka RF (1981) Lateral solute segregation during unidirectional solidification and inter-face instabilities during unidirectional solidification of a binary alloy. J Cryst Growth 54:167

16. Griffin PR, Motkef S (1989) Influence of nonsteady gravity on natural convection during microgravity solidification of semiconductors: part I. J Appl Microgravity Technol 2:121

17. Alexander JID, Ouazzani J, Rosenberger F (1989) Analysis of the low gravity tolerance of Bridgman–Stockbarger crystal growth: part I. Steady and impulse accelerations. J Cryst Growth 97:285

18. Naumann RJ (1994) Modeling flows and solute redistribution resulting from small transverse accelerations in Bridgman growth. J Cryst Growth 142:253

19. Derby JJ, Kwon Y, Pandy A, Sonda P, Yeckel A, June T, Müller G (2006) Developing quantitative, multiscale models for microgravity crystal growth. Ann N Y Acad Sci 1077:124

20. Pandy A, Yeckel A, Reed M, Szeles C, Hainke M, Müller G, Derby JJ (2005) Analysis of the growth of cadmium zinc telluride in an electrodynamic gradient freeze furnace via a self-consistent, multi-scale numerical model. J Cryst Growth 276:133

21. Derby JJ, Zhang N, Yeckel A (2013) Modeling insights on the melt growth of cadmium zinc telluride. J Cryst Growth 379:28

22. Duhanian N, Marin C, Abadie J, Chaudet M, Dieguez E, Duffar T (1997) Chemical segregation and crystal crucible interaction during the growth of Ga0.8In0.2Sb in space. Microgravity Sci Technol XI/4:187

23. Duhanian N, Duffar T, Marin C, Dieguez E, Garandet JP, Dantan P, Guiffant G (2005) Experimental study of the solid–liquid interface dynamics andchemical segregation in concentrated semiconductor alloy Bridgman growth. J Cryst Growth 275:422

24. Stelian C, Duffar T (2005) Numerical analysis of solute distribution and interface stabilization during experimental Bridgman growth of concentrated GaInSb alloys. J Cryst Growth 275:422

25. Stelian C, Duffar T (2005) Growth of concentrated GaInSb alloys with improved chemical homogeneity at low and variable pulling rates. J Cryst Growth 275:e585

26. Stelian C, Duffar T (2005) Modeling of a space experiment on Bridgman solidification of concentrated semiconductor alloy. J Cryst Growth 275:175

27. Nakamura S, Hibiya T, Kakimoto K, Imaishi N, Nishizawa S, Hirata A, Mukai K, Yoda S, Morita TS (1997) Temperature fluctuations of the Marangoni flow in a liquid bridge of molten silicon under microgravity on board the TR-IA-4 rocket. J Cryst Growth 186:85

28. Nakamura S, Hibiya T, Kakimoto K, Imaishi N, Yoda S, Nakamura T, Koyama M, Dold P, Benz KW (1999) Observation of periodic Marangoni convection in a molten silicon bridge on board the TR-IA-6 rocket. J Jpn Soc Microgravity Appl 16:99

29. Yano T, Nishino K, Kawamura H, Ueno I, Matsumoto S, Ohnishi M, Sakurai M (2011) Space experiment on the instability of Marangoni convection in large liquid bridge—MEIS-4: effect of Prandtl number. J Phys Conf Ser 327:012209

30. Duffar T, Serrano MD, Lerin L, Santailler JL (1999) Marangoni convective effect during crystal growth in space. Cryst Res Technol 34:457

31. Müller G, Neumann G, Matz H (1987) A two-Rayleigh-number model of buoyancy-driven convection in vertical melt growth configurations. J Cryst Growth 84:36

32. Schweizer M, Cröll A, Dold P, Kaiser T, Lichtensteiger M, Benz KW (1999) Measurement of temperature fluctuations and microscopic growth rates in a silicon floating zone under microgravity. J Cryst Growth 203:500

33. Lan CW, Chian JH (1999) Three-dimensional simulation of Marangoni flow and interfaces in floating-zone silicon crystal growth. J Cryst Growth 203:500

34. Selver R (2005) Experiments on the transition from the steady to the oscillatory marangoni convection of a floating-zone under various cold wall temperatures and various ambient air temperature effects. Microgravity Sci Technol 17:25

35. Rupp R, Müller G, Neumann G (1989) Three-dimensional time dependent modelling of the Marangoni convection in zone melting configurations for GaAs. J Cryst Growth 97:34

36. Kamotani Y, Matsumoto S, Yoda S (2007) Recent developments in oscillatory Marangoni convection. FDMP Fluid Dyn Mater Process 3:147

37. Cröll A, Kaiser T, Schweizer M, Danilewsky AN, Lauer S, Tegetmeier A, Benz KW (1998) Floating-zone and floating-solution-zone growth of GaSb under microgravity. J Cryst Growth 191:365
38. Nishinaga T, Ge P, Huo C, He J, Nakamura T (1997) Melt growth of striation and etch pit free GaSb under microgravity. J Cryst Growth 174:96
39. Tillberg E, Carlberg T (1990) Semi-confined Bridgman growth of germanium crystals in microgravity. J Cryst Growth 99:1265
40. Matthiesen DH, Wargo MJ, Motakef S, Carlson DJ, Nakos JS, Witt AF (1987) Dopant segregation during vertical Bridgman-Stockbarger growth with melt stabilization by strong axial magnetic-fields. J Cryst Growth 85:557
41. Kang J, Fukuda T (2000) Growth exploration of compositionally uniform bulk semiconductors under a high magnetic field of 80000 Gauss. Mater Sci Eng B 75:149
42. Yesilyurt S, Vujisic L, Motakef S, Szofran FR, Volz MP (1999) A numerical investigation of the effect of thermoelectromagnetic convection (TEMC) on the Bridgman growth of $Ge_{1-x}Si_x$. J Cryst Growth 207:278
43. Fripp AL, Debnam WJ, Rosch W, Chait A, Yao M, Szofran FR (2000) Melt stabilization of PbSnTe in a magnetic field. NASA Microgravity Mater Sci Conf
44. Cröll A, Szofran FR, Dold P, Benz KW, Lehoczky SL (1998) Floating-zone growth of silicon in magnetic fields II: strong static axial fields. J Cryst Growth 183:554
45. Baumgartl J, Müller G (1996) The use of magnetic fields for damping the action of gravity fluctuations (g-jitter) during crystal growth under microgravity. J Cryst Growth 169:582
46. Ma N, Walker JS (1997) Magnetic damping of buoyancy convection during semiconductor crystal growth in microgravity: spikes of residual acceleration. Phys Fluids 9:1182
47. Herrmann F, Müller G (1995) Growth of 20 mm diameter GaAs crystals by the Floating Zone technique with controlled as-vapour pressure under microgravity. J Cryst Growth 156:350
48. Fischer B, Friedrich J, Weimann H, Müller G (1999) The use of time-dependent magnetic fields for control of convective flows in melt growth configurations. J Cryst Growth 189:170
49. Barmin IV, Senchenkov AS, Greif A, Wunderwald U, Cröll A, Lyubimova T (2009) Application of rotating magnetic fields to crystal growth under microgravity (Experiments on Foton-M3). Magnetohydrodynamics 45:325
50. Feonychev A, Bondareva N (2002) Crystal growth under microgravity conditions with using of magnetic fields. In: 34th COSPAR scientific assembly, the second world space congress, Houston
51. Dold P, Benz KW (1997) Modification of fluid flow and heat transport in vertical Bridgman configurations by rotating magnetic fields. Cryst Res Technol 32:51
52. Lan CW, Tu CY (2002) Three-dimensional analysis of flow and segregation control by slow rotation for Bridgman crystal growth in microgravity. J Cryst Growth 237:1881
53. Lan CW (1999) Effects of ampoule rotation on flows and dopant segregation in vertical Bridgman crystal growth. J Cryst Growth 197:983
54. Lan CW (1999) Effects of centrifugal acceleration on the flows and segregation in vertical Bridgman crystal growth with steady ampoule rotation. J Cryst Growth 229:595
55. Churilov A, Ostrogorsky AG, Volz MP (2006) Solidification using a baffle in sealed ampoules: ground-based experiments. J Cryst Growth 295:20
56. Ostrogorsky AG (1990) Numerical simulations of single crystal growth by submerged heater method. J Cryst Growth 104:233
57. Ostrogorsky AG, Sell HJ, Scharl S, Müller G (1993) Convection and segregation during growth of Ge and InSb crystals by the submerged heater method. J Cryst Growth 128:201
58. Dutta PS, Ostrogorsky AG (1993) Segregation of tellurium in GaSb single crystals and associated diffusion coefficient in the solute layer. J Cryst Growth 197:749
59. Dutta PS, Ostrogorsky AG (2000) Segregation of Ga in Ge and InSb in GaSb. J Cryst Growth 217:360

60. Ostrogorsky AG, Marin C, Churilov A, Volz MP, Bonner WA, Duffar T (2008) Reproducible Te-doped InSb experiments in microgravity science glovebox at the International Space Station. J Cryst Growth 310:364
61. Ostrogorsky AG, Marin C, Volz MP, Duffar T (2009) Initial transient in Zn-dop ed InSb grown in microgravity. J Cryst Growth 311:3143
62. Cröll A, Volz MP (2009) Detached Bridgman growth—a standard crystal growth method with a new twist. MRS Bull 34:245
63. Yue JT, Voltmer FW (1975) Influence of gravity-free solidification on solute microsegregation. J Cryst Growth 29:329
64. Regel LL, Wilcox WR (1998) Detached solidification in microgravity—a review. Microgravity Sci Technol XI/4:152
65. Dold P, Szofran FR, Benz KW (2002) Detached growth of gallium doped germanium. J Cryst Growth 234:91
66. Schweizer M, Cobb SD, Volz MP, Szoke J, Szofran FR (2002) Defect density characterization of detached-grown germanium crystals. J Cryst Growth 235:161
67. Sylla L, Fauler A, Fiederle M, Duffar T, Dieguez E, Zanotti L, Zappettini A, Roosen G (2008) Dewetting during the crystal growth of (Cd, Zn)Te: in under microgravity. In: Nuclear science symposium conference record
68. Wang YZ, Regel LL, Wilcox WR (2000) Influence of contact angle, growth angle and melt surface tension on detached solidification of InSb. J Cryst Growth 209:175
69. Duffar T, Dusserre P, Picca F, Lacroix S, Giacometti N (2000) Bridgman growth without crucible contact using the dewetting phenomenon. J Cryst Growth 211:434
70. Palosz W, Volz MP, Cobb S, Motakef S, Szofran FR (2005) Detached growth of germanium by directional solidification. J Cryst Growth 277:104
71. Sangiorgi R, Muolo ML, Chatain D, Eustathopoulos N (1988) Wettability and work of adhesion of nonreactive liquid metals on silica. J Am Ceram Soc 71:742
72. Kaiser N, Cröll A, Szofran FR, Cobb SD, Volz M, Dold P, Benz KW (2001) Determination of wetting angles of germanium melts using the sessile drop technique. J Cryst Growth 231:448
73. Zemskov VS, Raukham MR, Barnim IV, Senchenkov AS, Shulpina IL, Sorokin LM (1983) Special features of solidification of alloyed single crystals of indium antimonide in zero gravity conditions. Fizika i Khimiya Obrabotky Materialov 17:56
74. Tegetmeier A (1995) Oberflächenspannung und Wachstumswinkel, zwei wichtige Parameter der Schmelzzonenzüchtung. Dissertation thesis, AlbertLudwigs- Universität
75. Shetty R, Balasubramanian R, Wilcox WR (1990) Surface tension and contact angle of molten semiconductor compounds: I. Cadmium telluride. J Cryst Growth 100:51
76. Mukai K, Yuan Z (2000) Wettability of ceramics with molten silicon at temperatures ranging from 1693 K to 1773 K. Mater Trans JIM 41:338
77. Duffar T, Abadie J (1996) Wetting of InSb melts on crucibles in weightlessness—results of the Texus 32/TEM 01–4 experiment. Microgravity Sci Technol IX
78. Harter I, Dusserre P, Duffar T, Nabot J-P, Eustathopoulos N (1993) Wetting of III–V melts on crucible materials. J Crys Growth 131:157
79. Wald FV (1981) Crystals: growth, properties and applications, vol 5. Springer, New York
80. Fiederle M, Duffar T, Garandet JP, Babentsov V, Fauler A, Benz KW, Dusserre P, Corregidor V, Dieguez E, Delaye P, Roosen G, Chevrier V, Launay JC (2004) J Cryst Growth 267:429
81. Yeckel A, Stelian C, Derby JJ (2009) Heat transfer, capillarity, and phase change during detached Bridgman crystal growth. In: Thermodynamics of phase changes (EUROTHERM-84), Namur, Belgium
82. Yeckel A, Daoutidis P, Derby JJ (2012) Stabilizing detached Bridgman melt crystal growth: proportional-integral feedback control. J Cryst Growth 356:33
83. Yeckel A, Daoutidis P, Derby JJ (2012) Stabilizing detached Bridgman melt crystal growth: model-based nonlinear feedback control. J Cryst Growth 361:16

84. Szofran FR, Cobb SD, Cröll A, Dold P, Kaiser N, Kerat U, Benz KW, Motakef S, Volz MP, Schweizer M, Vujisic L, Walker JS, Pettigrew P (2000) Reduction of defects in germanium-silicon (RDGS). In: NASAMSFC, Huntsville, AL, Science Requirements Document (SRD), 19–20 December 2000
85. Duffar T, Dusserre P, Giacometti N (2001) Growth of GaSb single crystals by an improved dewetting process. J Crys Growth 223:69
86. Mauk MG, Andreev VM (2003) GaSb-related materials for TPV cells. Semicond Sci Tehnol 18:S191
87. Wang XW, Lee H, Lan YC, Zhu GH, Joshi G, Wang DZ, Yang J, Muto AJ, Tang MY, Klatsky J, Song S, Dresselhaus MS, Chen G, Ren ZF (2008) Enhanced thermoelectric figure of merit in nanostructured n-type silicon germanium bulk alloy. Appl Phys Lett 93:193121
88. Kinoshita K, Yoda S (2011) Growth of homogeneous semiconductor mixed crystals by the traveling liquidus-zone method. J Cryst Growth 318:1026
89. Kinoshita K, Ogata Y, Adachi S, Koshikawa N, Yoda S (2005) Excellent compositional homogeneity in In0.3Ga0.7As crystals grown by the traveling liquids-zone (TLZ) method. Microgravity Sci Technol 16:71
90. Kinoshita K et al (2015) Growth of 2 inch $Si_{0.5}Ge_{0.5}$ bulk single crystals. Jpn J Appl Phys 54:04DH03
91. Kinoshita K et al (2012) Homogeneous $Si_{0.5}Ge_{0.5}$ bulk crystal growth as substrates for strained Ge thin films by the traveling liquidus-zone method. Thin Solid Films 520:3279
92. Kinoshita K et al (2011) Homogeneous SiGe crystal growth in microgravity by the travelling liquidus-zone method. J Phys Conf Ser 327:012017
93. Konoshita K et al (2015) Compositional uniformity of a $Si_{0.5}Ge_{0.5}$ crystal grown on board the International Space Station. J Cryst Growth 419:47
94. Abe K et al (2014) Numerical simulations of SiGe crystal growth by the traveling liquidus-zone method in a microgravity environment. J Cryst Growth 402:71
95. Tsuruta T, Hayakawa Y, Kumagawa M (1988) Effect of ultrasonic vibrations on the growth of $In_xGa_{1-x}Sb$ mixed-crystals. Jpn J Appl Phys 27:47
96. Tsuruta T, Hayakawa Y, Kumagawa M (1989) Effect of ultrasonic vibrations on the growth of $In_xGa_{1-x}Sb$ mixed-crystals (ii). Jpn J Appl Phys 28:36
97. Tsuruta T et al (1992) Effect of ultrasonic vibrations on the growth of $In_xGa_{1-x}Sb$ mixed-crystals (iii). Jpn J Appl Phys 31:23
98. Tanaka A et al (2000) Multi-step pulling of GaInSb bulk crystal from ternary solution. J Cryst Growth 209:625
99. Dutta PS, Ostrogorsky AG (1998) Suppression of cracks in $In_xGa_{1-x}Sb$ crystals through forced convection in the melt. J Cryst Growth 194:1
100. Kozhemyakin GN (2000) Indium inhomogeneity in $In_xGa_{1-x}Sb$ ternary crystals grown by floating crucible Czochralski method. J Cryst Growth 220:239
101. Stelian C et al (2005) Growth of concentrated GaInSb alloys with improved chemical, homogeneity at low and variable pulling rates. J Cryst Growth 283:124
102. Stelian C et al (2004) Solute segregation in directional solidification of GaInSb concentrated alloys under alternating magnetic fields. J Cryst Growth 266:207
103. Stelian C et al (2005) Bridgman growth of concentrated GaInSb alloys with improved compositional uniformity under alternating magnetic fields. J Cryst Growth 275:E1571
104. Yu D et al (2016) A review on InGaSb growth under microgravity and terrestrial conditions towards future crystal growth project using Chinese Recovery Satellite SJ-10. Microgravity Sci Technol 28:143
105. Murakami N et al (2008) Growth of homogeneous InGaSb ternary alloy semiconductors on InSb seed. J Cryst Growth 310:1433
106. Hayakawa Y et al (2008) In situ observation of composition profiles in the solution by X-ray penetration method. J Cryst Growth 310:1487

107. Rajesh G et al (2010) In-situ observations of dissolution process of GaSb into InSb melt by X-ray penetration method. J Cryst Growth 312:2677
108. Sakata K et al (2014) Thermal properties of molten InSb, GaSb, and $In_xGa_{1-x}Sb$ alloy semiconductor materials in preparation for crystal growth experiments on the International Space Station. Adv Space Res 53:689
109. Inatomi Y et al (2015) Growth of $In_xGa_{1-x}Sb$ alloy semiconductor at the International Space Station (ISS) and comparison with terrestrial experiments. npj Microgravity 1:15011
110. Kumar N et al (2016) Investigation of directionally solidified InGaSb ternary alloys from Ga and Sb faces of GaSb(111) under prolonged microgravity at the International Space Station. npj Microgravity 2:16026
111. Kazuhiko O et al (1997) Melt mixing of the 0.3In/0.7GaSb/0.3Sb solid combination by diffusion under microgravity. Jpn J Appl Phys 36:3613
112. Mirsandi H et al (2015) A numerical study on the growth process of InGaSb crystals under microgravity with interfacial kinetics. Microgravity Sci Technol 27:313–320

Wetting Behavior and Interfacial Characteristics of High Temperature Melts Under Microgravity

Zhangfu Yuan, Xiangtao Yu, Rongyue Wang, Bingsheng Xu and Likun Zang

Abstract The interfacial characteristics and microstructure of solder/substrate is an important indicator for welding performance. Due to weak gravity-induced convection under microgravity condition, Marangoni convection effect becomes more obvious, which significantly influences the interfacial characteristics and solidification microstructure of alloy by controlling bubble behavior and mass transfer in the melt. To obtain strong Marangoni convection, Sn–3.5Ag/Sn–17Bi–0.5Cu and Sn–3.5Ag/Sn–10Sb alloys with different surface tension are constructed. The alloys are solidified on Cu ring substrate under space microgravity condition (SJ-10 satellite) to study the reaction mechanism of melt/substrate interface and microstructure evolution of alloys based on the analysis of the structure of intermetallic compounds and element distribution in solidified alloy. The corresponding wetting experiment of Sn–3.5Ag and Sn–17Bi–0.5Cu or Sn–10Sb on the Cu ring ground is done under ground. The results indicate that, under microgravity condition, a large number of fine pores appear in the alloys, and a lot of scalloped or small rod-like crystals are formed at alloy/substrate interface. Many big rod-like crystals are also formed in the bulk of alloy. The elementary analysis results show that, after Sn–3.5Ag/Sn–17Bi–0.5Cu alloy being melted, Ag is evenly diffused and distributed in the bulk phase, Bi is also evenly diffused and block aggregation is formed, which is mainly clustered at the outer edge of interface layer at the bulk phase side. There are three layers of intermetallic compound at the interface of alloy and substrate, that is, the thin $Cu_{41}Sn_{11}$ transition layer near the substrate side, the thick Cu_3Sn layer in the middle, and Cu_6Sn_5 layer near the solder. Two layers of substances are formed on the ground. That is, the thin Cu_3Sn transition layer near the substrate side and the Cu_6Sn_5 layer near the solder. Similarly, after Sn–3.5Ag/Sn–Sb alloy being melted, interface reaction happens on the melt/substrate interface. And three layers of substances are also formed: the thin $Cu_{41}Sn_{11}$ layer near the substrate side, the Cu_3Sn in the middle side

Z. Yuan (✉) · X. Yu · R. Wang · L. Zang
Collaborative Innovation Center for Steel Technology,
University of Science and Technology, Beijing, China
e-mail: zfyuan@pku.edu.cn

Z. Yuan · B. Xu
College of Engineering, Peking University, Beijing, China

© Science Press and Springer Nature Singapore Pte Ltd. 2019
W. R. Hu and Q. Kang (eds.), *Physical Science Under Microgravity: Experiments on Board the SJ-10 Recoverable Satellite*, Research for Development,
https://doi.org/10.1007/978-981-13-1340-0_14

and the Cu_6Sn_5 layer near the solder side. The Sb element is more enriched in the alloy Sn in the other area, but in the internal rod-like crystal of bulk phase, Sn is not detected.

Keywords Microgravity solidification · Marangoni convection · Wetting behavior · Intermetallic compound layer · Sn-based alloy

1 Background

Solders, substrate materials and their interfacial reaction products play crucial roles in the reliability of the joint assemblies in microelectronic packages because they provide electrical, thermal and mechanical continuity in electronic assemblies [1, 2]. However, increasing environmental and health concerns over the toxicity of lead, combined with global legislation to limit or ban the use of Pb in manufactured products [3], has led to extensive research and develop works in lead-free solder materials [4–6]. Generally, the most prominent lead-free solders are likely to be tin-based multi-component alloys, with alloying elements such as Ag, Bi, Sb, and Cu, etc. Sn–3.5Ag solder has been broadly targeted as a potential candidate for replacement of Pb–Sn solder. Sn–3.5Ag solder possesses excellent mechanical properties and improved high-temperature resistance compared to Pb–Sn [7–10].

On the one hand, research on interface phenomena after melting of tin-based lead-free solder alloy under microgravity (including phase-change interface during melting, melt surface tension, wettability of container wall, interface phenomenon during solidification, and microstructure formation) not only reveals basic laws of metal melting and provides scientific basis for research on Marangoni convection phenomenon, but also has important significance for developing high-quality lead-free solder, preparing high-performance alloy materials, eliminating soldering and plating defects and improving plating process. At present, even though few researches have been performed on spreading behaviors and droplet shapes of alloy under microgravity [11], relevant research on above characteristics is an important prerequisite for developing a new alloy system. On the other hand, under the trend of continuous increase in electronic packaging requirements and miniaturization of electronic equipment, spontaneous growth of tin whisker is deemed as an important reason for internal short circuit or failure of electronic devices and causes serious effects on quality of electronic products. In particular, it also reduces the reliability of electronic products for long-term use in spatial environment. In highly vacuum spatial environment, spark caused due to short circuit of tin-based lead-free solder will result in tin atom gasification and even plasma arc. The latter can conduct heavy current, and eventually leads to catastrophic accidents. Currently, three American on-orbit satellites have completely failed since the launching of the Unha-7 satellite, and other major faults caused by solder defects and deficiencies also include danger to military aircrafts, Patriot missiles, F-15 radars and artificial cardiac pacemakers, etc. Therefore, it is necessary and urgent to effectively inhibit the growth of intermetallic compounds and improve reliability of microelectronic products through experimental research on wetting and spreading under microgravity and discussion on interface characteristics of tin-based lead-free solder.

In conclusion, this paper studies the Marangoni convection phenomenon of tin-based lead-free solder and discusses interface reaction laws and microstructure formation mechanism through microgravity experiments in space. This not only can theoretically fill the gap of material interface science in this field, but also provide theoretical guidance for actual development and application of lead-free soldering materials. Experimental research on melting of the tin coating and tin-based solder in bonding pad of the microelectronic circuit under microgravity and comparison with solder wettability in ground environment help to explain the effects of microgravity environment on spreading behaviors of molten solders and propose effective technical measures to avoid short circuit and whisker growth caused by open welds.

2 Research Items and Experimental Design

This research focused on experimental research on ground characteristics of tin-based alloy, preparations for ground support and design of microgravity experimental equipment and methods. Spreading and wetting research on tin-based alloy under microgravity was performed in two stages, including ground support preparations prior to space experiments, and wettability experiments of tin-based alloy of space satellites under microgravity.

Ground support experiments included design of centralized space experimental equipment and ground comparison experiment. Main research items included selection and determination of proper alloy system as the space experiment object and installation of tin-based alloy sample on the SJ-10 satellite experimental device [12]. Ground experiment helps to get an overall understanding of interfacial properties and wettability of selected alloy [13–16]. On this basis, the devices required for space experiments were designed, experiment methods and conditions were determined, the optimum experiment scheme was discussed, alloy surface tension and wettability test was performed, assistance in finishing ground and space comparison experiments were provided.

Microgravity experiment in space focuses on wettability experiment of tin-based alloy. The temperature rise experiment was completed in space through heating furnace, and the temperature was gradually decreased upon complete melting of samples until sample solidification. Conducted SEM scanning and EDS element analysis of the samples recovered from the satellite for the interface layer help to determine components of intermetallic compounds (IMCs) and the distribution of these components. In addition, the specialty of alloy interface reaction under microgravity could be determined, element movement laws could be analyzed, and effects of microgravity on spreading behaviors of molten solders through comparison with ground experiment results could be explained.

2.1 Selection of Experimental Materials

Based on space microgravity conditions and expected experiment objectives, an appropriate alloy system should be selected from the following binary alloy systems for research purpose, including Sn–Pb, Sn–Ag, Sn–Al, An–Bi, Sn–Cd, Sn–Cu, Sn–In, Sn–Ni, Sn–Pd, Sn–Pt, Sn–Sb, Sn–Tl and Sn–Zn, and surface tension at relevant melting points of pure metal is listed in Table 1.

In order to observe Marangoni convection phenomenon arising from differences in surface tension and realize the interface reaction between alloy and substrate, alloy density and surface tension shall be taken into full consideration. Specifically, under the condition that the system selected for the experiment satisfies microgravity conditions, alloy is kept at a melt status at a temperature ranging between 400 and 700 °C and has obvious difference in surface tension of different alloy systems, so as to restrict or avoid effects of molecular diffusion caused due to concentration difference on Marangoni convection phenomenon.

On these grounds, according to surface tension listed in the above table, alloy characteristics and phase diagram data, the binary alloy system with small and large surface tension at same temperature as the research object for space experiment was selected.

(1) Selection of alloy with small surface tension

    (a) Sn–Pb alloy: Pb is toxic, but wetting behaviors and interface phenomenon of Sn–Pb alloy system under microgravity are of significance to research on solder characteristics. However, due to limitations of space experiment positions, Sn–Pb alloy is not temporarily selected;

    (b) Sn–In alloy: as In and Sn have unobvious difference in surface tension at melting points, Sn–In alloy is not selected;

    (c) Sn–Tl alloy: as Tl is highly toxic, Sn–Tl alloy can be excluded;

    (d) Sn–Bi alloy: as Bi can significantly reduce surface tension of Sn and the foundation experiment is mature, Sn–Bi alloy can be selected as one of alternative alloys;

    (e) Sn–Bi alloy: as Sb can significantly reduce surface tension of Sn and the foundation experiment is mature, Sn–Sb can be selected.

In consideration of performance differences of above alloys in various aspects, we will select Sn–Bi and Sn–Sb alloys as alternative experiment materials.

Table 1 Surface tension at relevant melting points of pure metal (mN/m)

Metal	Sn	Ag	Al	Au	Cu	Ni	Pd
Surface tension	550	925	871	1145	1330	1796	1482
Metal	Pt	Zn	Pb	Bi	In	Sb	Tl
Surface tension	1763	817	457	390	556	371	459

(2) Selection of alloy with large surface tension

(a) Sn–Al alloy: Sn–Al phase diagram indicates that, when Sn content ranges 0–90%, the alloy is subject to solid-liquid equilibrium at a temperature of 400–650 °C. Therefore, Sn–Al alloy can be excluded;

(b) Sn–Au alloy: surface tension of Sn–Au alloy is unknown. The phase diagram indicates that, when Sn content ranges 0–50%, the alloy is kept at solid phase conditions at a temperature of 400–700 °C, and Sn–Au alloy cannot be completely melted. Therefore, Sn–Au alloy can be excluded;

(c) Sn–Cd alloy: the phase diagram indicates that the alloy can be selected, but its surface tension is unknown;

(d) Sn–Cu alloy: the surface tension of Cu is extremely high. Seen from the phase diagram, when Sn content exceeds 90%, Sn–Cu alloy can be completely melted within the temperature range, the substrate carrier is Cu in the space experiment as well. During the observation of Marangoni convection phenomenon under microgravity, as we cannot determine the source of Cu at the interface layer of alloy and substrate, Sn–Cu can be excluded;

(e) Sn–Ni alloy: seen from the phase diagram, the alloy is kept under solid phase conditions all the time at a temperature of 400–700 °C and cannot be completely melted, so Sn–Ni alloy can be excluded;

(f) Sn–Pd alloy: seen from the phase diagram, the alloy cannot satisfy requirements for experiment temperature;

(g) Sn–Pt alloy: seen from the phase diagram, Sn–Pt is inappropriate, and can be excluded;

(h) Sn–Pt alloy: Sn–Pt alloy has poor wetting performance, and presents non-wetting in Cu at a temperature of lower than 400 °C;

(i) Sn–Ag alloy: seen from the phase diagram, Sn–Ag alloy is accepted.

Optional alloy with small surface tension: (1) Sn–Bi and (2) Sn–Sb.
Optional alloy with large surface tension: (1) Sn–Ag; (2) Sn–Cd; and (3) Sn–Zn.

With comprehensive consideration of alloy system commonly used during welding process, theoretically, application characteristics of alloy in industry at the time of researching physical characteristics of the interface could be developed. Therefore, the combination of experiment materials is finally determined as ternary Sn–Bi–Cu alloy and binary Sn–Ag and Sn–Sb alloys which are most commonly used in daily life.

2.2 Sample Preparation

According to the mass ratio of the target alloy system, Sn–3.5Ag, Sn–17Bi–0.5Cu and Sn–10Sb samples were prepared by weighing the right amount of pure metal powdery raw materials of Sn, Ag, Cu and Bi, thoroughly stirring in the agate crucible, compressing by press for molding, placing the quartz tube containing samples into

WK-2 multi-functional medium frequency induction melting furnace for melting and casting molding.

In order to satisfy requirements for subsequent experiments, the copper-formed mold with inner diameter of 5 mm were selected, and solid steel cylindrical alloy bar with inner diameter of 5 mm was finally obtained. Before the wetting experiment, a cylindrical sample was cut by sample cutter with appropriate height of 3–5 mm, and the mass of experiment samples were kept about 0.7 g as required. Under ethanol wetting, samples were ground with abrasive paper smoothly and cleaned and dried by adopting the acetone-ultrasound method for standby use.

In order to avoid that excessively rough surface of the substrate hinders the measurement of the contact angle, it shall be ensured that the roughness of the substrate used in the experiment is consistent. Hence, the oxygen-free copper ring substrate shall be subject to multi-procedure cleaning before the experiment, including removal of dirt and oxide layer, and specific procedures are shown below:

Chemical degreasing → hot water cleaning → cold water cleaning → pre-pickling → cold water cleaning → chemical cleaning → cold water cleaning → mold release → cold water cleaning → acetone ultrasonic cleaning → drying with air blower;

Degreasing standards: NaOH 10–16 g/L, Na_2CO_3 18–26 g/L and Na_3PO_4 8–10 g/L(80–90 °C);

Cold water cleaning: deionized water;

Pre-pickling: H_2SO_4 100–140 mL/L, HCl 25–50 mL/L (room temperature) for 30–60 s;

Mold release: grind with abrasive paper (200#, 600#, 800#, 1000#, 1200#, 1500# and 2000#), and polish with 20, 10, 5, 2.5 and 1 μm diamond paste successively;

Chemical polishing agent: H_2O_2 400 mL/L, H_2SO_4 100 mL/L, CH_3COOH 60 mL/L, ethanol 40 mL/L, brightener 40 mL/L, gloss enhancer 0.5 g/L (30 °C), polishing frequency 6–20 s.

It shall be emphasized that substrate roughness is an important factor affecting experiment results. In order to ensure scientificity and systematism of the entire experiment and accuracy of experiment results, the abrasive paper shall be selected as per the permanent order during substrate grinding and polishing, and the uniform rotation speed and grinding and polishing direction were selected during procedure setting of the grinding and polishing machine. The polished substrate shall be detected by atomic force microscopy so as to ensure that internal surface roughness of the copper ring used in the experiment is consistent.

2.3 Sample Processing and Assembling

Space experimental programs and sample installation are shown in Fig. 1, which were according to requirements for recoverable satellites and multipurpose furnaces, process and assemble samples. Sn-based alloys with large and small surface tension

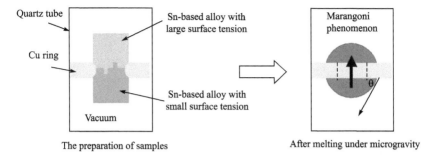

Fig. 1 Sample preparation scheme and schematic diagram for interface phenomenon under microgravity

were fixed to the cooper ring at the same time, and the cooper ring was fixed to the quartz tube wall.

The temperature was adjusted through heating furnace and heating procedures. As the temperature rose, alloy began to melt. When the temperature reached 800 K, two kinds of alloys were completely melted. Alloy elements presented a Marangoni convection trend under driving force caused due to surface tension differences, and the flow pattern was shown in the right figure of Fig. 1. The recycled samples should be analyzed, research on interface phenomenon and microstructure formation laws during solidification of tin-based alloy under microgravity (especially on Marangoni convection phenomenon arising from surface tension gradient with existence of element differences) should be performed, and the interface reaction between the alloy and Cu substrate should be analyzed.

(1) Sample processing and assembling

Sn, Cu, Ag and Sb powders with purity of 99.99% were selected, mixed and casted into a cylinder with a diameter of 6 mm by using the above mentioned method according to the fixed mix ratio, and finally Sn–3.5Ag, Sn–30Bi–0.5Cu and Sn–10Sb alloy samples were obtained. The samples were ground and polished with abrasive paper by layers, cleaned with acetone, and dried for standby use. A set of two ground samples was embedded into the ground Cu ring (as shown in Fig. 1), good contact between the sample and Cu substrate and sample firmness were ensured, and sample fallout under vibration caused by satellite launching would be avoided.

The quartz tube used to packaging samples is shown in Fig. 2. Hollow cylindrical quartz tube and boron nitride cushion at both ends were included, and quartz used in the sample experiment was provided by Shenyang Metal and was molded through machining. Before using, samples were placed in the oven at 50 °C for 2 h dewater treatment. Ground samples were packaged into the quartz tube under vacuum conditions.

(2) Requirements for ampoule packaging process and vacuum degree

In order to improve experiment accuracy, the inside of ampoule was in a high-vacuum state. As shown in Fig. 3, the ampoule was placed into the special heating fur-

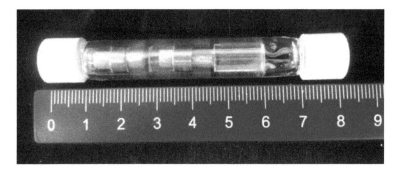

Fig. 2 Schematic diagram for samples packaging into quartz tube

Final seal position

Bottom of the ampoule

180

Fig. 3 Schematic diagram for packaging of vacuum type quartz ampoule

nace, under vacuum of 10^{-3} Pa and 0.5 barometric at a high purity argon atmosphere. The ampoule was heated and finally sintered and sealed at its neck. The length of the sintered quartz ampoule was 213 mm. After ampoule packaging, the surface was cleaned with detergent, immersed into aqua regia ($3HCl + HNO_3$) for about 8 h, and dried for standby use after flushing with purified water. Vibration isolation materials were prepared well to confirm that the samples have been placed in the crucible.

As the samples in the quartz tube shall be kept in a vacuum state, vacuumizing and subsequent sealing during packaging of quartz ampoule will result in formation of a gourd-like nozzle at one end of the ampoule. When fixing the ampoule, winding the nozzle with boron nitride can avoid breaking of quartz ampoule due to collision or vibration, and also avoid destruction of internal vacuum environment.

(3) Selection and assembling mode of vibration isolation piece inside the ampoule

In order to avoid damage to samples or ampoule due to external vibration, boron nitride and graphite materials were used to protect samples and vulnerable parts of the ampoule, and vibration isolation materials were placed into the quartz ampoule, as shown in Fig. 4. The vibration isolation pad in the ampoule shall be high-purity boron nitride pad (or graphite flake) + high-purity quartz fiber mat + metal sheet.

The structure of quartz ampoule is shown in Fig. 4, with a diameter of 16 mm and total length of 213 mm ($\leq$80 mm for samples).

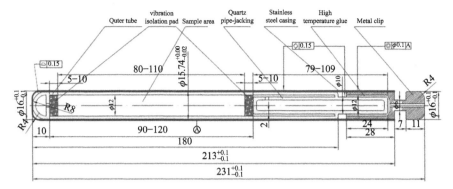

Fig. 4 Overall dimension design of quartz ampoule and distribution of vibration isolation materials

It should be noted that the axis of the quartz sleeve in the ampoule shall be kept on the same line so as to ensure uniform force bearing direction of samples, reduce effects of shear force in the vertical direction and reduce risks of sampling breaking due to vibration occurring during satellite launching or recovery. Tight contact between vibration isolation materials and quartz should be determined to reliably protect samples.

(4) Ampoule vibration experiment

After assembling the ampoule, in consideration of sample stability, the specific mechanical vibration experiment was performed before loading into the satellite so as to evaluate the ampoule assembling quality. In the vibration experiment, force bearing of samples during satellite launching and recovery were simulated to ensure that the ampoule was free of any damage and prevent fallout of samples embedded into the copper ring.

Ampoule vibration experiment was finished by Academy of Opto-electronics, Chinese Academy of Sciences, and experiment items include qualification grade random vibration experiment, qualification grade sine vibration experiment, qualification grade impact experiment, mainly referring to x-axis random vibration experiment, x-axis sine vibration experiment, y-axis random vibration experiment, y-axis sine vibration experiment, z-axis random vibration experiment, z-axis sine vibration experiment; x-axis impact experiment (3 times), y-axis impact experiment (3 times) and z-axis impact experiment (3 times).

Figure 5a, b indicate photo comparison before and after vibration and impact experiments.

During the shaking test before the experiment, the samples made no noise in the axial direction and slight noise in the radical direction. The shaking test after the vibration experiment indicates that the samples make slight noise in axial and radical directions. However, during the test after the impact experiment, the samples made noise in axial and radical directions. After removing the metal sleeve, the ampoule has good appearance. Vibration and impact experiments indicate that prepared ampoules

(a)

(b)

Fig. 5 Comparison diagram of ampoule samples before and after vibration and impact experiments

and samples meet experiment requirements, and subsequent satellite experiment research may be performed.

In conclusion, the entire embedded alloy sample has been packaged into the ampoule, the space multi-purpose materials synthesis furnace was to be assembled for space microgravity experiment.

2.4 Space Multi-purpose Materials Synthesis Furnace

As previously stated, the experiment equipment shall be space multi-purpose materials synthesis furnace. Overall effects for placing of the ampoule into the space multi-purpose materials synthesis furnace are shown in Fig. 6. The sample was placed under a temperature of 400–800 °C, and the furnace depth is 165.5 mm. S1 and S3 respectively represent temperature measuring points of two heating zones. The furnace bottom is sealed, and samples are accumulated in heating zones 1 and 2, with a heating length of 82 mm.

2.5 Experiment Process Under Microgravity

2.5.1 Experiment Conditions

According to overall requirements for SJ-10 satellite experiment in space and characteristics of alloy samples, two groups of experiment samples were assembled, and the position of alloy sample in the ampoule is shown in the figure. The internal quartz tube has outer diameter of 14 mm; the sample baffle has a thickness of 2 mm; the sample embedded into the copper ring has a height of 10 mm; and the cooper ring has a thickness of 2 mm (Fig. 7).

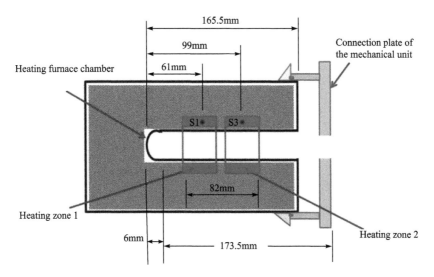

Fig. 6 Schematic diagram of overall effects for placing of ampoule into space multi-purpose materials synthesis furnace

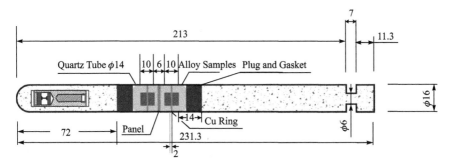

Fig. 7 Position distribution of space experiment samples in ampoule

2.5.2 Experiment Process

In order to observe Marangoni convection driven by surface tension, the experiment shall be performed under the condition that the alloy was completely melted and kept under a stable state, and experiment temperature curve and lifting process are shown in Fig. 8. The temperature of samples in the heating zone may reach 780 K, and the constant temperature shall be maintained for 1600 s for complete melting of samples. At the end of the experiment, the temperature drops at constant speed. When the furnace temperature reaches about 25 °C, the sample management unit of the heating furnace can lift samples at constant speed and remove them from the furnace. Samples were recovered at the time of satellite recovery after solidification, and relevant interface reaction analysis and test were performed.

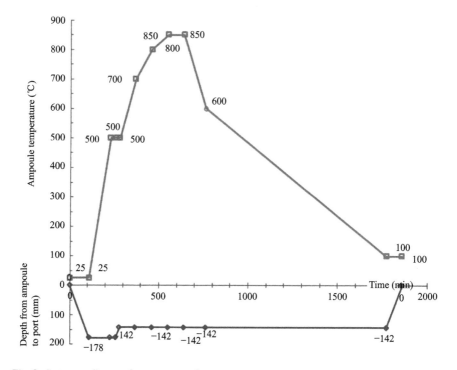

Fig. 8 Sequence diagram for space experiment

The total actual on-orbit experiment time reaches 192 h. The following table lists specific and detailed operation procedures for the experiment. The experiment was performed in the morning of the fifth day and lasted for 1893 min in total. The valid experiment time from heating to cooling was 1664 min, including 120 min for temperature rising to 500 °C, 52 min for maintaining temperature at 500 °C, 90 min for temperature rising from 500 to 700 °C, 90 min for temperature rising from 700 to 800 °C, 90 min for temperature rising from 800 to 850 °C, and 90 min for maintaining temperature at 850 °C. Subsequently, the temperature began to drop (Table 2).

The live ground monitoring diagram produced in the moving of orbiting satellite is shown in the figure below, which was used to monitor the real-time experiment progress and temperature field control conditions, and received the real-time actual temperature data feedback (Fig. 9).

Figures 10, 11 and 12 are the alloy morphology and position diagrams for equipment positioning observed through rays, the space experiment can be smoothly completed by measuring the position of alloy and fixing state. As for the melting morphology of the sample after experiment, it was almost the same with the sample before experiment. However, it was distinct from that on the ground, where the Cu ring would melt totally.

Table 2 On-orbit execution (experiment on working position 4, 09:48:35 (Day 5)–17:11:35 (Day 6))

S. no	On-orbit time	Execution action (or commands)	Duration (s)	Design value (typical value)	Remarks
1	Day 5, 09:48:35–09:49:35	Go upwards in reverse rotation under the action of motor 2 (limit switch)	60		Lift to the upper limit position
2	Day 5, 09:49:35–09:55:35	Rotate in the reverse direction under the action of motor 1 (limit switch)	360		Rotate to the reset position
3	Day 5, 09:55:35–10:01:35	Rotate in the reverse direction under the action of motor 1 (limit switch)	360		Rotate to working position 4
4	Day 5, 10:01:35–11:51:35	Go downwards in forward rotation under the action of motor 2 (limit switch)	6600		Samples are rapidly transferred into the position 178 at the bottom of the furnace within 110 min
5	Day 5, 11:51:35–13:51:35	PID2 closed-loop heating	7200	Temperature: room temperature—500 °C	The temperature rapidly rises from room temperature to 500 °C within 120 min
6	Day 5, 13:51:35–14:21:35	PID2 closed-loop heating	1800	Temperature: 500 °C	The temperature is maintained at 500 °C for 30 min
7	Day 5, 14:21:35–14:43:35	Go upwards in reverse rotation under the action of motor 2 (open loop); PID2 closed-loop heating	1320	Temperature: 500 °C	The ampoule upwardly moves to the second sample experiment position within 22 min; hold the temperature for 22 min at 500 °C
8	Day 5, 14:43:35–16:13:35	PID2 closed-loop heating	5400	Temperature: 500–700 °C	The temperature rises from 500 to 700 °C within 90 min
9	Day 5, 16:13:35–17:43:35	PID2 closed-loop heating	5400	Temperature: 700–800 °C	The temperature rises from 700 to 800 °C within 90 min
10	Day 5, 17:43:35–19:13:35	PID2 closed-loop heating	5400	Temperature: 800–850 °C	The temperature rises from 800 to 850 °C within 90 min

(continued)

Table 2 (continued)

S. no	On-orbit time	Execution action (or commands)	Duration (s)	Design value (typical value)	Remarks
11	Day 5, 19:13:35–20:43:35	PID2 closed-loop temperature preservation	5400	Temperature: 850 °C	Hold the temperature for 90 min at 850 °C
12	Day 5, 20:43:35–22:43:35	PID2 closed-loop cooling	7200	Temperature: 850–600 °C	The temperature drops from 850 to 600 °C within 120 min
13	Day 5, 22:43:35–Day 6, 15:35:35	PID closing stage	60,720	Temperature: 600 °C–room temperature	600 °C natural cooling stage
14	Day 6, 15:35:35–17:01:35	Go upwards in reverse rotation under the action of motor 2 (limit switch); PID closed	5160		The sample is rapidly drawn out of furnace hearth within 86 min; naturally cooled
15	Day 6, 17:01:35–17:02:35	Go upwards in reverse rotation under the action of motor 2 (limit switch)	60		Judging whether the upper limit switch works normally
16	Day 6, 17:02:35–17:03:35	Go upwards in reverse rotation under the action of motor 2 (open loop)	60		Go upwards by 15,000 steps under the action of lifting motor (open loop)
17	Day 6, 17:03:35–17:04:35	Go downwards in forward rotation under the action of motor 2 (open loop)	60		Go downwards by 15,000 steps under the action of lifting motor (open loop)
18	Day 6, 17:04:35–17:05:35	Go upwards in reverse rotation under the action of motor 2 (limit switch)	60		Judging whether the upper limit switch works normally
19	Day 6, 17:05:35–17:11:35	Rotate in the reverse direction under the action of motor 1 (limit switch)	360		Return to the reset state
20	Day 6, 17:11:35–18:40:00	Stand-by			Stand-by
21	Day 6, 18:40:00–Day 7, 03:00:00	Shutdown			Shutdown

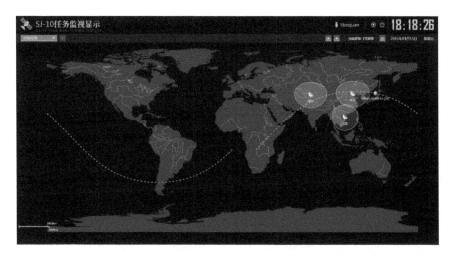

Fig. 9 Ground monitoring chart for moving trajectory of satellite

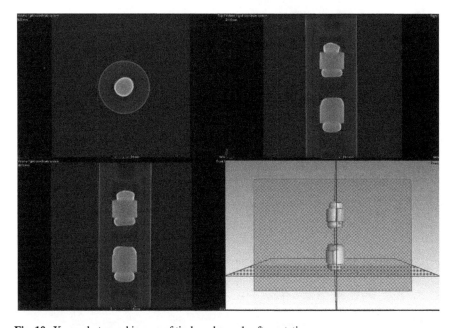

Fig. 10 X-ray photographic map of tin-based sample after rotating

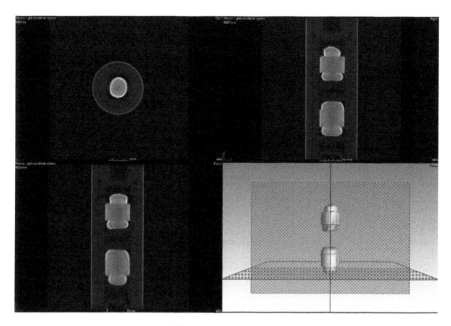

Fig. 11 X-ray photographic map of tin-based sample after rotating 45°

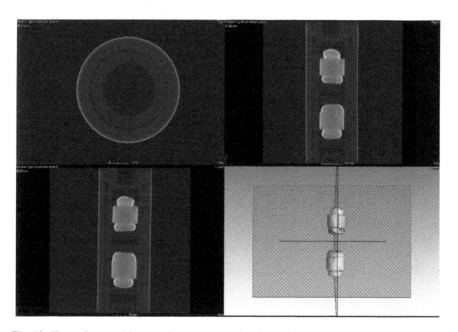

Fig. 12 X-ray photographic map of tin-based sample after rotating 90°

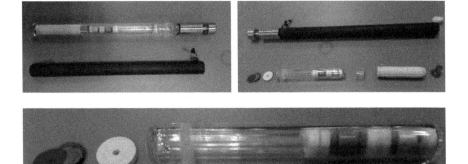

Fig. 13 Recovery and dismantling diagram of ampoule sample

The multi-function furnace was opened for sampling of high-temperature melt of SJ10 satellite re-entry module in Institute of Physics, Chinese Academy of Sciences on April 20, 2016. On April 26, 2016, 6 ampoules were opened in Shanghai Institute of Ceramics, Chinese Academy of Sciences, and Fig. 13 is the sample photo of ampoule No. 4 and the figure below is the profile photo of two groups of tin-based alloy samples in this experiment.

2.6 Testing and Analysis of Recovered Sample

After the completion of experiment, the samples were recovered from the satellite, cut along the vertical section of two types of alloy in the sample. Then the samples were ground, polished and etched. The element distribution at the interface of these two types of alloy was observed through SEM and EDS scanning, the reaction mechanism of alloy and the flow law of metallic elements were discussed.

Specifically, the solidified and mix-connected solder alloy and substrate sample was cut with sample cutting machine, mount the sample by mounting machine with phenolic plastic, ground from coarse to fine on the grinding machine with abrasive paper in difference particle sizes after cooling, and made mechanical polishing by polishing machine with canvas and flannelette. In such case, the polishing liquid was made of diamond polishing agent and water. After polishing, the sample surface was washed with deionized water, dehydrated with absolute ethyl alcohol, etched with etching agent (made of 10 g $FeCl_3$, 2 mL concentrated hydrochloric acid and 100 mL distilled water) for about 1 min. The residual etching agent on the sample surface was washed with plenty of water, the surface was cleaned again with absolute ethyl alcohol, and dried with cold air for future interface observation and analysis.

2.6.1 Optical Microscope (OM)

The optical microscope used in the experiment is produced by Olympus, with the model of BX51 M and magnification of 50–1000X. It is equipped with advanced software, which not only can conquer the impact of uneven sample surface on focusing and imaging, but also can accurately display the morphologic features at the interface.

With wave front aberration control technology applied, being equipped with image processing software, the optical microscope has good imaging effect on the surface morphology of samples to be measured. It can analyze the roughness of the substrate surface, and completely display the uneven interface morphology of sample surface in different depths of field by gathering groups of photos and through depth of field treatment. In addition, it can collage groups of photos and maximize the display of overall morphology of samples at the maximum magnification, so it is a significant tool for studying the formation of intermetallic compound and measuring the thickness of interface layer. Some of recovered sample are predicted in Fig. 14 by optical microscope.

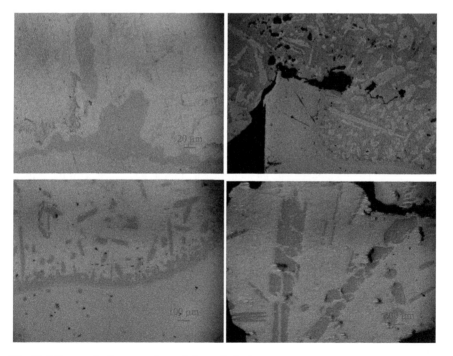

Fig. 14 Micrograph of recovered sample by optical microscope

2.6.2   Scanning Electron Microscope (SEM) and Energy Dispersive Spectrometer (EDS)

The formation and distribution of intermetallic compound at the interface of solder and substrate were tested with the help of the electron scanning microscope (model: KYKY2800B) of the research group. Figure 15 shows the morphology of the samples by using the SEM. The KYKY2800B scanning electron microscope mainly consists of three parts, namely, the electron optical system, including the electron emission gun, electromagnetic lens and scanning coil, the sample room and the gathering, processing and displaying system of signal from the sample room. The specific working process of electron scanning microscope is as follows: the tungsten filament on top of the electron beam gun excites electron beams under high pressure, which finally irradiates the sample surface through centering and accepting two times of focusing after anode acceleration. Scan line by line on the sample surface, and excite plenty of physical signals, including secondary electron, backscattered electron and X-ray signals, and the signal strength varies with the surface features. Excite the secondary electron signal formed on the sample surface to be received by the electronic detector and amplified by the annunciator, and finally display the actual morphology of sample on the display.

Features of KYKY2800B electron scanning microscope include:

(1) High magnification: wide range of variation in magnification (from several times to tens of thousands of times), with the magnification process and magnification being continuously adjustable;

(2) High resolution: the resolution of secondary electron image being 3.0 nm (tungsten filament);

(3) Large depth of field: the image being large in depth of field, and rough surface with great relief being directly observable;

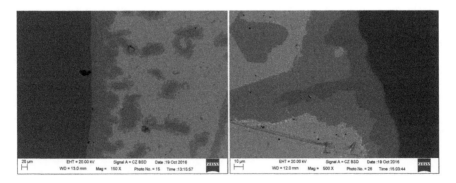

Fig. 15 Electronograph of recovered sample

(4) Good fidelity: the actual morphology of sample being fully displayable;
(5) Simple preparation of sample. Block or power samples can be directly placed under the electron microscope for observation without processing or with slight processing, which is simple relative to transmission electron microscope (TEM) in sample preparation.

Energy dispersive spectrometer (EDS) is used together with the electron scanning microscope to give quantitative and qualitative analysis to the sample using the characteristic X-ray produced by bombarding the sample with backscattered electron gathered by spectroscopy probe. The energy dispersive spectrometer is equipped with G7 micro-area energy dispersive spectrometer analysis system (Noran System Seven-NS7), which is mainly featured by the intelligent qualitative analysis program, intelligent quantitative analysis program, quantitative analysis correction algorithm, quantitative surface distribution, and line scanning multi-point automatic analysis. The composition and proportional relation of interface product are analyzed by point, surface and line scanning, so as to determine the intermetallic compound and promote the thermodynamics and dynamics analysis of interface reaction.

Qualitative analysis principle of energy dispersive spectrometer: the frequency of characteristic X-ray excited may vary from elements, and testing the energy (frequency) of photons can realize the qualitative analysis of elements. Quantitative analysis principle of energy dispersive spectrometer: measuring the X-ray strength of elements in the sample to be measured, comparing them with those of elements in the standard sample and making calculations can approximately obtain the percentage of elements in the respective samples.

Analysis features of energy dispersive spectrometer:

(1) Rapid qualitative and quantitative analysis can be made to all the elements included in the sample, including Be–U;
(2) If the magnification is not required, the scanning results can be obtained at a low magnification;
(3) The qualitative analysis is high in accuracy, the X-ray collection rate is high, and the test limit is 0.1%;
(4) The quantitative analysis is low in accuracy, but the quantitative analysis error in overlapping peak-free major element of medium atomic number is about 2%.

The use of electron scanning microscope and energy dispersive spectrometer are useful for analyzing the formation mechanism of interface layer.

3 Experiment Results Analysis

3.1 Contrast Analysis of Interface Reaction Between Ground and Microgravity Conditions

3.1.1 Analysis of Interface Morphology and Composition of Sn–3.5Ag/Sn–17Bi–0.5Cu on Cu Ring Substrate

Figure 16 is the SEM images for morphology of upper, middle and lower parts of alloy Sn–3.5Ag/Sn–30Bi–0.5Cu. A continuous interface layer is on top of the upper part of sample where obvious rod-like crystal appears. On the contrary, no continuous interface layer is on top of the lower part of sample where rod-like crystal appears to be less than that in the upper part. The sample is fully distributed with a plenty of fine air-holes, which is caused by the reason that the bubbles cannot escape from the sample in a timely manner under microgravity conditions and remain in it.

Seen from Figs. 16, 17 and 18, Cu substrate is corroded by the alloy, and the corrosion varies from alloy parts. The corrosion at the upper part of interface between Cu and alloy is more serious than that at the lower part, low arc-shaped corrosion part appears at the edge of Cu substrate, and scalloped crystal appears at the interface. The corrosion is not obvious at the lower part of interface, the shape of crystal at the interface gradually changes from scalloped to flat, and rod-like crystal appears at the edge of Cu substrate.

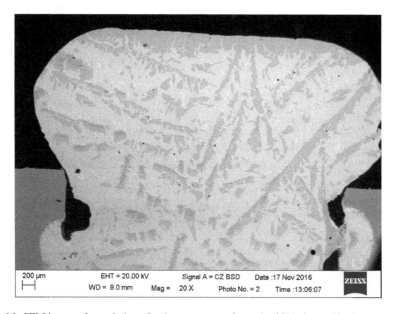

Fig. 16 SEM image of morphology for the upper part of sample (20X after etching)

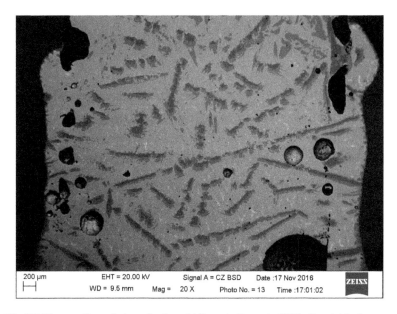

Fig. 17 SEM image of morphology for the middle part of sample (20X after etching)

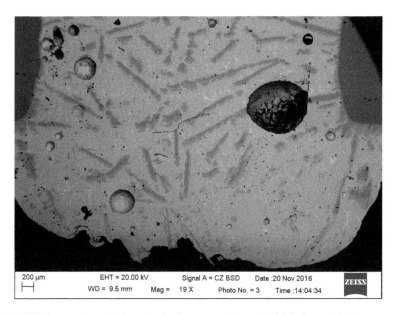

Fig. 18 SEM image of morphology for the lower part of sample (20X after etching)

It can be seen that a continuous intermetallic compound (IMC) reaction layer forms at the interface layer of upper and lower parts, which are different with substances on both sides in color and distributed in an arch shape, which proves that interface reaction has happened between solder and substrate, and columnar material has formed in the alloy. The composition and thickness of reaction layer is another significant factor affecting the quality of solder. Therefore, it is necessary to analyze the morphology and specific composition of reaction product by EDS and other interface characterization means.

Figure 19 is the line scanning results of interface reaction layer along the transverse route running through the solder and substrate, and it can be determined that chemical reaction has happened at the interface layer of solder and substrate. It can be preliminarily determined from the 20X image, compound of Cu and Sn, as well as compound of An and Sn have formed at the interface layer, which is consistent with the experiment results on the ground.

Analyze the surface distribution scanning results of metallic elements such as Cu, Sn, Ag and Bi at magnification of 600X respectively. It can be seen that Cu is mainly distributed on the substrate side and extends to the internal of solder alloy from the interface layer, which suggests that the reaction Cu is formed by the dissolution and diffusion of substrate, and correspondingly, Sn is distributed between the interface and bulk phase, which suggests that Sn has participated in the interface reaction,

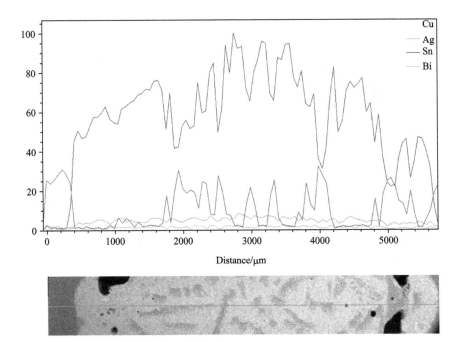

Fig. 19 Line scanning results at the interface around the system interface reaction layer (20X)

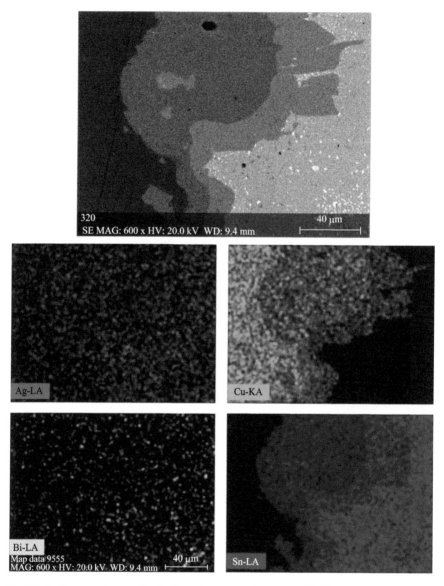

Fig. 20 Surface scanning results of alloy substrate reaction system (600X)

Ag is mainly distributed in bulk phase, which proves that a plenty of Ag_3Sn has formed in the solder, and Bi has not participated in the chemical reaction of interface (Fig. 20).

Through the analysis of energy dispersive spectrometer results and phase diagram relations after magnifying the arch-shaped interface at the left side by 600X, it can be seen that interface reaction has happened on the Cu substrate and a thick Cu_6Sn_5 layer near the substrate side and a transition layer at the alloy side have been formed as shown in the figure. According to the relevant ground experiment results, two compounds can be produced after the reaction between Cu and Sn, namely, Cu_6Sn_5 and Cu_3Sn. The specific reaction process is as follows: when Cu contacts the liquid Sn, Cu will rapidly dissolute, which makes the concentration of Cu in the liquid alloy at the interface between the liquid alloy and Cu increase, and this dissolution is unbalance; chemical reaction happens between Cu and metastable liquid alloy, and its driving force is stronger than that of Cu_3Sn, which makes Cu_6Sn_5 rapidly nucleates and grows on the Cu substrate. The Cu_6Sn_5 phase separates rapidly at the interface of liquid alloy and Cu, and it quickly grows laterally and finally grows to be a continuous compound layer after nucleation. The grain boundary diffusion of Cu in the Cu_6Sn_5 layer and the subsequent interstitial diffusion in Sn accelerate the growth of Cu_6Sn_5 phase. Cu atom, through grain boundary diffusion and interstitial diffusion in the Cu_6Sn_5 layer, reaches the Sn/Cu_6Sn_5 interface and reacts with Sn so that the Cu_6Sn_5 compound layer is thickened. As the Cu_3Sn phase is thinner, it is further magnified to determine its existence (Fig. 21).

Figure 22 is the line scanning results around the interface reaction layer after magnifying by 3500X. Through the analysis of energy dispersive spectrometer results and phase diagram relations, it can be seen that interface reaction has happened on the Cu substrate and three layers of substances have been produced as shown in the figure, namely, the thin Cu_xSn_y transition layer near the substrate side, the thick Cu_3Sn layer and Cu_6Sn_5 near the alloy side. Both the Cu_6Sn_5 compound layer and the Cu_3Sn transition layer can be seen in the figure. Cu_3Sn grows by consuming Cu_6Sn_5 at the Cu_6Sn_5/Cu_3Sn interface, but Sn atom can directly reacts with Cu atom to produce Cu_3Sn at the Cu_3Sn/Cu interface. In such case, the Cu_3Sn/Cu interface will move to the Cu substrate side, causing Cu corrosion.

The bright dotted substance in the alloy is Bi, which segregates at the alloy side, and mainly clusters at the edge of interface at the bulk phase side. Obvious pure Bi peak appears at the interface near the alloy side where almost no Cu exists, which

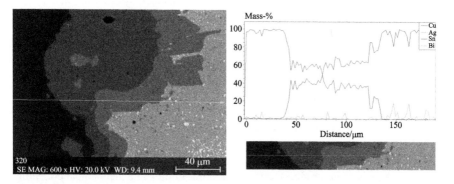

Fig. 21 Line scanning results at the interface around the interface reaction layer (600X)

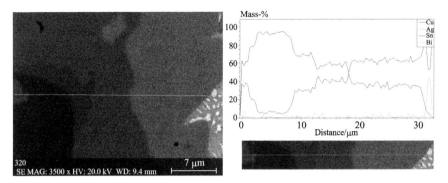

Fig. 22 Line scanning results at the interface around the interface reaction layer (3500X)

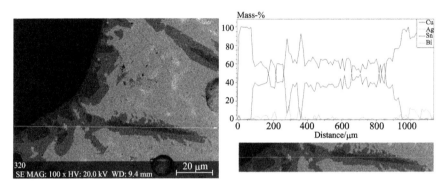

Fig. 23 Line scanning results at the interface around the internal substances of alloy (100X)

proves that Bi-rich layer exists. However, the Sn peak at the interface near the Cu substrate side increases, which is mainly due to the reaction of substrate Cu with Sn in the alloy. The appearance of Cu–Sn reaction layer is also the main reason for the appearance of Bi-rich layer at the interface, which is caused by the reason that Cu does not react with Bi and the Sn in Sn–Bi alloy is consumed by the reaction between Sn and substrate Cu, so a plenty of redundant Bi elements are discharged near the substrate side and a pure Bi layer area is formed at the secondary near the substrate side.

Additionally, Fig. 23 is the line scanning results around the interface reaction layer and rod-like crystal in the alloy after magnifying by 100X. Through the analysis of energy dispersive spectrometer results and phase diagram relations, it can be seen that the interface product is a two-layer intermetallic compound consisting of scalloped Cu_6Sn_5 near the alloy side and laminar Cu_3Sn near the Cu side. In addition to the interface layer around the substrate, the other layer partially passes through the bottom layer and extends to the bulk phase where part of the second interface is cracked due to the wear produced during the online cutting and mechanical polishing of solidified sample.

3.1.2 Analysis of Interface Morphology and Composition of Sn–3.5Ag/Sn–10Sb on Cu Ring Substrate

Figure 24 is the SEM images for interface morphology of upper, middle and lower parts of alloy Sn–3.5Ag/Sn–10Sb when the interface between Cu and substrate is magnified by 20X after reaction of them. It can be seen from the figure that the corrosion effects of Sn–3.5Ag and Sn–10Sb on the Cu substrate are similar, with a corrosion thickness of about 1000 um, which is more than one time thicker than that of Sn–3.5Ag/Sn–17Bi–0.5Cu on the Cu ring substrate (the thickest part is about 400 um). The rod-like crystal at the corrosion interface is small, but the rod-like crystals formed on top of the sample, on the alloy in the central of the sample, as well as on the bottom of the sample are big. Openings are clearly seen in the sample, especially at the interface layer of substrate and alloy, continuous and regular openings appear (Figs. 25 and 26).

It can be seen from Fig. 27 that the morphology of alloy formed outside the copper substrate covers the top area of the entire sample. According to the analysis of line scanning results, the intermetallic compound Cu_3Sn grows to the surface in columnar crystal shape and passes through the entire alloy system; the intermetallic compound is in a double-layer structure and a light Cu_6Sn_5 layer is covered at the outer edge of columnar intermetallic compound Cu_3Sn.

Figure 28 is the surface scanning results of reaction system of the right alloy substrate. Analyze the surface distribution scanning results of metallic elements such as Cu, Sn, Ag and Sb at magnification of 25X respectively. It can be seen that

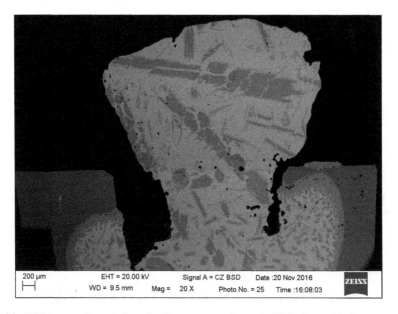

Fig. 24 SEM image of morphology for the upper part of sample (20X after etching)

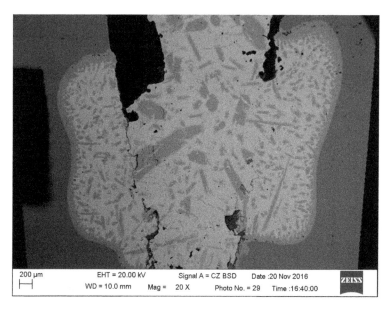

Fig. 25 SEM image of morphology for the middle part of sample (20X after etching)

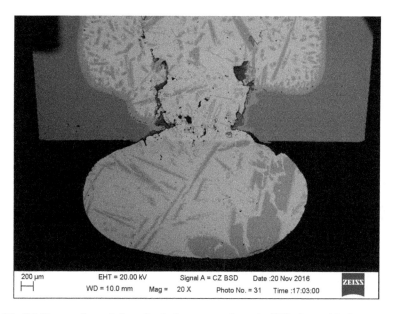

Fig. 26 SEM image of morphology for the lower part of sample (20X after etching)

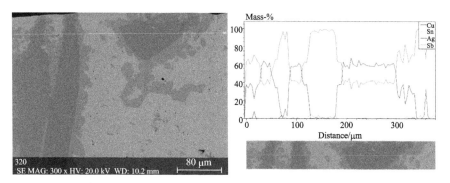

Fig. 27 SEM image of morphology on top of the sample (300X after etching)

Cu is mainly distributed on the substrate side and extends to the internal of solder alloy from the interface layer, which suggests that the reaction Cu is formed by the dissolution and diffusion of substrate, and correspondingly, Sn is distributed between the interface and bulk phase, which suggests that Sn has participated in the interface reaction. Ag is evenly distributed in the bulk phase, and Sn is mainly distributed in the bulk phase. However, Ag rather than Sb is detected in the rod-like crystal of bulk phase

Figure 29 is the line scanning results around the interface reaction layer after magnifying by 350X. Through the analysis of energy dispersive spectrometer results and phase diagram relations, it can be seen that interface reaction has happened on the Cu substrate and three layers of substances have been produced as shown in the figure, namely, the thin Cu_xSn_y layer near the substrate side, the relatively thick Cu_3Sn layer, and the Cu_6Sn_5 layer near the alloy. The enrichment of Sb in the alloy Sn is larger than that in other places.

Point scanning is made to the sample to further determine the specific composition of the thin Cu_xSn_y layer distributed between the Cu substrate and Cu_3Sn. Figure 30 the point scanning results of the internal alloy after magnifying by 1000X, and the first interface layer is mainly distributed by Cu and Sn, with few Ag. According to the quantitative analysis results of elements, Atom% (Cu): Atom% (Sn) = 79.62: 20.28, and the element is proved to be the interface compound $Cu_{41}Sn_{11}$, which is different from the results of ground experiment (Fig. 31).

3.2 The Interface Reaction Dynamics Analysis

Chemical reaction equations at four different interphases can be expressed as follows. The interfacial reactions were also a factor affecting the atom flux, as supported by both interface shifting and growth of IMC layers. Note that the elements in square brackets denote the diffusing species. The supply of Cu and Sn atoms to their

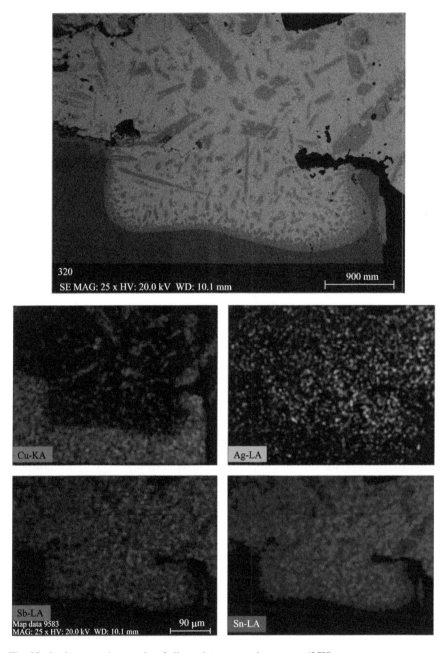

Fig. 28 Surface scanning results of alloy substrate reaction system (25X)

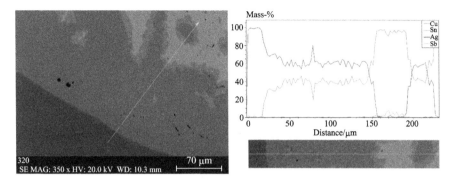

Fig. 29 Line scanning results at the interface around the interface reaction layer (350X)

E1	AN	Series	unn. C [wt. %]	norm. C [wt. %]	Atom. C [at. %]	Error [wt. %]
Cu	29	K-series	52.98	67.67	79.62	1.5
Ag	47	L-series	0.11	0.14	0.09	0.0
Sn	50	L-series	25.21	32.20	20.28	0.8
Sb	51	L-series	0.00	0.00	0.00	0.0

Total: 78.30   100.00   100.00

Fig. 30 Point scanning results of internal alloy (1000X)

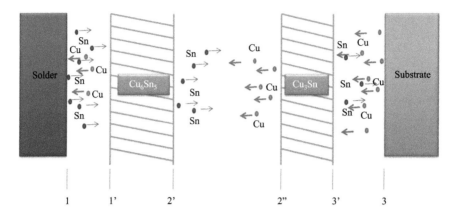

Fig. 31 The interface reaction mechanism of Sn–Ag–Cu–x alloy and Cu substrate (x = Bi or Sb)

corresponding interfaces was composed of two parts, released from Cu substrate or solder, and the decomposition of IMC phase.

At the solder/Cu_6Sn_5 interface:

$$5Sn + 6[Cu] \rightarrow Cu_6Sn_5 \tag{1}$$

The growth of Cu_6Sn_5 on the solder side depends on the availability of Cu atoms in the solder and the amount of Cu atoms that can diffuse to the solder/Cu_6Sn_5 interface from the Cu substrate. The solder/Cu_6Sn_5 interface moves toward the solder matrix.

At the Cu_6Sn_5/Cu_3Sn interface (the interface migrates toward the Cu substrate):

$$Cu_3Sn \rightarrow 9[Cu] + Cu_6Sn_5 \tag{2}$$

$$2Cu_3Sn + 3[Sn] \rightarrow Cu_6Sn_5 \tag{3}$$

The surplus diffuse-in Sn atoms would react with the Cu_3Sn layer to form the Cu_6Sn_5 layer. It is equivalent to saying that Cu_3Sn layer was consumed by the growth of Cu_6Sn_5 layer.

At the Cu_6Sn_5/Cu_3Sn interface (the interface moves toward the solder matrix):

$$Cu_6Sn_5 + 9[Cu] \rightarrow 5Cu_3Sn \tag{4}$$

$$Cu_6Sn_5 \rightarrow 3[Sn] + 2Cu_3Sn \tag{5}$$

After the Cu atoms arrive at the Cu_3Sn/Cu_6Sn_5 interface by diffusion through the grain boundaries of the Cu_3Sn layer. They react with Cu_6Sn_5 to form Cu_3Sn by consuming Cu_6Sn_5. This growth mechanism causes the Cu_6Sn_5/Cu_3Sn interface to move towards the Cu_6Sn_5 side.

At the $Cu_3Sn/Cu_{41}Sn_{11}$ interface:

$$3[Cu] + [Sn] \rightarrow Cu_3Sn \tag{6}$$

The Sn atoms diffuse through the Cu_6Sn_5 and Cu_3Sn layers, and react with Cu atoms from the Cu substrate at the $Cu_3Sn/Cu_{41}Sn_{11}$ interface to move towards the Cu substrate or dissolution of Cu.

At the $Cu_{41}Sn_{11}/Cu$ interface:

$$3Cu + [Sn] \rightarrow Cu_{41}Sn_{11} \tag{7}$$

The Sn atoms diffuse through the Cu_6Sn_5 and Cu_3Sn layers, and react with Cu at the $Cu_{41}Sn_{11}/Cu$ interface to move towards the Cu substrate or dissolution of Cu.

At the initial stage, when Cu contacts the liquid Sn, Cu will rapidly dissolute, which makes the Cu atoms supersaturated in solder matrix. Then, Cu_6Sn_5 is formed first by the chemical reaction between Cu and metastable liquid alloy as its driving force is stronger than that of Cu_3Sn and Cu_6Sn_5 has a lower activation energy compared with Cu_3Sn phase. Cu atom, through grain boundary diffusion and interstitial diffusion in the Cu_6Sn_5 layer, reaches the Sn/Cu_6Sn_5 interface and reacts with Sn by Eq. (1), so that the Cu_6Sn_5 compound layer is thickened. Thirdly, Cu_3Sn layer grow on both sides by Eqs. (4), (5) and (6), make the interface moves toward the solder matrix and the Cu substrate. In some conditions, $Cu_{41}Sn_{11}$ was formed by Eq. (7). Finally, the thick layer would act as the diffusion barrier, retarding the diffusion of Cu and Sn. According, the growth of IMC layer was suppressed, and the move of interphase interfaces was hindered. The final IMC layer is formed.

4 Summary

In the background of studying the interface phenomenon and microstructure formation rules of tin-based alloy during the solidification under microgravity, the paper focuses on the study and preparations of relevant ground supporting experiment. To observe the Marangoni flow due to surface tension caused by composition difference during the melting of tin-based alloy, the wetting experiment of Sn–3.5Ag and Sn–17Bi–0.5Cu or Sn–10Sb on the Cu ring is selected as the object of study to design the experiment scheme and process, which include discovering the sample composition and packaging method, assembly of ampoule used for space experiment, mechanical testing and tube furnace modification, helping to complete the space and ground contrast experiment. Based on this, the whole process of microgravity experiment is described in details, and SEM and EDS analysis are made to the recovered sample. The sample is distributed with numerous fine pores, and scalloped or small rod-like crystals are formed at the interface, and big rod-like crystals are formed in the alloy area. According to the elementary analysis results, after Sn–3.5Ag/Sn–17Bi–0.5Cu alloy melts under microgravity conditions, Ag is evenly diffused and distributed in the bulk phase, Bi is also evenly diffused and block aggregation is formed, which is mainly clustered at the outer edge of interface layer at the bulk phase side. Three layers of intermetallic compound is formed at the interface of alloy and substrate, namely, the thin Cu_xSn_y transition layer near the substrate side, the thick Cu_3Sn layer in the middle, and Cu_6Sn_5 layer near the solder. However, only two layers of substances are formed on the ground, namely, the thin Cu_3Sn transition layer near the substrate side and the Cu_6Sn_5 layer near the solder. Similarly, after Sn–3.5Ag/Sn–Sb alloy melts, interface reaction has happened on the Cu substrate and three layers of substances have been produced, namely, the thin $Cu_{41}Sn_{11}$ layer near the substrate side, the Cu_3Sn in the middle and the Cu_6Sn_5 layer near the solder side. The enrichment of Sb in the alloy Sn is larger than that in other places, but Sn is not detected in the internal rod-like crystal of bulk phase.

References

1. Tu KN (2007) Solder joint technology: materials, properties, and reliability
2. Tu KN (2011) Reliability challenges in 3D IC packaging technology. Microelectron Reliab 51(3):517–523
3. Fournelle RA (2003) Lead-free solders and processing issues in microelectronics. JOM 55(6):49
4. Kang SK (2002) Developments in lead-free solders and soldering technology. JOM 54(6):25
5. Frear DR, Jang JW, Lin JK et al (2001) Pb-free solders for flip-chip interconnects. JOM 53(6):28–33
6. Chiang HW, Chen JY, Chen MC et al (2006) Reliability testing of WLCSP lead-free solder joints. J Electron Mater 35(5):1032–1040
7. Frear DR, Vianco PT (1994) Intermetallic growth and mechanical behavior of low and high melting temperature solder alloys. Metall Mater Trans A 25(7):1509–1523
8. Plumbridge WJ, Gagg CR (2000) The mechanical properties of lead-containing and lead-free solders—meeting the environmental challenge. Proc Inst Mech Eng, Part L: J Mater Des Appl 214(3):153–161
9. Abtew M, Selvaduray G (2000) Lead-free solders in microelectronics. Mater Sci Eng: R: Rep 27(5):95–141
10. Ochoa F, Williams JJ, Chawla N (2003) Effects of cooling rate on the microstructure and tensile behavior of a Sn-3.5 wt.% Ag solder. J Electr Mater 32(12):1414–1420
11. Hu WR, Zhao JF, Long M et al (2014) Space program SJ-10 of microgravity research. Microgravity Sci Technol 26(3):159–169
12. Xu B, Chen J, Yuan Z et al (2016) Spreading dynamics and interfacial characteristics of Sn–3.0Ag–0.5Cu–xBi Melting on Cu Substrates. Microgravity Sci Technol 1–8
13. Zang LK, Yuan ZF, Zhan YP et al (2012) Spreading kinetics of a Sn–30Bi–0.5 Cu alloy on a Cu substrate. Chin Sci Bull 57(6):682–686
14. Zang L, Yuan Z, Zhu Y et al (2012) Spreading process and interfacial characteristic of Sn–17Bi–0.5 Cu/Ni at temperatures ranging from 523 K to 673 K. Colloids Surf A: PhysChemical Eng Asp 414:57–65
15. Zang L, Yuan Z, Zhao H et al (2009) Wettability of molten Sn–Bi–Cu solder on Cu substrate. Mater Lett 63(23):2067–2069
16. Paul A (2004) The Kirkendall effect in solid state diffusion. Laboratory of Materials and Interface Chemistry